Scientific Practice

Scientific Practice

Theories and Stories of Doing Physics

Edited by
Jed Z. Buchwald

The University of Chicago Press
Chicago and London

Jed Z. Buchwald is director of the Dibner Institute for the History of Science and Technology at MIT and the author of *From Maxwell to Microphysics*, *The Rise of the Wave Theory of Light*, and *The Creation of Scientific Effects*, all published by the University of Chicago Press.

The University of Chicago Press, Chicago 60637
The University of Chicago Press, Ltd., London

Printed in the United States of America
04 03 02 01 00 99 98 97 96 95 1 2 3 4 5

ISBN: 0-226-07889-2 (cloth)
ISBN: 0-226-07890-6 (paper)

Library of Congress Cataloging-in-Publication Data

Scientific practice : theories and stories of doing physics/edited by Jed Z. Buchwald.
p. cm.
Includes bibliographical references and index.
1. Physics—Experiments—Methodology. 2. Physics—Experiments—History. 3. Physics—Philosophy. 4. Science—Philosophy.
I. Buchwald, Jed Z.
QC33. S33 1995
530'.0724—dc20 94-49715
CIP

∞The paper used in this publication meets the minimum requirements of the American National Standard for Information Sciences—Permanence of Paper for Printed Library Materials, ANSI Z39.48-1984.

Contents

FIGURES

Analytical Contents

I. Theories

2. Context and Constraints

Peter L. Galison explains a constraint as a boundary beyond which inquirers find it difficult, if not unreasonable, to pass. Constraints are found at every level of theory and experiment—in enduring conservation laws, in common but transient knowledge about plastics, or in the material and social structure of laboratories. Because they are contingent (and local in time and subject matter), they may be short or long lived. By tracking the quasi-autonomous clusters of practice that pick out the different subcultures of physics (e.g., instrument makers, theorists, and experimentalists) Galison argues that a radical change in experiment may accompany persisting and unaltered theory, or vice versa. Finally, from anthropological linguistics, he introduces recent notions of pidgins and creoles to analyze the ways in which these differing subcultures bind through local "trading." This picture of locally connected, quasi-autonomous subcultures within physics gives rise to what one might call contextual antirelativism; each piece is embedded in a particular culture, but the whole of physics does not split into disconnected frameworks.

3. Beyond Constraint: The Temporality of Practice and the Historicity of Knowledge

Andrew Pickering aims at recovering a robust sense of the objectivity of scientific knowledge. Like Galison he refuses to explain objectivity either by the shared standards of scientists or by their interests. Pickering emphasizes that objectivity is thoroughly temporal; that is, it does not arise from some features that are already there, so to speak, but emerges in scientific practice. This is one reason why he rejects Galison's reliance on constraints. An experimenter, he argues, begins with current cultural resources: apparatus, theories about how it works, and hypotheses about some aspect of nature. Any experimental investigation involves all of these, and commonly it will be met by "resistance": results that cannot be interpreted by the available hypotheses or apparatus that works, if at all, in apparently aberrant ways. These three resources are what he calls plastic, and any of them may be modified, in the "real time" of the experiment, without being subject to constraints that were in place before experimentation began. Pickering illustrates this process of establishing an accommodation between his three elements with an experiment on free quarks conducted from 1965 to 1980. He finds that the practical dialectic of resistance and accommodation is a satisfying way to conceptualize the otherness of nature, its independence of human activity and intelligence.

4. Experimenting in the Natural Sciences: A Philosophical Approach

Hans Radder argues for a revaluation of the almost invisible craftwork used in the natural sciences. It is demanded by the production and reproduction of experiments. He defines three distinct types of reproducibility. Reproducibility has meant that a result could be witnessed objectively by others, and thereby become established fact. In our times an even more potent social legitimation of the natural sciences has arisen. Reproducible experiments become experimental technologies that can be exploited outside the laboratory. Radder also emphasizes a fairly sharp distinction between the theoretical description and the performance of an experiment: people can be trained to perform an experiment while having no idea what it means. As a technology is moved outside the laboratory, increasingly many people can use it with less and less understanding of its theoretical basis, and yet be all the more committed to the technology itself.

5. Scientific Practice: The View from the Tabletop

Brian S. Baigrie examines three different perspectives on experimental practice, here represented by Galison, Pickering, and Radder. All three maintain that scientific practice involves many distinct types of activity that collaborate so as to propel scientific culture in new and sometimes unexpected directions. But as the preceding papers show, pronounced philosophical disagreements lie beyond this consensus. Baigrie accepts that a Durkheimian approach to the experimental sciences has a good deal of potential. Placing experiment in a very broad social framework of allied intellectual and practical activities, it leads to a framework like Galison's or, at an extreme, to that of Bruno Latour. From the point of view of small-scale experimentation itself, however, the most useful descriptions will refer to "resourcelike cultural elements" reminiscent of Max Weber's analysis of social formations. Instead of a Durkheimian vision of autonomous social forces acting outside the experiment, we find a place for the intentions of experimenters; their cultural resources are stamped with the goals and aims of scientists themselves.

6. Following Scientists through Society? Yes, but at Arm's Length!

Yves Gingras challenges the actor-network and "heterogeneous-engineering" approaches to science in action that have been urged by Michel Callon and Bruno Latour. After arguing that these approaches are neither clear nor consistent, he advances a sociological view of the microdynamics of research. Although taking full account of our present knowledge of small-scale interactions, this view adds constraints on scientific practice that cannot be detected at the microlevel. Thus there are (contra Pickering) *invariants* within the endless variations observed in case studies of controversies or of scientific practice.

II. Stories

7. Why Hertz Was Right about Cathode Rays

Jed Z. Buchwald presents Hertz's pioneering investigation of the new cathode rays in its own terms and thereby illustrates an important thesis about the historiography of experiments. Hertz took for granted that cathode rays were a new phenomenon. He showed that their behavior differed at almost every significant point from anything obtained using known electrical or magnetic devices. Whatever the rays might be, they were not electric currents. Within a decade J. J. Thomson had proved that they were beams of moving electrons. So we naturally ask, what did Hertz do wrong? Did his instruments confuse too many effects which at that time could not be analyzed, or did he make a mistake of procedure, or of analysis? Buchwald argues that these questions are improperly posed and that the experimental work of Hertz's day cannot be usefully described in terms of what we now know about cathode rays or of what "must have been" happening with Hertz's apparatus.

8. Is the Identification of Experimental Error Contextually Dependent? The Case of Kauffmann's Experiment and Its Varied Reception

Giora Hon proposes that the identification or recognition of error—in particular, experimental error—may be just as elusive as positive knowledge. Kaufmann's experiments claimed to refute both Lorentz's theory of the electron and Einstein's relativity theory. These theories seemed to confirm instead the classical analysis of Abraham and Bucherer. Hon presents the experiments in some detail and reports the reactions of Poincaré, Einstein, and Lorentz. Poincaré accepted the result but warned that there might be an error. Einstein praised Kaufmann but rejected the result without locating any error in the experiments. Lorentz vacillated: he needed fifteen years to reject the results outright.

9. Scientific Conclusions and Philosophical Arguments: An Inessential Tension

Margaret Morrison examines the roles played by philosophical and scientific considerations in assessing both theories and evidence. Some scholars take the late-nineteenth-century debate about atomism and kinetic theory to be primarily philosophical, about the existence of unobservable entities; others take the issues to be internal to the sciences themselves. Such a split is overrigid; it is not supple enough to account for evolving practices. Hertz's approach to electromagnetism suggests how philosophical and scientific criteria can be combined in an interactive model. Because of his normative philosophical standards Hertz was skeptical about his own experiments even though they were uncritically accepted by the British. His canons were nevertheless able to evolve in the light of new results, changing practices, and new ways to interpret evidence.

10. Where Experiments End: Tabletop Trials in Victorian Astronomy

Simon Schaffer tells how solar astronomy became experimental, overturning the dicta of Laplace and Comte, who located astronomy in a purely observational/mathematical nexus. The new technologies of photography and spectroscopy enabled new types of research centers to treat the sun like the earth. Two of the great Victorian debates—sunspot spectra (1860s) and solar energy (1880s)—expressly concerned the extent to which astronomy could be deemed experimental. These illustrations more generally show how the work of experiment demands close links between laboratories and other sites of knowledge and practice. These links involve, in a fundamental way, the mutual calibration of instruments from one type of site to another.

11. The Sturdy Protestants of Science: Larmor, Trouton, and the Earth's Motion through the Ether

Andrew Warwick writes of the final years of a network of Maxwellians, 1900–1910, to illustrate three points. First, what in retrospect seem like ad hoc hypotheses may in their day be vigorously defended by entire experimental programs. Second, an experimental test may be in effect coproduced by theoretician and experimenter, the former needed to create the evidential context in which the test can be designed and conducted.

Third, by studying the successful operation of the Maxwellian network when challenged, and its subsequent collapse, we can understand the changing economies within which a theory holds currency. The example is itself astonishing: the attempt of Joseph Larmor, theoretician, and Frederick Trouton, experimenter, to generate unlimited amounts of energy from the contraction of matter as the earth moves through the ether.

ONE

INTRODUCTION

Ian Hacking

TABLE TOP EXPERIMENTS WAS A SMALL CONFERENCE held in Toronto at the end of March 1990. Our pedantic subtitle made plain what it was all about: Philosophical and Historiographic Questions about Small-Scale Experimentation in the Physical Sciences. The previous decade had been rich in contributions to a burgeoning field: the social, historical, and philosophical study of experimentation. Theory had ceased to dominate historical and philosophical thinking about the sciences. A surprising number of contributions in this vein were about big science, and this was particularly true of some of the books that we liked best. There were, for example, Andrew Pickering's *Constructing Quarks* and Peter Galison's *How Experiments End,* both marvelous studies of high-energy physics. Then there was that remarkable revival of work on the scientific revolution of the seventeenth century—a field that one might well have thought exhausted because it had been the mainstay of history of science for two full generations of scholars. It was resuscitated by Simon Schaffer and Steven Shapin, who addressed the great absence in most previous work: namely, experiment and instrumentation. Their *Leviathan and the Air Pump* is nevertheless also a story of big science: the air pump was, as Rupert Hall observed, the cyclotron of its day. Air pumps did not fit on tables, and their cost in research and development was prodigious; only a very rich man who also had access to government funding could have contemplated the enterprise.

The books I have mentioned are not silent about smaller experiments. Parts of *Leviathan* can be read as a story of transition; what began small and cheap became big and expensive. *How Experiments End* tells how table-sized experiments, like that of Einstein with the gyrocompass, were followed by middle-sized work, like the cosmic-ray studies, and culminated in high-energy physics. Nevertheless, Galison's project was to characterize what is new and peculiar about the logic of demonstration used in big science. In organizing our workshop we thought that it was time to take a look at smaller and cheaper and what one might call more personal experiments in their own right. I should add that although we have learned an enormous amount from the several schools dedicated to social studies of science, we hoped that our meeting would focus more on the interactions between experimenters, instruments, apparatus, and different layers of theorizing than on the relationships between laboratories, editors, patrons, competitors, manufacturers, and the prosperous public. This was not because we thought the relationships were not central to science, but because we thought that the interactions between materials, people, and ideas were being submerged in the enthusiasm for social studies.

The work of our guests has created a surprisingly stable architecture of its own. That's not to say that Jed Buchwald hasn't done an immense amount of editing, suggesting, and, sometimes, cajoling. Nevertheless, the house built itself; he did the cabinet work. Among the things that neither he nor I had planned was that the final product would so tidily divide into two parts: one group of essays that are general in scope, primarily philosophical in their type of reflection, and another group, of the same size, providing a series of equally thoughtful case studies of experimentation. We did expect that some themes popular a decade or so ago would have become invisible by 1990–topics such as "scientific realism" or "rationality versus relativism." We also expected that our contributors would take for granted that standards and practices in the sciences are various and "local"—and that we would not find this recent buzzword used as if it were the name of a new discovery. We live in a time when the disunities among the sciences are more familiar to historians, sociologists, and philosophers than are the (equally important) unities. One thing that we did not plan was that the case studies would be so curiously restricted in time to the late nineteenth and early twentieth centuries. They are bounded by Schaffer's look at the astronomical observatory on the one hand and by Hon's analysis of Kaufmann's attempts to shore up the ether at the other. Hertz is at the center, not only as an experimenter but also as a leading and influential methodologist of science. We were a little ahead of our time, in 1990, in seeing Hertz loom so large. The centennial of his death, 1994, was marked by many scholarly celebrations, some of which brought out the impact of Hertz on philosophical thought: on the immortals, such as Ludwig Wittgenstein, and on the mortals, Hertz's immediate colleagues in physics. Morrison's paper may indeed be read as an opening contribution to those festivities, although I am sure that she had no such idea in mind when she presented her paper. She was addressing what will continue to be central to any historical study of Hertz and his colleagues, namely, how scientific and philosophical principles are to be distinguished and how they are to be brought to bear on specific issues. Those of us who work in an analytic tradition of philosophy tend to suppose that normative, analytical, and metaphysical assumptions always trump scientific conclusions by limiting possibilities. Scientists sometimes imagine that the normative prescriptions can always be abandoned in the face of experience. Hertz was far subtler than either side in that type of debate. Morrison sensitively draws out how the two types of constraint are not independent boundary conditions but are instead interactive and jointly evolving.

The collection printed here does, quite inadvertently, honor the old prescription of symmetry demanded in bygone years by the Edinburgh "strong programme in the sociology of sciences" associated with Barry Barnes and David Bloor. In this volume we have detailed examinations of both failures and successes, in almost equal numbers. Neither is privileged here.

On the failure side, Warwick tells us about Larmor's fierce determination not only to preserve the ether theoretically but also, through his alliance with Trouton, the Irish experimenter, to harness the motion of the earth through the ether in order to generate endless clean energy. This was a turn-of-the-century dream of something almost as good as the perpetual motion that nineteenth-century thermodynamics had proved to be impossible—if only the giant rotating electrical condensers could be built. It was the cold-fusion fantasy of the day. Hon writes of what went wrong with Kaufmann's work, and how Einstein, for example, remained sublimely indifferent to it. Einstein did, at the end of his celebrated relativity paper, suggest three experimental tests, but he also had a great deal to do with the enthusiasm for theory during the first half of our century, because he was usually right when the experimenters were wrong. Worse, he kept on predicting phenomena that he rather confidently said could not be observed and took an impish delight in the fact that they could not be observed. It took decades until observers or experimenters would get their timid revenge, apparently detecting what he had so confidently thought was real but undetectable. I think not only of the upshot of Hon's story, but also of one I've told elsewhere, about that newly fashionable branch of observational astronomy, gravitational lenses, which Einstein puckishly hoped would be forever undetected (Hacking, 1989a).

The case studies here are restricted not only in time but also in space: they are German and British. Worse, they are delimited in topic to what could well be called post-Maxwellian experimental science. The only qualification to be added to that characterization is modest: Schaffer's study of the construction of the instruments for the new astronomy. Even there, Maxwell looms quite large. We could rearrange and supplement the second half of the book, giving it an Irving Stone title such as "Clerk Maxwell: The Triumph and the Decay." The man died; his name and his equations became immortal; his program flourished for a while but then became senescent. One of the happy by-products of this collection is the clear demonstration of the different things that Maxwell's ideas meant to different people and of the entirely distinct and indeed conflicting groups that evolved from his electrodynamical conceptions. That is fascinating for historians, but for philosophers it is a bit embarrassing, for we seem to have far "too slender a diet of examples," restricted in time, place, and topic, assuredly rich in some essential vitamins and minerals but deficient, perhaps, in others.

The philosophical essays, in contrast, mention experiments from all over the European shop, from the early days of experimental science (Baigrie) to Italian quark searches of the 1970s (Pickering). Although I have no hesitation in calling the first group of essays philosophical, three of the five contributors have made a reputation for themselves with their historico-sociological studies of big science, Galison and Pickering with the books already mentioned and Gingras with his sociological and institutional studies of, for example, the decisions about building a Tokamak (nuclear fusion) project in Quebec. I think that among our authors, no one has been more directly concerned than Gingras with the interplay between power and industrial-strength science. None has known better the ways in which networks and alliances are built up in the course of deciding where substantial chunks of national treasure are to be invested in research, development, and prestige. So it is important to have his careful rejection of the "network" approach to science studies urged by Michel Callon and Bruno Latour. This essay can well be read as a companion piece to Schaffer's "The Eighteenth Brumaire of Bruno Latour" (1991). Gingras's objection begins with the fact that "network theory" denies the distinctions that we all do make and must make in describing scientific activity, be it big or small, public or private.

I see things a bit differently from Gingras, probably because I am at a greater distance than he from the disagreements among different practitioners of "science studies." That is the present pseudoneutral name for examinations of science that are very different from any old-style sociology but still roughly think of themselves as sociological, sometimes even being conducted on the margins of university sociology departments. Latour, Pickering, and Gingras alike do science studies, but they do not agree on how to do them. Many of the papers collected in Pickering's *Science as Practice and Culture* (1992) focus on just these differences of approach; that book of essays is perhaps the best place to find out about the present state of play.

I would like to put in a word for Latour's perspective since he is not here to speak for himself. I see Latour and Callon as forging new categories, new distinctions, deliberately intending to challenge and replace old ones. That is particularly clear in a 1991 essay by Latour about the Shapin and Schaffer air-pump book. He called it "Not Post-modern—Amodern," and it is now most easily found, in expanded form, in Latour 1993. Latour does, as Gingras says, challenge the usual contrasts made between the natural and the social, but his aim is not wanton destruction. He argues that the separation is itself a historical artifact that we do not need. The year of the Armada, Hobbes said, gave birth to two monsters, "fear and myself." Perhaps that very year, Latour might continue, gave birth to two more monsters, Nature and Society. Hobbes we remember as the student of Society. (What had he to do with nature, until we read Shapin and Schaffer?) Boyle we remember as the student of Nature. (He had nothing to do with society, had he? Until we're reminded that he was a cofounder of that paragon and veritable begetter of scientific societies, the Royal Society of London.) Latour contends that the separation of Nature and Society is the characteristic malaise of modernism. It is our task, in trying to understand our 1990s world, which is inextricably social and scientific, to start all over again. To replace a Hegelian metaphor by a better one, we don't want *Aufhebung* but *Abschaffung*.

To say that Latour is abolishing old distinctions in favor of new ones is not to dissent from Gingras's criticisms of Latour and Callon but to profit from the way Gingras so clearly attacks one set of targets. That leaves us free to look at another aspect of the work of those two authors. We are now better placed to learn from their critique (here in the Kantian sense of the word) of the way we understand our entire intellectual and interventionist relationship to our sociomaterial environment.

Such pacifist observations, laying blessings upon both houses, are not really my forte, but the most serious contretemps in the book bids me continue in this vein. Galison and Pickering genuinely disagree, even if the common decencies of scholarship have led both of them to mute their differences in print. Baigrie offers one analysis of what is at issue between them. He sees the two as defining two fundamentally different ways in which to study science. He sees Galison as Durkheimian and concerned with autonomous social forces that act upon scientists and their communities, establishing constraints beyond which inquirers will not reasonably pass. He sees Pickering as Weberian, preoccupied by the use that individuals make of available cultural resources and the ways in which they adapt them. I do not see the two as exclusive. My point is not to make peace; up to a point, I quite like my friends to disagree with each other. I find a lot to learn from these two that I would not have noticed had they agreed with each other.

Baigrie's characterization of Galison as Durkheimian and Pickering as Weberian is slightly mischievous. Pickering would contend that he is, if anything, more truly modern than Latour. Baigrie implies that the disagreements between Pickering and Galison are paradigmatically "modern": what could be a better paradigm of the modern than the confrontation of those two unifying and systematizing giants of sociology, Durkheim and Weber? I believe that neither Pickering nor Galison is happy with the comparisons drawn. Yet I would use an even larger—and thoroughly "modern"—canvas to sketch what is happening.

Their disagreement reminds me, even to a matter of words, of one between William James and F. C. S. Schiller, the spokesman for pragmatism in Britain. Note that this was an in-house dispute; forgive me if I see Galison and Pickering in the same light. Their house is science studies, deeply infused with the 1980s version of pragmatism. Most people would regard Pickering as more radical than Galison, as most would regard Schiller as more radical than James. When Bertrand Russell wrote on pragmatism, what he really could not bear was that Schiller said that truth is made up by people: maximal social constructivism, as the labelers among us would now label it. But Schiller, perhaps just because he was the more radical pragmatist, was more aware than James of the need to insist that we cannot make up anything. We are "resisted." James, as sociable in his philosophy as in his daily life, thought of the "constraints" on belief as being primarily determined by our interactions with other people, our past, our traditions, and our yearnings (see, e.g., James 1907, chap. 7). And so it is with Pickering and Galison, and I think it no accident that they take issue over the very same words as vexed the two pragmatists a century ago. Pickering wants to know, what is the amazingly difficult business of getting an experiment to work? How do we defy the resistance of recalcitrant apparatus? We live (says Pickering's philosophy) in a world of total contingency, and yet something all too tangible, all too material, resists us. And he needs an account of that from within his philosophy, an account unwanted and unheeded by most species of "scientific realists." Galison wants to know something quite different, namely, the impositions on knowledge and action that arise, not from some unstructured story about "society," but from a very structured story about how various projects, techniques, cliques, and long-term investments of intellectual as well as material capital determine the organization of research and thereby constrain the very kinds of things that can be found out.

It seems to me that one can see Pickering's "resistance" and Galison's "constraints" in the several stories of experimentation that make up the second part of the book. Pickering regards apparatus, theories of how the apparatus works, and rival hypotheses about an aspect of nature as plastic resources that have to be remolded and accommodated to each other in the light of some general aims or goals of the scientist. Warwick's account of Larmor, Trouton, and the end of the Maxwellian network wonderfully exemplifies this. Each successive incident recounted by Warwick could be taken as an illustration of Pickering's type of analysis. Yet I can also read this story in Galison's way, of a network of researchers with certain rather fixed horizons engendered by both theoretical and instrumental traditions.

If we want an example of fundamental work in experimental physics for which Galison's type of analysis is appropriate, it is surely to be found in the events described by Buchwald. Hertz, Buchwald writes, was right: cathode rays are an entirely new type of phenomenon. And yet J. J. Thomson immediately proved they were beams of electrons. Hertz was operating within a set of Galisonian constraints. And what is important, from my point of view, is that these were not particularly intellectual constraints, or what Popper once called "the myth of the framework." What Buchwald shows us, among other things, is the way in which they were constraints that arose from the instrumental tradition followed by Hertz and his community.

Schaffer, in my opinion, adds essentially to the visions of both Pickering and Galison, for he is concerned with the creation and extension of an entirely new tradition, one that flew against all astronomical wisdom. Yes, there can be an astronomical laboratory, deploying a technology quite undreamt of by Laplace and Comte, spokesmen for the purity and autonomy of mathematical-plus-observational astronomy. Schaffer describes two events, separated by two decades, in which the experimental study of the solar system came into being. The "astronomical laboratory" had, nevertheless, to relate itself to other experimental practices and bodies of knowledge. It had to make its instruments part of an extended continuum of measuring devices familiar for terrestrial exploration. It had, in short, to calibrate its results to more-familiar ones, but how it would do that was very much a matter, as Pickering would put it, of exploiting and changing available cultural resources. In so doing, it fixed the kind of thing that solar astronomy could be and the kind of numbers with which it could deal, thereby constraining, in part by sheer material investment, the projects of all future investigators.

In the case of Kaufmann's seeming experimental validation of classical electromagnetic theory, Hon provides a good sense of what it was like to do the experiments. He emphasizes the fact that accuracy was critical because the differences between the predictions of competing theories were minute. But the major part of the paper is dedicated to interpretation. The physicists who responded to Kaufmann were theoreticians, not experimenters. There was no more modifying of the apparatus to be done, although there was rethinking of Pickering's second plastic resource, namely, understanding what happened in the doing of the experiment. But we cannot always reinterpret, reanalyze. Buchwald's account reminds us of the fact that although at the time of the experiment reinterpretation is a necessary activity, it is implausible and perhaps even improper for the historian to try to decide what was "really" happening in a piece of equipment that led a gifted experimenter to what we now take to be a mistaken conclusion. Hertz must be understood in terms of the cultural resources available to him.

It is very likely that neither Galison nor Pickering will welcome my eclecticism. Baigrie claims that when their views are situated by analogy with the incompatible sociologies of Durkheim and Weber, then we can see why each ought to reject my eclecticism. Reader, judge for yourself.

It should go without saying that our other authors did not write to expand upon the disagreements between Galison and Pickering with which this book opens. I have been putting the other papers to unintended ends. We find, in the accounts of experiment, an extraordinarily rich contribution to established questions in the history and philosophy of science. Although the second part of this collection is properly described as historical, philosophy is always close to hand. Thus we have Warwick, as philosopher, on ad hoc hypotheses and, as historian, on the specifically theoretical approaches of that great experimenter Larmor. Or Schaffer, as historian, on the emergence of experimental astronomy and, as philosopher, on the role of calibration in extending or creating a science. Or Hon, as historian, on the role of the Kaufmann experiment and, as philosopher, on the resilience or even indifference of theory to experiment.

We also, as is so often the case with conferences, have samplers from much larger innovative enterprises. Thus Hon's paper is a portion of an ongoing series of studies on the different kinds of experimental error, and Buchwald's is part of a major reanalysis of the work of Hertz. Likewise, Schaffer's use of calibration as a key analytic concept for understanding the emergence of new disciplinary matrices (as Kuhn called them) is a sampler from a larger tapestry which is also developed by his Cambridge colleague Nicholas Jardine (1991).

Hans Radder, whom I have not thus far mentioned, brings to the book a perspective rather unfamiliar to English- and French-speaking students of the sciences, namely, Habermas's early work on instrumental and experimental action. Radder's distinction between different types of reproducibility may clarify some of the other debates in the book. Thus his views on nonlocality and stability may agree with Galison's views about constraint, although the strongest statement one should make is that the stabilities guide practice; they do not determine it. Likewise, Radder finds a place for Pickering's notion of the coherence between material procedure, instrumental model, and phenomenal model in his own framework. He would describe that coherence as reproducibility by the original experimenter under a fixed interpretation. However, other types of reproducibility may well tie in with a more comprehensive notion of objectivity, of which Pickering's is only a special case.

It is good to have reproducibility—or kinds of reproducibility—restored to the philosophy of science. It became commonplace, in the recent years of enthusiasm for experiment, to say that nobody ever reproduces an experiment. That may have been an excellent antidote to the complacency about experimental life that was characteristic of the days when science studies consisted of the study of theory. But of course reproducibility and the ideal of replication are an integral part of what makes a scientific result legitimate. Hence it is important for Radder to distinguish, for the first time, between quite different types of reproducibility and to consider the role that they play in the norms of communication and acceptance. He deliberately twists our topic of tabletop experiments (which could become all too inward) so that it once again faces outward. He reminds us how we have now come to turn reproducible experiments into off-the-shelf technology, which many people—and certainly government agencies—now take as the central legitimation of science.

The difference that is important to Radder is an aspect of science not hitherto discussed—and certainly different from the evolution of small into big science. Indeed, as Baigrie says, most laboratory work is still tabletop work, and it never will become big science. What Radder notices is the increasingly intense desire for rapid transformation of pure experimental science into impeccable technology that creates its own consumers. We are undoubtedly in an era of science far different from that of Hertz, the centerpiece of the present collection of essays. Yet Hertz is a remarkably handy icon standing for what Radder points to, and what is elsewhere missing from this book. I mean the transition from a tabletop experiment to a tabletop tool tuned into all the rest of the technological and institutional infrastructure of the world. Indeed, what was the first household piece of technology derived from purely theoretical and theoretically motivated experimentation? My candidate would be the wireless. What is the first product of modern technology that will induce a subsistence-level family in the "third world," one with literally zero materialistic resources, to grow a cash crop in order to make its first major purchase? A transistor radio.

I

Theories

Two

Context and Constraints

Peter L. Galison

1. Constraints

At the root of many historical-philosophical debates over science are two conflicting intuitions. In the first, science is divided into what I will call "island empires," each of which encompasses a relatively consistent assembly of experimental and theoretical procedures and results. Variously designated as frameworks, conceptual schemes, and paradigms, each such islet of knowledge supports its own language; taken as a whole, this picture of fragmented science is taken as a blow against a realist, and for a relativist, metaphysics. The best-known historical-philosophical account of this type is Kuhn's *The Structure of Scientific Revolutions,* where he argues that science divides itself into units grouped around exemplary problem solutions. The island of experiment and theory in Newton's physics stands forever isolated from that of experiment and theory in Einstein's special relativity. But Kuhn was not the first or the last to divide science into such islands: the intuition behind his work shares much, as I will argue below, with Carnap's "frameworks" of the 1950s and with later sociological work of the 1970s and 1980s. It is an enormously compelling vision of science—not only for philosophers but for historians. Without a doubt it helped to dismantle the grand but flawed program of logical positivism that had held such sway for several decades in the history and philosophy of science.[1]

In the end, however, I want to argue against the island empire picture of science for three reasons. Historiographically, the assumption that theory and experiment change of a piece strikes me as unwarranted. Cultural historians have made it clear over recent years that we cannot view

1. On different schemes of periodization in history of science and their debt to underlying philosophies of science, see Galison 1988a; on the cultural links between the logical positivists and the modernism of the 1920s, see Galison 1990a, 1993.

revolutionary France, for example, as if the different sectors of that society responded to political dislocations in the same way or at the same pace. On similar grounds, we can learn to view the different sectors of a scientific culture as having their own rhythms of development and styles of argumentation. Sociologically, the homogenization of the physics community (for example) into a single entity with a single set of commitments is supportable only by focusing attention on a single, narrow band of theorists, while simultaneously subordinating the other cultures constituted by experimentalists, engineers, technicians, and others who make up the community. Philosophically, the picture of theories as embodying full languages which then must be translated *globally* if commensurability is to follow seems to me in error; we need models of *local coordination,* not global translation, among theories, among experiments, and between experiment and theory.

In articulating the distinctiveness of these subcultures, I find the notion of constraints helpful; constraints enter the discussion because they mark the endpoints of scientific inquiry, the boundaries beyond which inquirers within the community find it unreasonable to pass. Such borders may be at the lofty plane of conservation laws and symmetries or buried in the common knowledge of plastics, metals, and computer chips. Constraints may last for decades or only persist for the duration of an experimental run. They play a pivotal role in epistemological discussions, because the establishment of an experimental result closes or aims at closing skeptical challenges. On the power of such challenges rides much of the debate over the nature of demonstration in scientific practice: Is the apparatus functioning? Are the assumed theories themselves in question? Does it matter if they are? Is the environment responsible for a given signal? Are the observers themselves creating artifacts? An argument for a given effect must block such objections by invoking constraints, not in the logically closed form of a mathematical proof, but sufficiently unto the day. The central historical-philosophical questions therefore are: how do these constraints arise, what sustains them, how do they act, and what makes them fall?

In experimental particle physics at mid–twentieth century, constraints took many forms. In some parts of the community, only sharp and complete photographic images from a cloud chamber could halt debate. For these advocates of "golden events," statistical demonstrations with Geiger-Müller counters would never close off objections that some unknown process was occurring between points where the counters registered a passing particle. For other physicists, the photographs themselves were suspect. Because of the small number of pictures in early cloud-chamber work, the photographs could never persuade Geiger counter physicists that their results were not flukes—unlikely artifacts of unknown origin. There are many other such boundaries to the field of the reasonable and the persuasive. How many witnesses and what kind of witnesses are needed to certify a laboratory phenomenon (Shapin and Schaffer 1985)? Can (or must) fossils count as evidence for stratigraphic sequencing (Rudwick 1985)? When are statistics allowed in physics experiments as part of an argument for a new effect? When is statistical significance required—and why? Is it enough to produce an elementary interaction and to explain a complex one by theoretical exposition—or is it necessary to mimic complexity in the laboratory? How do inferences from many levels interact—from the widest-ranging theories to practical rules of thumb? Taken together such questions (posed by many recent authors, including Daston, Hacking, Hesse, Porter, Rudwick, Schaffer, Shapin, Cartwright, myself, and others) have begun to constitute a field of inquiry by articulating what delimits the acceptable in laboratory demonstration. Talk of constraint strikes me as useful because it allows us to speak about the domain of allowable actions without referring to an external scientific method valid always and everywhere *and* to avoid a representation of laboratory practices as relying purely on extrinsic forces to silence skepticism. And most importantly, constraints have histories—the view that the ability to mimic is a prerequisite to understanding has a location and a history; photography as a constitutive element of objective representation has specific origins; conservation laws have advocates and opponents in their rise and demise.

In this openly speculative set of reflections, I will proceed as follows. First, I return to the historians' use of constraint—particularly that of the Annalistes and their cultural-historical successors[2]—and argue that the notion of constraint plays an exceedingly important and substantive part of their explanatory framework. As part of this line of inquiry, I will argue that the Annales school's conception of constraint comes in large measure from two sources: from the French sociological tradition (especially Durkheim) and from a materialistic focus on technology, topography, and power. Next, I turn to the ways in which what I term "multiple constraints" function in physics, blurring the boundary between the old categories of the "internal" and the "external." "Multiple" here has two senses. Clusters of constraints carve out different subcultures of physics, distinguishing the standards of demonstration of experiment, for example, from those of theory. But "multiple" also refers to the situation of these constraints and designates the diversity of their origin. An acceptable move in laboratory practice is bounded simultaneously by strictures from highly heterogeneous sources ranging from the properties of plastic insu-

2. As discussed further in Galison 1987, I find the work of Carlo Ginzburg particularly helpful in his use of cultural constraints.

lators to the economics of circuit design; I argue against any reductionist theory of their origin.

Using this language of subcultures and constraints, I then argue both historically and analytically that there is striking continuity between the "framework" ontology of Carnap and of Kuhn and one particular strand of social constructivism. Against this view, I would suggest that the language of constraints offers a more plausible picture of science as an *intercalated* enterprise, in which several, often quite disparate subcultures follow their own pace and practices, and with their own often conflicting views on the laws, objects, and demonstrations appropriate to their science. Freed from the presumption that experiment and theory move as one, we can shift our attention toward the complex and highly local means by which cultures in contact can coordinate their actions and beliefs—even in the absence of shared global views about meaning.

• • •

IN THE HEYDAY of the Annales school, the most famous use of the notion of constraint came, no doubt, from the restrictive power of "obstacles" presented by the terrain itself on the social, economic, religious, and cultural structure of the mountainous regions of the Mediterranean area. The Church, for example, could never secure its domination among the peaks: "It was only in places where its actions could be persistently reinforced that the Church was able to tame and evangelize these herdsmen and independent peasants" (Braudel 1972, 34). Similarly, "the Berbers of North Africa, protected by the mountain peaks, were still hardly at all, or very imperfectly, won over to Mohammed" (34). Even when the mountains were penetrated by one of the dominant sects, the rough terrain sheltered continuing resistance—militarily, culturally, and religiously. Sometimes the belief systems that opposed the new religions were vestiges of traditional practices, sometimes conscious revolt. Even linguistically the mountains remained unconquerable. In Roman times the territories of North Africa and Spain never adopted Latin as their principal language (34). But if the mountains were "backward" in their institutional (or anti-institutional) structure, they were then, and remained, persistent enclaves of freedom: "It is only in the lowlands that one finds a close-knit, stifling society, a prebendal clergy, a haughty aristocracy, and an efficient system of justice. The hills were the refuge of liberty, democracy, and peasant republic" (40).

In the hills, steep terrain, rough weather, bad roads, and rocky, isolated agricultural plots combined to form "obstacles" to feudal society, which consequently could not take hold as "a political, economic, and social system, and as an instrument of justice" (Braudel 1972, 38). Said another way, mountains do not determine a certain social form, but gross deviations from a "peasant democracy" encounter very real resistance (Braudel 1950). Here Braudel dilates the notion of obstacle explicitly to include the technical, social, and administrative. "Think," he writes, "of the difficulty of breaking certain geographical frameworks [*cadres*], certain biological realities, certain limits on productivity, or such and such spiritual constraints: mental frameworks [*cadres mentaux*] also are prisons of the *longue durée*" (Braudel 1958, 731). These ideas do historical work for Braudel; indeed, they are fundamental to his whole conception of the growth of material life in the early modern era. As such they are not "glosses" on a previously inscribed, neutral "history"; they prescribe the kind of history that gets written: floods, plows, pumps, and horses become subjects of a history where otherwise they were not. Consider Braudel's example of Chinese rice production in *Capitalism and Material Life, 1400–1800,* where he contends that the cultivation of this crop demands large amounts of land and extensive irrigation and drainage systems, all of which require a large, coordinated workforce:

> All in all an enormous concentration of work, human capital and careful adaptation was involved. Even then nothing would have held together if the broad lines of this irrigation system had not been firmly integrated and supervised from above. This implies a stable society, state authority and incessant large-scale works. . . . It also implied the concentration of villages, as much because of the collective requirements of irrigation as because of the dangerousness of the Chinese countryside. (Braudel 1973, 101)

Thus where rice production is the dominant form of agriculture we find the same kind of "obstacle" effect, this time from "state authority." Attempts to breach the basic, centralized, authoritarian structure collapse the vital irrigation and drainage systems and lead to food shortages and diseases. While there remains a degree of freedom in choosing a societal form among the rice paddies, the imposition of a peasant democracy of the kind found in the hills of Tuscany would meet strong resistance. What at first glance appears to be simply a crop choice reveals itself to bring with it a set of "obstacles" which guide and restrict culture in the realms of government, religion, and societal organization. Braudel *never* argued that it is *impossible* for a Chinese rice-growing community to shed authoritarian political structure; instead he pointed to the obstacles such a move encountered because it threatened to destabilize the massive irrigation and drainage systems in an environment where such a collapse would jeopardize food and health. Nor are these constraints permanent features of the universe: they had origins, and within the scope of Braudel's work they have ends. But the rhythm of their change is slow with respect to the pace of political action, wars, and treatises.

Braudel and his fellow *Annalistes'* debt to sociology is deep; on theoretical grounds, Braudel owed a great deal to Durkheim. In particular, his cluster of analytic tools—obstacle, resistance, and constraint—derives in large measure directly from Durkheim. The historical constraints (or "envelope" in the mathematical sense) that Braudel spoke of diachronically are precisely the sociological limits that Durkheim studied synchronically. Their function in the sociologists' work is virtually identical to their function in Braudel's work:

> [Certain] types of conduct or thought are not only external to the individual but are, moreover, endowed with coercive power, by virtue of which they impose themselves upon him independent of his individual will. Of course, when I fully consent and conform to them this *constraint* is felt only slightly, if at all, and is therefore unnecessary. But it is, nonetheless, an intrinsic characteristic of these facts, the proof thereof being that it asserts itself as soon as I attempt to resist it. (Durkheim [1895] 1964, 2, emphasis added)

That this notion of constraint is essential to Durkheim is clear. In his work, it serves as a demarcation criterion for the discipline of sociology itself.

> We thus arrive at the point where we can formulate and delimit in a precise way the domain of sociology. It comprises only a limited group of phenomena. A social fact is to be recognized by the power of external coercion which it exercises or is capable of exercising over individuals, and the presence of this power may be recognized in its turn either by the existence of some specific sanction or by the resistance offered against every individual effort that tends to violate it. (Durkheim [1895] 1964, 10)

Paying for goods in the wrong currency or responding in the wrong language would instantly confront someone with the kind of resistance it would be impossible to ignore. The Annales school transformed the sociological category by giving constraints a specific historical context—and often (but not always) a material foundation.

2. Multiple Constraints

I like the language of constraints for two reasons. First, it offers one the beginning of a historical vocabulary that leaves scientific practice neither utterly divorced from its cultural context nor relegated to a mere puppet of other forces. The language of constraints allows one to consider a scientific problem setting in its own terms—ones that cut across an after-the-fact classification into the "internal" and the "external." In quantum chromodynamics, many of the theory's predictions could not be examined by analytic means but could only be studied by computer simulation. In some parts of the particle physics community this was not an adequate demonstration. How should this constraint be viewed? Was it "internal" because it involved some technical objections to the simulation method? Or was it "external" because resistance to nonanalytic solutions was fundamentally tied to accepted attitudes about the role and function of computers within the hierarchy of knowledge?

Challenging the bubble chamber in 1955, for example, meant having to confront not only the machine's multiple tests and applications, it also meant challenging the intimately related cryogenic practices. These technologies were fundamentally implicated in a range of technical systems from the recently tested wet hydrogen bomb to jet propulsion systems. So, when Luis Alvarez announced the production of a new phenomenon such as muon-induced fusion or the cascade zero, he owed more than his machines to the AEC (they were brought up from the H-bomb site at Eniwetok). Constraints on skeptical challenges about, say, the thermodynamics of liquid hydrogen, issued from processes at once "purely" scientific and manifestly "applied"; technological reliability was part and parcel of physical demonstration. Even much of the staff carried over from hydrogen bomb to hydrogen chamber. The bubble-chamber argument for the cascade zero thus was in part *constituted* by knowledge from weaponeering. Constraints will not yield to a post hoc division into the scientific, the technological, and the social.

Here is a metaphor. In mathematics it is common to search for a function or class of functions that will satisfy many constraints simultaneously. Sometimes the function will be underdetermined and a whole class of objects may fit the bill. Sometimes the function will be well defined—unique. In still other situations the constraints may leave no solution at all. Physicists (experimentalists or theorists) typically find themselves confronted by a set of constraints but with no clear notion of whether the solution will be many, one, or nonexistent. Moreover, the constraints themselves come from a variety of domains, and these domains are not neatly sorted; action depends on navigating within this heterogeneous set.

That these various constraints act in concert and must be considered simultaneously is manifest to the actors who must confront them: only historians and philosophers can ignore their joint presence. Listen to A. K. Mann, a prominent high-energy experimentalist, as he introduced a workshop entitled Collider Detectors: Present Capabilities and Future Possibilities:

> The goal of our field is easily stated in a general way: it is to reach higher center of mass energies and higher luminosities while employing more sensitive and more versatile event detectors, all in order to probe more deeply into the physics of elementary particles. The obstacles to achieving this goal are equally apparent. Escalating costs of construction and operation of our

> facilities limit alternatives and force us to make hard choices among those alternatives. The necessity to be highly selective in the choice of facilities, in conjunction with the need for increased manpower concentrations to build accelerators and mount experiments, leads to complex social problems within the science. As the frontier is removed ever further, serious technical difficulties and limitations arise. Finally, competition, much of which is usually healthy, now manifests itself with greater intensity on a regional basis within our country and also on an international scale. (Mann 1990, 3)

Such admixtures of what might be classified as "social" and cognitive occur all the way down the line into the very fabric of experimental practice. David Nygren, the inventor of the time projection chamber, mused at this same collider conference on one of the proposed detectors, Čerenkov counters: "It is easy to nurture technological fantasies marrying advanced photocathodes with micro-channel plate electron multipliers and CCD [charged coupling device] imaging arrays in a single device, yet such devices are not likely to be developed with the usual financial resources devoted to instrumentation for high energy physics" (Nygren et al. 1990, 62). He could—and did—take certain steps in the direction of exploring such integrated CCD/microchannel technologies, but each move became more and more costly until the direction was abandoned. In the modern laboratory technology and physics cannot be isolated one from the other; and in the environment of massive data and signal processing the practical and the pure are not divisible. When a physical theory is questioned because it does not lead to calculable predictions using analytic techniques, is this internal? Is it external? How do we treat resistance to a physical theory (such as quantum chromodynamics) if the theory remains intractable analytically but is open in restricted cases to computer simulation? Is the evolution of supercomputer hardware and software for other purposes (cryptography, weapons design, meteorology, economic forecasting) internal or external?

Recently Steven Shapin and Simon Schaffer have suggested that science embraces a complex of technologies from the literary to the mechanical (Shapin and Schaffer 1985). I would argue that this expansive notion of a technology could be taken further by trying to determine the boundary of each such set of practices. What is it that marks the limits of action within a particular technology? The literary technology invoked by Boyle and his followers would not admit an intervention in the style of his competitors. Such limits are fully visible as technological practices get inscribed in the complex weave of computer programs such as the computer-aided design programs that now govern so much of the design of objects ranging from cars and airplanes to chips, computers, and electronics.

An architect working at a computer console uses a light pencil to move a structural element into a roof support; immediately, software designed by a structural engineer flashes a message, rejecting the move as dangerous or in violation of a building code.[3] Or the software warns that a particular wall placement calls for a nonstandard part and offers a slight variation that can employ the cheaper and more easily installed standard piece. Aesthetic, structural, legal, and subcontractural constraints shape one another; design proceeds simultaneously from top to bottom and from bottom to top. From the moment the architect begins to sketch plans on a console, the computer interweaves, in an integrated fashion, constraints ranging from the avoidance of roof collapse to the cost of plywood forms for poured concrete. In computer-aided electronic circuit design (Ackland et al. 1985), a designer exploring variations of a device encounters constraints not only when the disposition of components violates production capabilities (such as demanding an overly dense configuration of connections) but also when it fails to meet electronic standards (of noise and loss) or demands economically unfeasible architecture (such as being too large). Flipping back and forth between laws of nature and the laws of humans, the computer tries to satisfy multiple constraints. It is possible—indeed usual—for the solution offered to be neither the best from a *purely physical* nor the best from a *purely economic* standpoint. Both as an example and as a metaphor, computer-aided design makes the boundaries of action visible; constraints either block or temporarily complicate transgression. From within the depths of a computer code, hidden from view and written by others, come obstacles blocking the path of a free walk in the space of design.

The fact that computer-aided design works on a machine merely throws in relief what is true throughout scientific practice. Intertwined constraints like these shape the experimental physicist's work as well (Galison 1988a, 1988b, 1988c). In 1984, the high-energy physics community was preoccupied with basic design decisions about detectors for the Superconducting Supercollider. One group listed a set of reasons for moving the magnet outside the calorimeter. The first reason was that there was "[n]o deterioration of calorimeter performance with respect to energy resolution, shower spreading or, especially, hermeticity" (Galison 1988c). Examined out of context, a part of this list would seem to provide grist for the internalist's mill: the decision to move the magnet was made because of technical demands associated with the search for heavy-quark decay. But a second constraint on the same list came from the cost and

3. I would like to thank Martin Fischer for several helpful discussions about computer-aided design. See also Tatum and Fischer 1990; Fischer 1990.

availability of uranium within the whole network of military and nuclear-industrial production and consumption. On the basis of *this* argument, taken in isolation, the externalist could claim that detector design "followed" from military and industrial demands. Yet a third constraint issued from technique itself: engineering simplification would result from moving the magnet out of the calorimeter. And a technological determinist would smile with satisfaction. But none of these reductive schemes—internalism, externalism, or technological determinism—captures the physicists' routine experience of working within a multiplicity of constraints, any one of which can be violated, but at a cost. Grasping the constraints of technology, economics, and engineering (or of other branches of science and culture) and understanding the costs of their violation are, or ought to be, central to an analysis of the closing of experimental argument.

3. Construction

Examples such as those I have just given might wrongly suggest that the function of constraints is purely negative; it is not. Or these examples might imply that the constraints bind the experimentalist's laboratory or the engineer's blueprints but not the theorist's chalkboard. False again. To a large extent, and across many domains of science, constraints are constitutive of the positive research program. They create a problem domain, giving it shape, structure, and direction. In the following remarks, the theoretical physicist Steven Weinberg describes his reaction to renormalizability, the metatheoretical constraint that demands that a theory make predictions to all orders of accuracy with a fixed and finite set of parameters as input:

> I learned about renormalization theory as a graduate student, mostly by reading Dyson's papers. From the beginning it seemed to me to be a wonderful thing that very few quantum field theories are renormalizable. Limitations of this sort are, after all, what we most *want;* not mathematical methods which can make sense out of an infinite variety of physically irrelevant theories, but methods which carry constraints, because these constraints may point the way toward the one true theory. In particular, I was very impressed by the fact that [quantum electrodynamics] could in a sense be *derived* from symmetry principles and the constraint of renormalizability; the only Lorentz invariant and gauge invariant renormalizable Lagrangian for photons and electrons is precisely the original Dirac Lagrangian of [quantum electrodynamics]. (Weinberg 1980, 1213)

One need not subscribe to Weinberg's faith in the existence, or even the approximation, of "one true theory" in order to recognize the tremendous positive role of the theoretical constraint in defining the field of inquiry. This positive role appears most directly in the *derivation* of quantum electronics; it is in just such cases that constraint becomes a constructive element of argumentation. Upon accepting renormalization as a constraint, whole classes of theories are suddenly ruled out of court, where before they were serious contenders. And among these ruled-out theories are many, such as the previously standard theory of the weak interactions, that had been enormously successful in summarizing, organizing, and predicting subatomic interactions.[4]

The inscription of constraints into the tools of daily practice serves as an indicator of their acceptance. For example, renormalizable theories cannot contain terms with dimensional coupling constraints; with time, theoretical-particle physicists either stopped writing such terms down even as *candidates* for theoretical terms or else developed arguments as to why they should be insignificantly small. Other theoretical constraints arose in different contexts: restrictions against nonrelativistic terms, against terms breaking symmetries or combinations of symmetry. Bit by bit, theorists used the graphics and nomenclature of symbolic representation to carve out a subculture of physics. Consequences of theoretical moves no longer needed to be drawn explicitly by derivation. As four-vectors replaced explicit time and space variables, equations came to wear their relativistic invariance on their sleeves. Notation itself became a technology. Diagrammatic representation—of electrical circuits, of logic modules, or of Feynman diagrams—structured combinations by routinized conventions. It becomes as hard for a particle theorist to write down a term that "accidentally" violates relativity (or parity or probability) as it would be for an electrical engineer to leave an exposed high-voltage lead near a delicate circuit. Standardization works in part by rendering the deviant obvious. The danger of such violation is internalized and, as Braudel would have it, enters into the mental framework that bounds the possible.

Quite evidently, constraint clusters in theoretical physics are not the same as constraint clusters in experimental physics or in instrument making. And in each subculture constraints attach differently to the wider culture in which they are embedded. Each subculture has to work out means of exchange not only with other subcultures but also with the wider world within which they both lie.

The issue of the relevance of cultural embedding is crucial, for at stake is a vision of history and philosophy of science. Let me proceed by example. About thirty years ago, Thomas Kuhn and Paul Feyerabend opened a tremendously interesting avenue of inquiry by pointing to the

4. An extensive discussion of the origin of the electroweak theory can be found in Pickering 1984b.

disjuncture in meaning that occurs between successive scientific theories when they are separated by the great gulf of scientific revolution, such as that between Ptolemy and Copernicus or between Newton and Einstein. Kuhn extended this "meaning incommensurability" (Ian Hacking's phrase) to the myriad of smaller mutations that occur between theories such as cathode-ray theory before and after it was discovered that cathode rays could emit X rays. The continent of science was cut into thousands of isolated islands, each domain ruled by its own sovereign paradigm. With the focus on meaning change across a change of theory, attention naturally turned away from the by then discredited positivist dominion of observation and away from experiment toward theory.

Years ago, in his book *The Structure of Science,* Ernest Nagel insisted that experimental law "has a life of its own, not contingent on the continued life of any particular theory" that might justify or explain it (Nagel 1961, 87). By this statement, he meant that there was an observational substrate that functioned as the foundation of a hierarchy of levels, passing through bridge principles and culminating in theory. High theory might readjust the meaning of terms within the experimental claim, but there must be a determinate sense to the law that is statable without reference to the theory. When Ian Hacking revived the slogan "experimentation has a life of its own" in *Representing and Intervening* (Hacking 1983, 150), he had in mind something quite different: that experimentation has a momentum and a motivation that are not the slaves of theory; experiments have goals that range widely from the standard of merely checking theories (though they certainly do that too). Experimenters do their work to explore new domains of phenomena, to try out new equipment, to gain added precision for physical constants, and much else besides. On the basis of the robustness of experimentation and the objects that it studies, Hacking concluded that though he could not accept a realism about theories—the entities posited in high-level accounts of physics, biology, and astrophysics—he could (contra van Fraassen's revised instrumentalism; see van Fraassen 1980) plump for a realism about experimental entities. The ability to manipulate physical objects such as electrons and positrons leaves us in an altogether different position from the one we are in when facing the full panoply of relations implicit in a theory such as quantum electrodynamics.

But there is a third sense to the striking motto "experimentation has a life of its own"—a literal sense. The life associated with experimentation is not the life affixed to theorizing. Relations to industry, to material objects, and to standards of right reasoning, methods of argumentation, and techniques of apprenticeship are all different. The life of an experimenter involves knowing the properties and costs and societal location of materials. In the 1990s a young precision bubble-chamber experimenter, for example, must know not only about heavy-quark decays but about the tough transparent plastic lexane—about the manufacturers that make it, about its optical qualities, its tensile, shear, and temperature-resistant strength. In short the experimenter must be fluent with the many properties that make lexane ideal for military canopies on jet fighters; properties, after all, which constituted its original reason for existence. A colliding-beam experimenter has to learn not only about rare decays and Higgs searches but about computer and electronic problems that are shared by electronics engineers preparing radiation hardening for fighting on a nuclear battlefield. The intellectual and social world of the experimenter is different from that of the theorist. It takes place in a different physical setting, with different standards of acceptable argumentation.

4. Social Construction

Standards of demonstration are historically determined; they frequently (but not always) have roots in the wider culture; their specification gives us reason to treat science as an intercalated field of subcultures. What follows? Three theses do not: foundationalism, relativism, and what one might call plasticism.

If one accepts that the standards of demonstration not only shift between fields but within fields over time, it becomes immensely difficult to think of a fixed method that has been or ever could be characteristic of science as a whole. A statistical argument for a physical phenomenon was, as Ian Hacking would put it, not even a candidate for being true or false much before the nineteenth century. In the twentieth century it frequently occurs that a single photograph can persuade the physics community of the existence of a new kind of force or particle; such a demonstration was simply unthinkable before the cloud chamber and its various descendant technologies.[5]

Of the many types of relativism, I want to address only one: the claim that the absence of a single, absolute foundation is an argument for the fragmentation of science. Since each epoch carries different standards of evidence and demonstration, this reasoning goes, each can be evaluated only on its own isolated terms. According to both Kuhn and Feyerabend, Einstein's relativity is irremediably isolated from Newtonian mechanics; only a punster or a superfast translator could believe that their common use of terms like "space," "energy," and "time" makes them commensurable.

5. The idea of standards of demonstration has much in common with the idea of styles of reasoning developed by A. C. Crombie, Ian Hacking, and Arnold Davidson. See Hacking 1986 for the discussion of this idea and references to Crombie; see also Davidson 1987, n.d.

Their argument goes something like this. Newtonian and Einsteinian physics use terms like "time" and "space" differently; for Einstein, these terms rest on an operational foundation, whereas for Lorentz and Poincaré, the terms have no such operational foundation. For Lorentz and Poincaré, mass as a concept would ultimately be replaced by a fully articulated electromagnetic theory; for the Einstein of 1905, mechanics and electrodynamics retained their independent existence, neither reduced to the other. Because of this fundamental disagreement, any attempt to compare the theories by means of experiment must necessarily be only illusory (Kuhn 1970, 102).

To understand the original appeal of the Kuhn-Feyerabend view, it is necessary to recall their real target: positivist sense-data language. Indeed, for Kuhn the *only* alternative to radical incommensurability is a third term, a separate, extratheoretical language to which, for example, both pre- and post-Einsteinian physics could be reduced:

> The point by point comparison of two successive theories demands a language into which at least the empirical consequences of both can be translated without loss or change. . . . Ideally the primitive vocabulary of such a language would consist of pure sense-datum terms plus syntactic connectives. Philosophers have now abandoned hope of achieving any such ideal, but many of them continue to assume that theories can be compared by recourse to a basic vocabulary consisting entirely of words which are attached to nature in ways that are unproblematic and, to the extent necessary, independent of theory. (Kuhn 1976, 266)

Kuhn goes on to argue, surely correctly, that such a "basic" vocabulary cannot exist in an "unproblematic" and totally theory-"independent" fashion. To this part of his argument I am sympathetic: no historicizing account would claim that there was a basic vocabulary with all that it implies about a building up from eternally valid starting points grounded in experience or intuition. It is hard to say what "unproblematic" means here; surely no set of activities in the laboratory could ever be immune to skeptical doubt. Nothing rides (either for the positivist philosopher of 1960 or for the scientist) on the absolute indubitability of *any* aspect of scientific practice. So the crux of the incommensurability argument centers on independence, about whether laboratory practice is independent in relevant ways from theoretical practice. But even here, in this paradigmatic instance of paradigm change, the history of experimentation will not cooperate with the idea of radical disjuncture. Poincaré, Einstein, and Lorentz all corresponded with the experimenters; all thought the experiment pertinent to the evaluation of their theories. They retained (or established) a *local* coordination between terms such as "velocity," "momentum," and "mass" and the magnetic and electric fields of Kaufmann's and Bucherer's experiments. Though Einstein and Poncaré suspected systematic error in some of the experiments, neither fundamentally had any problem coordinating his own or the other's predictions with the experimental results. Without a *global* agreement on the goals and ultimate principles of physics, and without a neutral observation language, there could still be a *local* coordination between actions in the laboratory and theories on paper. I have introduced the term "trading zone" to capture this region (conceptual and spatial) of partial coordination."[6]

The dependence of observation on theory lies at the root of Kuhnian relativism; its corollary has been a historiography that puts theory first. But in the face of so much recent work in the history of experimentation, the theory-dominated account has had to retreat. In the place of the "theory-first" position, a more recent view has arisen that acknowledges the role of experiment but only as an unindicted co-conspirator with theory. Experiment and theory are essentially "plastic" resources that physicists manipulate until they conform to each other. It is no miracle that two colliding pieces of clay present conformal surfaces; they distend one another until it is so.[7]

For many of the reasons I have given earlier, this view seems no more tenable than that presented by the theory-firsters. Standards of demonstration, classes of objects, types of interactions, all differ in the two communities. Put this way, it becomes clear that the traditional opposition between theory and practice is misplaced. The point is not that theory is opposed to practice—there is a theoretical practice on an exact par with

6. "Whereas in the first phase Kaufmann had . . . tested one single theory in which he appeared to believe, namely, Abraham's theory, in the second phase his task was to discriminate between the various theories that had been suggested during the years 1904–1905. Yet, his experimental method was essentially the same as in his early experiments" (Hon 1985, 304). See also Hon, chap. 8, this volume. I first discussed these issues in Galison 1989, 1990b. These issues are discussed in greater depth in Galison n.d.

7. Recently, A. Pickering has dropped his emphasis of the ways in which theory stabilizes experiment and begun to speak of the "interactive stabilization of practices and beliefs" (1986). This represents a constructive step toward a more symmetric treatment of the experimental and theoretical communities but still stops short of granting each the quasi autonomy it often exhibits. Also in this paper Pickering argues that constructivism should distance itself from global holism and subscribe to a local holism, this latter (oxymoronic) term designates the linked nature of experiment and theory for a single experiment, not for a worldview. My argument is that even this is not sufficient; at a given time when theorists find themselves in the midst of great storms in their practices and beliefs, the quasi-autonomous experimentalists and instrument makers often find themselves, on the contrary, in the midst of calm weather. Conversely, at the moments of enormous dislocations in instrumental practices (development of the cloud chamber or the cyclotron), theorists saw nothing but continuity.

experimental practice—but rather that scientific work is composed of a whole variety of constraint clusters. The location and nature of these constraint clusters are changeable: in the cosmic-ray physics of the 1930s, instrument makers and experimenters were in many instances one and the same; in particle physics during the 1980s, instrument makers, computer programmers, and experimenters often drifted apart into quasi-autonomous communities. Unraveling the relationships between these different aspects of knowledge production is jointly a philosophical and historical task: there lies the changing nature of evidence and argument.

My view of the task of history and philosophy of science thus differs fundamentally from the legacy of the 1960s. That legacy prescribes inquiry along the lines followed so closely by Pickering and others: identify a leading theoretical problem solution (the paradigm) and look for the crisis, breakdown, and freestanding replacement to it. My understanding of the task undertaken by the new history and philosophy of science is rather different. Our present goal is to follow the activities of scientific subcultures in their differing contexts. And the way to do that is to understand the multiple constraints that instrument makers, theorists, technicians, and experimentalists face as they confront the technological, economic, physics, and manufacturing difficulties in making decisions about the assembly of scientific arguments.

Taken out of cultural context, provided with the dominant single-line narrative of theory, it is possible to see physics as fragmenting into cohesive chunks, isolated one from the other. Described this way the scientific enterprise has room for progress, truth, ontology, and realism only within a single "conceptual scheme." In this respect one particular strand of social constructivism, represented forcefully by Pickering, is utterly Kuhnian—and, I will argue next, entirely consonant with some of the oldest and deepest commitments of the Vienna Circle. Such a brief historical venture into the history of recent philosophy will, I hope, pay off by further underlining what I consider to be the spuriousness of adding the island empire view to a historical-philosophical commitment to contextualism. Said differently, I would argue that contextualism need not pass through a framework conception of science to relativism.[8]

8. Andrew Pickering has criticized Kuhn and Feyerabend for trying to hold the line at "frame relativism" (something like what I mean by framework relativism); Pickering thinks that even objectivity-within-a-frame is a lost cause ("there is no reason in the world to think that the line can be held at objectivity-within-a-frame"). See Pickering 1990a, 690. Elsewhere Pickering tries to defend a "local holism" between experiment and theory taken synchronically (within the conduct of an experiment), whereas I would argue that even within small-scale experimentation there are frequently quasi-independent clusters of instrumental, theoretical, and experimental practice.

5. Positivism, Antipositivism, and Constructivism

From their very first pronouncements Feyerabend, Kuhn, Hanson, and the other antipositivists of the 1960s defined themselves in opposition to the Vienna Circle, particularly to Rudolf Carnap. There is of course a rather striking circumstance that has usually been handled by invoking "irony": Kuhn's *Structure of Scientific Revolutions,* undoubtedly the best-known and most-influential work of antipositivism, appeared as a major contribution to the logical positivists' volume 2 of the *Encyclopedia of Unified Science.* One recent work put it this way:

> Yet one more irony: Kuhn (1970)—*The Structure of Scientific Revolutions*—which has proved so crucial in stimulating studies that undermined the notions of science central to the Vienna Circle, was originally published as part of . . . *Foundations of the Unity of Science.* The lists of editors and members of the organizational and advisory committees for this series, provided on the back of the title page . . . constitute a useful guide to who was who among the logical positivists. History moves in mysterious ways. (Harding 1986, 250)

History does take unexpected turns; here, I will argue, it did not.[9] On many grounds the Kuhnian antipositivism is only awkwardly extricable from positivism—especially Carnap's, from which it emerged. But let us step back to understand in somewhat more detail the links between the Unity of Science movement and the history of science.

Already in 1936, Charles Morris, the most active advocate of the Vienna Circle in America and an editor (with Carnap and Neurath) of the *Encyclopedia,* was keen to get George Sarton (one of the earliest and most vigorous boosters of the history of science) on the permanent committee of the movement.[10] Sarton's participation was probably encouraged by the growing closeness between Harvard and the Unity of Science movement, as Harvard president James Conant pressed for the first big Unity of Science meeting to be held at his university.[11] To Morris, Sarton offered the possibility of grounding the Unity of Science movement with a sympathetic history of science. It was an eminently sensible choice: Sarton was a

9. Since the time that this paper was presented (March 1990), a fine article by Reisch has appeared that also discusses the Kuhn-Carnap interaction. See George A. Reisch, "Did Kuhn Kill Logical Empiricism?" *Philosophy of Science* 58, no. 2 (June 1991): 264–77.

10. Morris to Neurath, 26 Jan. 1936, Unity of Science Movement Records, Department of Special Collections, University of Chicago Library, all rights reserved (hereafter abbreviated USMR), box 1, folder 15.

11. Morris to Neurath, 5 Jan. 1937, USMR, box 1, folder 16; Morris to Conant, 18 May 1937, USMR, box 1, folder 16—accepts Conant's invitation.

positivist (though more Comtean than logical). But in the end, Sarton declined for lack of time, disappointing Morris, who very much wanted the history of science in the second volume of the *Encyclopedia:* "you would have seen the whole larger significance of the history of science for a comprehensive science of science program."[12] In his place, Sarton offered the services of his assistant, I. Bernard Cohen. Cohen, too, eventually withdrew, and it was then that the task fell to a third Harvard historian of science, Thomas Kuhn.

In the literature of the philosophy of science Rudolf Carnap has played an irreplaceable role as the foil against which Kuhn is positioned. In the simplistic reading to which his texts are frequently subject. Carnap is depicted as an advocate of a naive foundationalism who thought theories could be built through induction on the basis of indubitable sense data. It is a difficult thesis to sustain, since from at least the early 1930s Carnap was a thoroughgoing conventionalist.[13] Far from opposing Kuhn's view, Carnap found himself entirely in sympathy with Kuhn's opposition to a monotonic progressivism based on a goal-*directed* movement of science toward truth. Nor should this surprise us. For decades Carnap had switched freely between different starting points for the construction of scientific knowledge, and he had held a deep suspicion toward the ultimate validity of any given theoretical representation.

More specifically, Carnap insisted that all scientific questions about the reality of entities were necessarily dependent on a linguistic framework; either this framework has a natural place for talk about a given entity or it does not. But Carnap added that

> it is a practical, not a theoretical question; it is the question of whether or not to accept the new linguistic forms. The acceptance cannot be judged as being either true or false because it is not an assertion. It can only be judged as being more or less expedient, fruitful, conducive to the aim for which the language is intended. Judgments of this kind supply the motivation for the decision of accepting or rejecting the kind of entities. (Carnap 1952, 219)

So important is this distinction between questions "internal" and "external" to a framework that Carnap quite uncharacteristically breaks with his usual style to insert "a brief historical remark." Questions challenging the framework from outside ("external questions") have no cognitive content, and this was fundamental to the Vienna Circle's earliest and most

12. Morris to Sarton, 14 Jan. 1939, USMR, box 1, folder 18; also see Morris to Neurath, 15 Dec. 1938, USMR, box 1, folder 17.

13. For more sophisticated readings of the early Carnap, see the excellent work of Richard Creath (1990, 1987), Friedman 1987, Richardson 1992.

characteristic writings. It was a view expressed not only in Carnap's own 1928 *Pseudoproblems in Philosophy* but also in *Positivism and Realism,* written by the leader of the Circle, Moritz Schlick, and perhaps most importantly by Ludwig Wittgenstein, who utterly and influentially rejected theses about the world's reality or irreality (Carnap 1952, 220). As Carnap emphasized in April 1960, some two years before Kuhn finished his *Structure,* the aspect of Kuhn's planned work that brought questions back to these linguistic frameworks appealed to him a great deal:

> I believe that the planned monograph will be a valuable contribution to the *Encyclopedia.* I am myself very much interested in the problems which you intend to deal with. . . . Among many other items I liked your emphasis on the new conceptual frameworks which are proposed in revolutions in science, and, on their basis, the posing of new questions, not only answers to old problems.[14]

Kuhn's own reaction to the appearance of his work in the *Encyclopedia* was a bit diffident, but never in his correspondence with Carnap or Morris was his hesitation grounded in the *Encyclopedia*'s objectives. Rather, Kuhn's concern centered on the declining influence of the journal in a world in which logical positivism could no longer carry the readers it had in the 1930s. As he put it to Morris, "Twenty years ago the circulation of the *Encyclopedia of Unified Science* would have been ideal for my purposes. Today, I am not nearly so clear that it will enable me to reach all of the audience I, perhaps optimistically, hope for."[15]

With the book typescript completed, Carnap took private reading notes on the manuscript in his horrendous private shorthand. The date is 27 April 1962, and it is quite evident that nothing in the completed manuscript disappointed him at the level of "truth" or "progress": "not 'we are coming closer to the truth,'" Carnap inscribed, "instead: we are improving an instrument." He continued:

> *I think:* So it is also the *task of the development of an inductive logic.* Not everyone is looking for the one perfect system, or at least a stepwise approach zeroing in towards it; instead we really have an effort to improve a solution. . . . So it was with the [basic?[16]] system which was an important step. And now I am looking for others. And I see already for languages with

14. Carnap to Kuhn, 12 Apr. 1960, 088-47-08, University of Pittsburgh Carnap Collection (hereafter abbreviated UPCC).

15. Kuhn to Charles Morris, 13 Oct. 1960, 088-47-07, UPCC.

16. The word immediately preceding "system" is illegible, according to Gerald Heverly, who is the world expert on reading Carnap's private shorthand. Richard Creath has suggested to me that "basic" might be this missing word, as that would fit with Carnap's concerns at this time.

> one relation . . . a first, simple solution; and a better solution; and I have ideas for possible further refinements.[17]

Just as in the evolution of species, Carnap adds, it is not evident that the third in a series will be in the direction of the second; it is precisely like this in the evolution of theories.

To Kuhn, Carnap was reassuring; he was "especially gratified" that he and Charles Morris could publish the *Structure* in the *Encyclopedia.* "I am convinced," Carnap wrote Kuhn the day after taking his private notes,

> that your ideas will be very stimulating for all those who are interested in the nature of scientific theories and especially the causes and forms of their changes. I found very illuminating the parallel you draw with Darwinian evolution: just as Darwin gave up the earlier idea that the evolution was directed towards [a] predetermined goal, men as the perfect organism, and saw it as a process of improvement in natural selection, you emphasize that the development of theories is not directed toward the perfect true theory, but is a process of improvement of an instrument. In my own work on inductive logic in recent years I have come to a similar idea: that my work and that of a few friends in the step for step solution of problems should not be regarded as leading to "the ideal system," but rather as a step for step improvement of an instrument. Before I read your manuscript I would not have put it in just those words. But your formulations and clarifications by examples and also your analogy with Darwin's theory helped me to see clearer what I had in mind.[18]

In the above remarks by Carnap about Kuhn's work, it is clear that Carnap is espousing an instrumentalist stance not only with respect to scientific theories but reflexively, toward his own ideas on confirmation as well. From these remarks there can be little doubt that he held no objection to three aspects of *The Structure of Scientific Revolutions:* Kuhn's endorsement of the schema of "new conceptual frameworks," Kuhn's rejection of any ultimate truth, and Kuhn's denial of linear progress.

Kuhn's break with Carnap occurs on other levels: Kuhn's insistence on the role of textbooks as providers of exemplary problem solutions; his application of Wittgenstein's critique of rule governance in the application of these exemplars; his interest in the crisis that attends the collapse of one paradigm and its succession by another; and the strong focus on *theory* rather than *observation* as the starting point of scientific activity. But the division of science into fragments (Carnap's linguistic frameworks, Kuhn's paradigms), each with its own ontology, remained intact.

17. Carnap notes, 27 Apr. 1962, RC 082-03-01:ir/2, UPCC.
18. Carnap to Kuhn, 28 Apr. 1962, 088-47-01, UPCC.

Equally important, for our purposes, is that the picture of island empires of science carried with it a powerful, and extremely fruitful, historiography. It structured a style of narration that has been enormously stimulating in any number of arenas. Among sociologists of scientific knowledge, the Kuhnian picture took hold at a deep level. The notion of Kuhnian incommensurability explicitly shaped H. M. Collins and T. J. Pinch's examination of paranormal science.[19] In her fine historical-philosophical study, *Crystals, Fabrics, and Fields* (1976), Donna Haraway exploited the Kuhnian model to organize the history of developmental biology into two periods, the mechanistic and the organicist; each carried metaphors and problem solutions constitutive of a paradigm. One linguistic framework yields to another, and a new set of experimental entities is ushered into place. In Marc de Mey's work of 1981 on Harvey's discovery of circulation of the blood, the Kuhnian picture of a paradigm shift is used to elucidate the transition that takes place from the anatomy of Galen to that of Harvey. Throughout this literature, it is repeatedly emphasized that it is theory that drives change. This is also the narrative structure of Andrew Pickering's work on charm and color in high-energy physics.[20] Such concerns motivate Pickering in his *Constructing Quarks* to drive a wedge of Kuhnian incommensurability between the self-sufficient island of "old physics" and another of "new physics."

> Kuhn's argument was that if scientific knowledge were a cultural product then different scientific communities . . . would inhabit different worlds. . . . One striking consequence of this hypothesis is that the theories proper to different worlds would be immune to the kind of testing envisaged in the "scientist's account"; they would be, in philosophical language, incommensurable. The reason for this is that each theory would appear tenable in its own phenomenal domain, but false or irrelevant outside it. There would be no extra-cultural facts against which the empirical adequacy of different theories could be impartially measured. . . . I have analyzed the emergence of the new physics against the background of the old, and manifestations of incommensurability are stamped across the transition between the two regimes. The old and new physics constituted, in Kuhn's sense, distinct and disjoint worlds. (Pickering 1984b, 407, 409)

19. See Collins and Pinch 1982, chap. 1 and the introduction for discussion of Kuhn's incommensurability and its function within a sociology of science.

20. "It is clear that as theoretical models become more securely tied within the whole network of theory they can play an increasing role in the 'stabilisation' of data. Furthermore it can, I believe, be argued that this role extends far beyond that of simply removing the ambiguity inherent in data of poor statistics, ultimately to the level of stabilising experimental techniques, methods and procedures" (Pickering 1981c, 130–31). In support of the view that theory stabilizes data and that this is what produces a coherent network of theory and experiment, Pickering cites (131) S. Shapin, H. Collins, T. Pinch, and others.

What in Kuhn's writing appears as a stricture against "pure sense datum terms" outside theory now appears in Pickering as a stricture against "extra-cultural facts." Both views underline the impossibility of a phenomenal domain that crosses the theoretical divide.

Philosophically, each of these framework narratives supports a small-scale relativism on the grounds that the isolated island empires exist in an impassible ocean. But it may well be that the impassability of the seas appears as such only because the experimental and instrumental navigators who daily cross them are given no place in the history. For the men and women who built the spark chambers used in the discovery of the psi and tau at the Stanford Linear Accelerator Center's (SLAC) colliding-beam facility (Stanford Positron-Electron Asymmetric Rings [SPEAR]) 1974 did not mark a great year of revolution; it was a time of virtually complete continuity with efforts begun over a decade earlier. Indeed, for most of the SLAC experimentalists, the "November Revolution" brought more prominence—but their commitment to the methods and strategies of colliding-beam physics, resonance hunting, and the search for detectors that could measure the entire volume around a collision was long-standing and formed the heart of the old physics and continued long into the new.

Experiments that came from SPEAR certainly made contact with both the experimentalists' "old" and their "new" worlds of physics. Does that mean that experimental and instrumental practices were "extracultural" or "extratheoretical"? Of course not. The colliding-beam facilities were deeply embedded in the technical and scientific cultures of their time, but they are not neatly classifiable on one or the other side of the 1974 gulf. If one takes seriously the quasi autonomy of these different subcultures, the history or, more precisely, the histories of the period look very different.[21]

In the standard account it is very apparent that the dominance of theory is important: in word choice, narrative structure, and in the silences. "The penetration of experimental practice by [the theory of quantum chromodynamics] constituted the last act in the establishment of the new physics" (Pickering 1984b, 272). The trope of classical theater, with its concentrated narrative, linear development, focus, and resolution is a *choice* of how to do a certain kind of science study. Breaking the history of particle physics incommensurably in 1974 has many benefits, but it

21. Pickering's account of the discovery of the tau at SPEAR emphasizes the theorists' crisis (the "mini-R-crisis") and spends relatively little time following those SLAC experimentalists who were not particularly interested in what the theorists had to say about the tau. This is perfectly understandable: Pickering's goal in *Constructing Quarks* is to refute the view that phenomena *determine* or fix theoretical structures. But for more on the tau seen from the experimental and instrumental side, see Treitel 1987, 1986.

comes at a price: the exclusion of other voices, the voices of those experimentalists whose practices did not mutate, those technicians whose machines did not change, those Monte Carlo programmers whose goals were constant—along with the broader domain of culture in which techniques in particle physics are shared by those in industry and in the military.

6. The Trading Zone

Carnap's modernist epic, *Der logische Aufbau der Welt,* is usually translated as *The Logical Structure of the World;* a better translation would be *The Logical Construction of the World.*[22] And it was a construction. From elementary observations one builds up to high-level theories; the work served as a model for the kind of fact-grounded reasoning that would sustain the scientificity of the Vienna Circle's program. Together with allied movements, such as Bridgman's operationalism, the positivists shaped the conduct of the history of science by focusing case histories on what they considered the essential feature of science: its grounding in observation. The corresponding antipositivist text, *The Structure of Scientific Revolutions,* was, as is evident from the reading many historians and sociologists gave it, understood to have inverted this hierarchical relation. As one commentator put it, "Operations and manipulations, [Kuhn] feels, are determined by the paradigm and nothing could be practically done in a laboratory without one. A pure observation language as the basis of science exactly inverts the order of things" (Haraway 1976, 7–8). Whereas the positivist "order of things" sets observation as the ground and theory as the superstructure built upon it, the antipositivist order renders the paradigm as basic, and the observations only exist insofar as they emerge within the paradigm. Ontology, epistemology, and nomology are embodied within the paradigm; the solved problem is prior to its observational instantiations.

Each hierarchical order dictates a historiography: Conant's case studies of experimental science are privileged for the positivists; the theory-first studies are favored by the antipositivists. (It is in this manner that one should understand Hanson's [1963] study arguing that the positron appeared on photographs but was not "seen" because of theoretical predilections.) The antipositivists taught both the historical and the philosophical community a crucial lesson: observations do not follow a simply progressive, unbroken trajectory. But once this is acknowledged, the antipositivists march in a direction I am loath to follow: they assume that

22. For more on Carnap's *Aufbau* in the wider context of cultural modernism, see Galison 1990a, 1993.

the breaks in experimentation will necessarily occur synchronically with breaks in theory. It is as if brick masons stacked their bricks, instead of intercalating them—a break at one level of the wall would cascade catastrophically through the whole structure.[23] The received view of paradigms is just this: experiment and theory are "stack periodized," if you will. As one sympathetic commentator put it: "In actual historical cases, as a new paradigm arises so there is an associated transformation of the entire conceptual fabric. What has to be evaluated are two alternative frameworks of discourse and activity. There is a reconstruction of the whole pattern of both" (Barnes 1982, 67).

An alternative vision of history renounces the assumption that the conceptual fabric tears as one. In history, more broadly conceived than the history of science, such an alternative is by now so deeply embedded in historical practice that it goes without saying. But it was not always so. For many of us it is in living memory that arguments were fought for a history of women, a history of workers, a history of illiterate societies that would not periodize and structure themselves through the history of high politics and war.

Here, however, lies a paradox—or at least an apparent one. If the histories of experiment, theory, and even instrumentation are truly distinct sociologically (with internal exchange of reprints, their own summer schools and training procedures), then there is good reason to consider them to be distinct communities. Furthermore, if it is quite common for these different communities to maintain different views about the status of what objects there are, how we learn about them, and how they interact, then no quasi-mechanical model of intercalated bricks will solve the problem of understanding the felt continuity within the scientific community. In other words, if the different subcultures are themselves incommensurable one with the other, the incommensurability problem only gets worse.

For the Vienna Circle and their contacts (Carnap, Wittgenstein) and for the antipositivists (including Kuhn, Feyerabend, Hanson), Gestalt psychology provided a guiding metaphor for the understanding of conceptual change. Perception, after all, was considered to be, not only metaphorically, but also literally at the center of observation. But when it comes to the encounter of different symbolic systems and the problems of translation between frameworks, paradigms, conceptual schemes, and the like, a focus on the psychology of individual perception strikes me as an extremely awkward point of analytical reference. What actually happens when different cultures encounter one another and need to communi-

23. On the problem of co-periodization of the different subcultures of physics see Galison 1987, 1988a.

cate? Quite frequently—as a whole field of anthropological linguists has shown—what results is not a full-scale oscillation back and forth between different natural languages. Instead, those in contact forge foreigner talk, pidgins, and creoles—languages that colliding groups fabricate on the boundary between two cultures.[24]

The processes by which these contact languages form are diverse, but in general reduced and regularized syntax may be borrowed from one group and an abbreviated lexicon from the other. Sometimes a pidgin will relexify: key words from one language will be substituted for words in another. As these trading languages form and grow, they facilitate the exchange of goods and actions. Above all, a trading language is characterized by a *local* agreement about the use of terms, even where there is global disagreement about the corresponding word usage from the two natural languages.

Linguists find most of these pidgins and creoles clustered around coastal islands and other natural boundary points, for it is there that cultures encounter one another. The important point here is that the trading language, and the exchanges that accompany and support it, can exist in extraordinary complexity despite the absence of any consonance of meaning between the two cultures.

Here lies a whole domain of investigation left out of a history predicated on island empires condemned to perpetual isolation. How do the high-energy physicists working on radiation hardening for Superconducting Supercollider detectors speak to the avionics personnel from Livermore and Lockheed? Can we understand Monte Carlo programs and computer simulations more generally as providing a trading language between experimenters and theorists? When J. D. Bjorken and Sidney Drell wrote their famous text in quantum field theory, they actually produced two separate volumes. One was a text in quantum field theory for theorists; the other, self-consciously and explicitly, was addressed to experimentalists. In effect, by eliminating exceptions and keeping the arguments focused on experimental phenomena, they produced a pidginized quantum field theory.

As in culture contact, we expect friction as the different symbolic systems interact; understanding the heat generated at such contact points helps explain who and what the systems actually carve out of the cultural world. For example, it is worth attending to the ways in which particle theorist Sheldon Glashow presents his relation to mathematics in his preface to a book by Howard Georgi, *Lie Algebras in Particle Physics:* "Here, the reader will find no narcissistic cry of mathematics for mathematics'

24. This analysis of scientific pidgins and "trading zones" is discussed far more extensively in Galison n.d.

sake. To the contrary, we physicist-chauvinist-pigs regard mathematics as the mere handmaiden of physics. Flights of mathematical fancy are tolerated only insofar as they are tethered to observable physical phenomena" (Glashow 1982, xvii). Georgi reduces proofs to a minimum, and the style of the book manifestly violates the theorem, lemma, proof sequence of mathematical texts. Subjects are omitted (such as noncompact groups) that would be a centerpiece of any mathematical exposition of the subject, and examples occasionally must stand in for full proofs. In a variety of ways a pidgin has been created from the field of Lie algebras, and the "handmaiden" speaks a new tongue. By seizing portions of the mathematics, truncating its internal structure, and renaming many elements within it, the physicists have created a language at the boundary of the two domains. Nor was Glashow's attitude toward mathematics unique at Harvard in this period. Sidney Coleman's steady emphasis from the mid-sixties into the 1980s was away from "untethered" mathematical structure. "This lecture will be devoted to explaining functional integration and its connection with field theory. Our approach will be, from a mathematical viewpoint, despicable. Nothing will be proved; everything will be done by analogy, formal manipulation of ill-defined (and sometimes divergent) quantities, and handwaving. I hope that this will at least give you an idea of what is going on and teach you to manipulate functional integrals; if you want a deeper understanding, you must go elsewhere" (Coleman 1985, 145). Pidginized mathematics is locally defined for local concerns: while it offers a medium of exchange between mathematics and physics, it is ripped helter-skelter from the full body of mathematical technique and theory in which it originally had meaning. Now, exceptions and rigor removed, it serves other purposes.

Importantly, the physicists split the mathematical fragments from the place they occupy within mathematics. Nothing in the mathematics itself, Georgi insists, should seduce the physicist to a conclusion more pleasing to the mathematician than satisfactory to the physicist. As a cautionary tale, Georgi points to the example of one particular mathematical group, *SU(3)*, the most successful application of group structure to physics during the 1960s. The application of this technique involved approximating three quark masses by zero, a move justified by the smallness of their masses and the difference in masses with respect to the "natural" scale of mass associated with the theory. With the discovery of a fourth quark, the natural mathematical generalization would have been from *SU(3)* to *SU(4)*. Georgi not only disparages this move, he elevates his disparagement to a general lesson:

> There is a more general moral here. A symmetry principle should not be an end in itself. Sometimes the physics of a problem is so complicated that sym-

> metry arguments are the only practical means of extracting information about the system. Then, by all means use them. But, do not stop looking for an explicit dynamical scheme that makes more detailed calculation possible. Symmetry is a tool that should be used to determine the underlying dynamics, which must in turn explain the success (or failure) of the symmetry arguments. Group theory is a useful technique, but it is no substitute for physics. (1982, 155)

An utterly different attitude toward mathematics pervades the string theorists of the 1980s and 1990s; instead of "despising" "untethered" and "narcissistic" mathematics, physicists and mathematicians have worked together to create a language with sufficient structure and expanse for people to accomplish more than merely piecemeal coordination. In the literature on contact languages such expanded pidgins are dubbed "creoles," languages created frequently as children grow up within a pidgin and press its expressive capacity far beyond their parents' trading language. Similar phenomena undoubtedly occur as fields come together: biology and physics; biology and chemistry; chemistry and physics. Within each, participants carve out a trading zone, an arena of interaction marked not only by joint papers, conferences, and faculty appointments but by a realignment in the direction of research. The production of formal pidgins and creoles occurs in a myriad of spheres, recognizable, for instance, in textbook form by titles such as *Statistics for Social Scientists, Electronics for Physicists, Topology for Physicists.* Perhaps these ideas might be useful outside the natural sciences: Theodore Porter argues that quantification itself can usefully be considered a trading zone within the social sciences (1990, 1992a, 1992b).

We have come this far: physics cannot be considered a homogeneous enterprise, unified either from below by observation or from above by theory. Instead, diverse constraint clusters move according to largely autonomous tempos, with their own breaks and continuities. But the laboratory itself is the site of an immensely complex set of trading relationships, and it is here in the everyday exchange between tabletop and blackboard that contact languages play an ever-present role in crossing the boundaries thought forbidden by theorists of incommensurability and in linking their parent subcultures to the culture around them.

7. Context, Constraint, and Construction

Let me summarize the view presented here, in four steps.

Argument I. I am against any account of the standards of argumentation that is socially reductionist, for two reasons. First, for the actors themselves, movement between "social" and "cognitive" constraints is fluid;

and second, because the constraints interact: they are not artificially separable. We can think of a new theory or laboratory move as the simultaneous satisfaction of multiple constraints, instead of privileging an old-fashioned internal or external story of scientific work.

Argument II. Historiographically, it seems to me utterly unwarranted to defend the view that theoretical and experimental practices change in lockstep. This is a purely contingent matter, and the case against a stack periodization (and for an intercalated one) is grounded in the numerous studies made by many people in the history, philosophy, and sociology of science over the last decade. From air pumps to bubble chambers, from eighteenth-century electrical experiments to parity violation, the dynamics of the workbench conform only piecewise to theoretical ideas. Indeed, in retrospect, the defense of the stack periodization scheme relies on an artifact of the history of philosophy of science and its relation to a fledgling history of science: antipositivists, attacking sense-data reductionism, argued that theory came first and historians dutifully ferreted out the desired case studies. What we have learned is that the subcultures of physics all have their breaks and continuities; the strength of the whole derives, as in a cable, from the intercalated finite strands of these various finite traditions.

Argument III. On the basis of a history of science predicated on stack periodization, and a philosophy of science built on linguistic holism, the antipositivists and a certain sector of social constructivists have defended what one might call a framework relativism, science partitioned into island empires. In some quarters the view that standards of demonstration have historically specific origins has been considered an argument for framework relativism. After all, this line goes, if there is no universal model of scientific argumentation, then the only criteria of demonstration must be internal to each of a myriad of frameworks. But I have argued here that a thoroughgoing contextualism works in exactly the opposite direction. In a world in which the solution to scientific questions is often simultaneously the solution to problems in the wider culture, it becomes more, not less, difficult to see scientific knowledge as carved neatly into isolated frameworks. Contextualism works against framework relativism.

Argument IV. Again reacting against the positivists' defense of finite neutral observation languages, the antipositivists have handed down a picture of language use in science that poorly fits scientific practice. We do not need a view of a finite, unchanging observation language. Nor do we want a holistic account in which paradigms, frameworks, or conceptual schemes embody whole languages, and these whole languages confront one another in unison across paradigmatic divides. Instead, our attention should be local and historical, focused on the foreigner talk, pidgins, and creoles that bind the subcultures of a discipline to one another and to the larger culture in which they are embedded. The relation of theory to experiment or of theory to mathematics or of theory to instrumentation is given through the local coordination of action and belief, even in the absence of global agreement.

My real quarrel, then, is with the central metaphor of science as an amalgam of unrelated, internally homogenized frameworks handed down from Carnap and the positivists to Kuhn and the antipositivists, and then on to one strand of constructivism. Instead I see science, or at least physics, as a collection of finite bits, put together piecewise—and it is just this disorder that holds it together. Carpenters know that a homogeneous board is far less strong than a laminated one: where one layer cracks, another holds. A cable gains its strength not by passing a single strand down its center but by the twisted, mutual support of thousands of finite fibers with differing endpoints.

By according the different sectors of scientific activity partial autonomy, the picture of even single disciplines like particle physics is immensely complexified. Adding the cultural embedding seems at first to militate for a fragile picture of science, in which at every moment a scientific discipline threatens to tumble into isolated worlds. But this has it backward. Far from obliterating the interconnectedness of science, the intercalation of a myriad of scientific subcultures is precisely what lies behind the strength and coherence of the endeavor.

Acknowledgments

I would like to thank Jed Buchwald, Richard Creath, Ian Hacking, Caroline A. Jones, Timothy Lenoir, Betty Smocovitis, and David Stump for helpful conversations and Nani Clow and Jean Titilah for manuscript preparation. Gerald Heverly at the Pittsburgh Archives for Scientific Philosophy (PASP) kindly assisted both in locating archival material and in transcribing it. Material from the PASP Carnap Collection is quoted by permission of the University of Pittsburgh. All rights reserved. Material from the Unity of Science Movement Records is quoted by permission of the Department of Special Collections, University of Chicago (UC) Library. All rights reserved. Support for this work came in part from the Andrew W. Mellon Foundation, the Center for Advanced Study in the Behavioral Sciences, and the Presidential Young Investigator Award, National Science Foundation grant 8911508.

THREE

BEYOND CONSTRAINT: THE TEMPORALITY OF PRACTICE AND THE HISTORICITY OF KNOWLEDGE

Andrew Pickering

ONCE UPON A TIME it went without saying that thinking about science meant contemplating scientific knowledge.[1] Nowadays, few people would agree. It is clear that to make sense of science one has to think about both scientific knowledge *and* the practice with which it engages. This realization lies behind the shift in history, philosophy, and

1. Given their common emphasis on "constraint," readers might be puzzled about the relationship between Peter Galison's contribution to this volume and my own. I presented this paper (except for minor subsequent revisions such as this note) at the conference held in Toronto in March 1990 on which the book is based. Peter Galison gave a paper there on an unrelated topic (a chapter from his next book) but made some remarks in response to mine which have, I suppose, developed over time into his "Context and Constraints." He does not react there to the points I made in Toronto but multiplies his usage of "constraint" and seeks to assimilate my work, especially *Constructing Quarks,* to his "theory-first/island-empires" reading of Carnap and Kuhn. This assimilation proceeds by undocumented assertions concerning my work, and I hope that the reader will be suspicious of it. It is quite misleading to say, for example, "[that] the dominance of theory is important in [my] account is everywhere apparent" (Galison draft, 21). It is likewise misleading to make remarks such as "they assume that the breaks in experimentation will necessarily occur synchronically with breaks in theory" (22) and "If the histories of experiment, theory, and even instrumentation are truly distinct sociologically . . . then there is good reason to consider them to be distinct communities" (23) as if they applied critically to my book. In fact, I made it clear there that experimenters and theorists constituted different communities and that I thought that the history of particle physics should be understood in terms of the making and breaking of symbiotic relations between very diverse experimental and theoretical research traditions. And I did not suggest that in such processes theorists necessarily hold the initiative. There is no quarrel between Galison and myself on this issue, and I am happy to remark that in his recent work he has explored the linkages between some subcommunities very interestingly. It remains the fact that the "new physics" of the 1980s—experimental, instrumental, and theoretical—did turn out to be organized in very important ways around the quark–gauge theory worldview, but this is a fact of history, and complaints about it should be directed to the physicists, not to me.

sociology of science toward studying scientific practice that has become a self-conscious movement during the past decade.[2] Nevertheless, I have the feeling that we have yet to get our money's worth from this practical turn. The problem is that the old vocabularies and problematics of science-as-knowledge continue to haunt us. We document practice, but for want of a proper analytic framework, we fail to foreground the new and important insights to be gained there. Instead we simply reinforce the positions we thought we had abandoned. One symptom of this malaise is, I think, the prevalence of the concept of "constraint" in discussions of practice. This concept, I believe, marks one of the points where the vocabulary of science-as-knowledge runs out, and where thought accordingly stops. To get out of this impasse, I will try here to come to grips with the genuine feature of doing science that constraint-talk names: its difficulty. Not coincidentally, to do so I find it necessary to develop an analytic framework for thinking about the temporality of practice—its constitutive unfolding in time—an aspect of science that the old vocabulary of science-as-knowledge definitively erases.[3] My arguments about temporality are, in a sense, technical ones, designed just to help us see and think about this central aspect of science. But I also want to develop an epistemological argument, to illustrate what might be achieved in the analysis of practice. I will try to bring new life to a debate which, in its classic form, has bored everyone to death: the debate over the objectivity, or lack of it, of scientific knowledge. My suggestion is that a serious analysis of the temporality of practice points to a kind of historicist understanding of scientific knowledge that cuts across the old clichés of objectivism and cultural relativism in interesting ways.

I can begin biographically, with my book *Constructing Quarks* (1984b). Its reception induced in me a profound allergic reaction to the word "constraint." Suddenly, "constraint" seemed to be everywhere: the first essay review of the book was entitled "Constraints on Construction" (Gingras and Schweber 1986), Peter Galison devoted an entire chapter of his *How Experiments End* to "constraint" and made me the villain of his discussion (1987, chap. 5), two philosophers from St. Louis later savaged

2. Probably the precipitating factor was the appearance of the first ethnographic studies of laboratory life: Latour and Woolgar 1979; Fleck 1979; Knorr-Cetina 1981; Collins 1985, chap. 3; and Lynch 1985. Though the two are connected, it is useful, I think (Pickering 1992), to distinguish between studies of practice and the sociology of scientific knowledge tradition, which, at its inception at least, lay largely within the science-as-knowledge orbit: Barnes 1974; Bloor 1976; Shapin 1982.

3. The classic instrument for this erasure is the logical-empiricist distinction between the contexts of discovery and justification in science. In terms of this crude dichotomy, the context of discovery is where practice unfolds in time but the context of justification is where the knowledge is.

me in similar terms (Roth and Barrett 1990; my reply is Pickering 1990a), and so on. Looking at things a bit more dispassionately, I realized that I was not alone in getting the constraint treatment; anyone claiming to be a constructivist, relativist, or sociologist of scientific knowledge came in for the same.[4]

It is instructive to ponder this outburst of constraint-talk in the second half of the 1980s. The thought behind it is, I think, this. People like me are held to want to reduce science to the play of some purely social variable, usually "interest." This is then held to be equivalent to the belief that interest overrules all else—in other words, scientists just say whatever they wish about nature. Constructivism is thus understood to assert not just a form of cultural relativism but something much more specific, a self-fulfilling wish- or desire-relativism. And through this series of displacements, the critics of constructivism put themselves in a position to say something terribly sensible as if it were news. Scientific practice is difficult, they say sagely, to much nodding of heads; scientists cannot just believe what they like, and we can show it. There follows a list of examples in which scientists emerged from struggles with the world believing something other than what they had originally hoped or expected. These struggles are then glossed into the language of constraint, which sounds impressively technical, and constructivism and relativism have been safely disposed of.

Now, to speak for myself, I could protest that since a single flirtation in my youth (Pickering 1981c), I have deliberately avoided any species of interest-talk. I could also protest that I have never anywhere denied that doing science is indeed hard work, nor have I spoken in defense of wish-relativism. But, just to be agreeable, I want to make a concession instead. It is that in *Constructing Quarks,* and in my other writings of the time, I did not deal in any systematic way with *why* scientific practice is difficult.

4. Even though the language of constraint is not uniformly foregrounded, the common strategy for reproving constructivists is continued in Rudwick 1985, Richards 1987, Giere 1988, and Hull 1988. At the Toronto meeting where this paper was first presented, Yves Gingras presented a paper entitled "Choosing and Constructing a Measuring Apparatus: Strategies and Constraints"; in the abstract to his contribution, Simon Schaffer promised to show how "the knowledge produced in these inquiries into the instruments itself generates knowledge in and constraints on the experiment proper"; and the organizers referred centrally to "constraint" in three of their "questions" circulated in advance to contributors. Since writing this paper in March 1990 I have learned more about the origins of constraint-talk. One of the most powerful sources is, as Galison (chap. 2, this volume) proclaims, the Durkheimian tradition in sociology. Pickering 1993 is a critique of traditional sociology, Durkheimian and otherwise, and contains further critical discussion of the language of "limits" and "constraints." I had better confess that my own writing has not always been free from constraint: see Pickering 1981a.

To be honest, I did not then know how to do it. However, to stop being agreeable, I want to say that the critics of constructivism were in the same boat and, worse, they are still in it. I think the anticonstructivist, antirelativist rhetoric of constraint gets one nowhere in understanding what the difficulties of scientific practice are. Further, the rhetoric functions as a kind of placebo, just the kind of block to inquiry that marks the exhaustion of the vocabulary of science-as-knowledge in the face of aspects of scientific practice that have hitherto been ignored.

Before I take the discussion further it will help, I think, to make a distinction between two species of constraint-talk. We can call these species "traditional" and "postmodern," and Martin Rudwick's defense of the last chapter of *The Great Devonian Controversy* (1985) can serve to exemplify the former. Replying to the constructivist critique of his commentary on how the controversy was closed, he had this to say:

> The fundamental issue, however, is whether the cumulative [empirical] base effectively *constrained* the actors . . . towards the (almost) consensual resolution of the controversy. My conclusion that it did . . . demonstrated to Harry Collins that I had "not quite managed to go native." He, Trevor Pinch and . . . Roy Porter all commented on the problem posed by the two lightweight or "marginal" actors (Weaver and Williams) who obstinately declined to join the consensus. . . . [But as] his criticism of Weaver shows, the judgment that Williams was *wrong* (or at least unwise) to rely on such an area is a judgement based on tacit communal standards to which Williams himself claimed to adhere. (1988, 157)

What is happening here is that Rudwick is glossing his own admirable account of science-as-practice in terms of the tired vocabulary of the traditional philosophy of science-as-knowledge, of a "cumulative [empirical] base" and "communal standards" vaguely understood as "constraints." Here, then, is a clear instance of the general problem. Historians, philosophers, and sociologists of science have started seriously to examine practice, but the understandings that emerge continue to be couched in quite inadequate vocabularies inherited from the days when it was permitted to talk about scientific knowledge as though practice were irrelevant. I have criticized this traditional species of constraint-talk elsewhere, and for the remainder of this essay I want to concentrate on the second, postmodern, variant, which is exemplified, for me, in Peter Galison's analyses.[5]

5. Traditional constraint-talk is simply one variant of the logical-empiricist appeal to epistemic rules or norms as controls upon the agency of scientists. In Pickering 1991 I argue that one can understand the extension of scientific culture without such appeal, and in Pick-

Galison's *How Experiments End* (1987) has many similarities with *The Great Devonian.* It combines a fascinating set of studies of scientific practice with a commentary designed to refute wish-relativism (imputed to me and my ancestors) through a vocabulary of constraint. But while Rudwick locates constraint in the old categories of "evidence" and "standards," Galison finds them somewhere new and interesting. Much of chapter 5 of *How Experiments End* is devoted to a discussion of three categories of constraint that Galison, following Fernand Braudel, calls long-, middle-, and short-term constraints, and for Galison, these constraints relate to substantive elements of scientific culture. An example of a long-term constraint would be the "metaphysical commitments" of scientists to principles like the conservation of energy (Galison 1987, 246); a short-term constraint might be the scientist's "faith in a type of [experimental] device—even in a particular piece of apparatus" (254), the faith that this bubble chamber is working now, say. Galison's idea, then, is that commitments to the conservation of energy, or to this bubble chamber today, constrain the arguments that scientists can produce and hence structure knowledge production, possibly even against their own wishes or desires. These substantive constraints are the reasons why scientists cannot believe just whatever they like.[6]

Here I can make two positive comments. First, under his rubric of constraint, Galison directs attention to what I call the *topology* of scientific culture. His lists of constraints exemplify the patchy, interrupted, and heterogeneous nature of scientific culture (this is the sense in which his approach can be described as postmodern). They provide a partial map of cultural landmarks, of a space in which all sorts of moves can be made, all sorts of configurations constructed. To see scientists as moving around in a kaleidoscopic culture is, I am sure, a major step forward in the analysis of scientific practice, a step that just cannot be made within the unitary and monolithic images of scientific culture that feature in traditional discussions of science. Second, Galison is also right, I think, in associating the difficulty of scientific practice somehow with that of moving around in a complex cultural space. I still think, though, that his articulation of this difficulty falters when it adopts the language of constraint.

What does Galison tell us about what he means by "constraint"? Not much, actually. His most explicit statement is: "I want to use the notion

---

ering and Stephanides 1992 we argue that even explicitly formulated epistemic rules are subject to emergent and discontinuous transformation in practice. Arguments concerning the engagement of epistemic rules with practice, and the vocabulary of "constraints" and "limits" in general, are developed at greater length in Pickering 1995.

6. Galison 1988b and chapter 2, this volume, elaborates upon the neo-Durkheimianism of Galison 1987, but nothing in the later writings constitutes a response to what follows.

of constraint the way historians often do—to designate obstacles that while restrictive are not absolutely rigid" (1987, 257). From this I understand the idea to be that cultural elements can be bent around a bit in scientific practice, but not too much, and that this inflexibility is what inhibits the free play of interests in knowledge production. Constraints somehow set limits to and *contain* practice, an idea expressed more graphically in Braudel's discussion, which Galison cites, of "the difficulties of breaking out of certain . . . prisons of the *longue durée*" (246 n. 3). This sequence of displacements—difficulty to constraint to restriction to the prison—is, as far as I can make out, characteristic of a whole genre of thought on the nature of cultural practice.[7] And, not to beat about the bush, I think it is a mistake. Scientific practice is difficult—right—but the move to prisons and restrictions is a red herring, for two reasons. First, it encourages us to inspect cultural elements for some property that limits practice; second, more specifically, it encourages us to think that this property is already and enduringly there in the elements in question. In contrast, I am of the opinion that cultural elements have no such inherent cagelike properties, and that whatever obstacles do indeed arise in practice are not "already there" to begin with; instead they genuinely emerge in time. To see this, we have to do what has proved just about impossible in history, philosophy, and sociology of science. We have to take the temporality of practice seriously—meaning we have to try to come to grips with and find a framework for the analysis of practice as it happens.

1. The Temporality of Practice

We need an example to think about, and I will discuss the history of a program of tabletop experiments in modern physics. The experiments in question are the searches for isolated quarks conducted by the Italian physicist Giacomo Morpurgo between 1965 and 1980 (Pickering 1981b, 1989b, 1990a). Though there is much more to say on the topic, the tem-

7. Gingras and Schweber likewise appeal to the "rigidity" of culture in their analysis of "Constraints on Construction": "Pickering's frequent statements about the possibility of alternative theories are based on the assumption that it is *easy* to construct a theory to fit any set of data . . . we believe that the rigidity of the network is much greater than Pickering allows, and this rigidity makes it difficult to construct alternative theories" (1986, 379–80). Galison cites Carlo Ginzburg's *The Cheese and the Worms: The Cosmos of a Sixteenth-Century Miller* (1980) with approval (1987, 255); there, in a passage that Galison does not quote, one learns: "In the eyes of his fellows, Menocchio [the miller] was a man somewhat different from others. But this distinctiveness had very definite limits. As with language, culture offers to the individual a horizon of latent possibilities—a flexible and invisible cage in which he can exercise his own conditional liberty" (xii). I think one should start worrying when one finds oneself talking about invisible rubber cages with "very definite limits."

porality of Morpurgo's practice can be immediately highlighted by noting its orientation toward future *goals;* Morpurgo continually aimed at the creation of a cultural configuration that did not presently exist. As I have discussed it before, Morpurgo's goal in each of the experiments that made up his program was to bring together three disparate cultural elements: a material procedure (assembling and running a piece of apparatus), an interpretive model (a theoretical understanding of how the apparatus functioned), and a phenomenal model (a theoretical understanding of the phenomenon under investigation). In Morpurgo's case, the material procedure was a scaled-up version of the oil-drop experiment with which Millikan had demonstrated the quantization of electric charge in units of the charge on the electron, *e*. The interpretive model was derived from the laws of classical electrostatics. And, actually, Morpurgo was prepared to entertain just two phenomenal models: either all of the charge in the universe was quantized in units of *e,* or third-integral charges, 1/3 or 2/3*e,* also existed. Third-integral charge was the distinctive property postulated for quarks, so experimental configuration of Morpurgo's second phenomenal model would correspond to the discovery of isolated quarks.

Now, listing these three elements of Morpurgo's practice brings us back to the question of the topology of scientific culture I mentioned above; these elements map the heterogeneous space in which Morpurgo worked. Further, they correspond well, I believe, to what Galison calls the constraints on scientific knowledge production. So now I can begin to make clearer what I have against that way of speaking. Think first about Morpurgo's material procedure. This, I want to say, was clearly and explicitly *modeled* on Millikan's oil-drop experiment. Morpurgo looked at and thought about the classic version of Millikan's experiment and then sought to modify it so as to optimize his chances of finding third-integral charges. And here I need to emphasize one fundamental aspect of modeling, namely, that it is an *open-ended* process having no determinate destination (Kuhn 1970; Barnes 1982). There were an indefinite number of ways in which Morpurgo could extend Millikan's example. And, in case I am accused of making philosophical points of no historical substance, I can note that Morpurgo did indeed explore all sorts of extensions in the real time of his experiment. He first sought to suspend particles of graphite (his equivalent of Millikan's oil drops) in a liquid in order to observe their response to an applied electric field and hence to determine their charges. Later he suspended them in a magnetic field; later still he suspended iron cylinders in a differently contrived magnetic field. And within the magnetic-suspension experiments many particular variations in material procedure were tried. What I am trying to get at here is that if one examines Morpurgo's material practice closely enough, metaphors of constraint—"obstacles that while restrictive are not absolutely rigid," the prison—seem both artificial and misleading. I can see no payoff from applying them in this instance; I cannot see how they can do any work. The model of Millikan's experiment should, I think, be seen instead as a primary *resource*—an "exemplar" in Kuhn's terminology—creatively and open-endedly extended in Morpurgo's practice, the very medium of his practice, rather than a constraint upon it. One might, of course, say that Morpurgo was constrained by his initial choice of a model in the sense that if he had adopted some other model, his practice would have been different; he might, for example, have ended up looking for quarks in a big accelerator experiment, as many other physicists did—and there is some point to this locution. But it is, at the least, out of balance. It puts all the weight on the negative and limiting aspects of the choice and gives none at all to the positive, constructive aspect of model as resource. I also think it conceptualizes the negative badly—a point I will return to shortly.

Let me now discuss the other two elements of Morpurgo's practice, beginning with his interpretive model. Again following Millikan, Morpurgo's idea was to observe the response of test samples to known electric fields and to use the laws of classical electrostatics to translate his observations into conclusions about the charges of those samples. Electrostatics is an almost totally dead and codified field, and Morpurgo's apparatus was designed as a textbook instance for its application, so one might imagine that, in this respect at least, prison- and constraint-talk must be appropriate. Not so. A large part of Morpurgo's practice consisted in the continual generation, revision, and rejection of a whole series of electrostatic interpretive models. Here, again, electrostatics is better thought of as a resource for creative and open-ended practice. Finally, what about Morpurgo's pair of phenomenal models? In this instance I have to concede something to the constraint theorists. All through his experimental program, Morpurgo never wavered in his conviction that the only findings that might count were integral or third-integral charges. This conviction was, I would say, something more than a constraint, understood as an obstacle "that while restrictive" is "not absolutely rigid." It was an absolutely rigid *fixed point* in Morpurgo's practice.

Does this concession give the game back to the constraint theorists? I don't think so. First, one can note that even if one element of Morpurgo's practice functioned like an absolute constraint, the others did not. There is something left to think about there. Second, the fixity of Morpurgo's phenomenal models is untypical of experimental practice in general. It is rare that the expected outcome of an experiment is limited to a couple of discrete and disjoint possibilities. Third, and this is the important point, there is still something odd going on here from the perspective of constraint, namely, that for more than a half a century leading up to Morpurgo's experiment the constraint had been *different*. From the days of

Millikan until 1964, the fixed constraint on all experiments involving charged particles was simply that all charges were integral. It was only with the quark proposal of Murray Gell-Mann and George Zweig in 1964 that third-integral charges came on the scene. So, constraints, even absolute and inflexible ones, can change radically and discontinuously, and it escapes me how such change can be conceptualized and analyzed within the language of constraint itself.[8] I cannot see how "constraint" can speak to the requisite dynamics. Alternatively, within the language of resources and modeling that I am recommending, such conceptualization and analysis are possible (Pickering 1984b).

To see practice as a process of open-ended modeling, then, points up a difficulty in the notion of constraint: to identify cultural elements as constraints is precisely to lose sight of the openness of their future extension. Conversely, however, what vanishes in the discussion of openness is the genuine attribute of practice that constraint-talk gestures toward, the very difficulty of practice. Doing science is hard work, and what I have said so far about openness does not do justice to that observation. But if one recognizes the openness of modeling, one has made some progress in thinking about where the difficulty of practice resides. As I earlier asserted, it cannot be situated *within* cultural elements as constraint-talk suggests. It must emerge, I claim, in the gaps *between* such elements as extended in modeling. Further, I claim, this emergence is irredeemably tied to the real time of practice. These remarks take us back to the topology of practice and further into its temporality. To see what is at stake we need to think about what lies between future goals and their practical achievement; back to Morpurgo.

Morpurgo began his program of quark-search experiments with a relatively well defined goal. In topological terms, he envisaged contriving, as I indicated, an *association,* a bringing together, of the disparate elements of his practice—his material procedure and his interpretive and phenomenal models—the first two of which were transformed and projected into the future through a process of modeling. His material procedure, translated through his interpretive model, should underwrite one of his two phenomenal models. What I now want to say is that the difficulty of achieving this goal lay in the making of such an association. It is crucial to note here that while modeling is truly an open-ended process, one can only, in real-time practice, extend a model in one direction at

8. Galison begins by stating: "The central historical-philosophical question therefore seems to me how these constraints arise, what sustains them, how do they act, and what makes them fall?" (draft, 2). But he does not address the first, second, and fourth of these questions in the rest of the paper.

once. The simplest case to think about is that of building a piece of apparatus bit by bit. Millikan's apparatus could be extended in an indefinite number of ways, but in building his own apparatus, Morpurgo could only explore these avenues one at a time. The same goes for Morpurgo's electrostatic interpretive models. And this is the point: there is no reason in the world to think that the products of such tentative and open-ended modeling processes will all fit together with one another as desired. Which is just what Morpurgo found. Repeatedly it turned out that his latest material procedure, performed on a particular piece of apparatus in a particular way and translated through his latest interpretive model, did not underwrite either of his phenomenal models. The standard finding of all of the versions of Morpurgo's experiment was that charge was not quantized in either integral or third-integral units, a finding which, I repeat, he was not prepared to accept.

What should we make of this? The message, I take it, is that instead of thinking about the difficulty of practice in terms of constraints within particular bits of culture, we should see it as arising in the interference of particular open-ended extensions of bits of culture within the context of goal-oriented practice (Tiles 1984). This interference constitutes a *resistance* to the kind of association of cultural elements that Morpurgo aimed at. So here is a concept, resistance, that, as promised, lies between cultural elements (creatively transformed) rather than in them, as constraint-talk would suggest. Resistances are *situated* with respect to particular projects, models, and extensions of models. That they arise, and what they look like, has to be understood in relation to those projects, models, and extensions. Further, let me emphasize that unlike constraint—which seems somehow "already there"—resistances *emerge* in the real time of practice: this extension, this way of doing modeling, does not fit with that extension in the context of this project. And, to emphasize one more point for future use, there is an irreducible element of chance and *contingency* in the emergence of resistance. By this I mean that I can think of no principled way of explaining why particular resistances surface at particular times within particular projects. I can find nothing better to say than that "it just happened" that Morpurgo's apparatus misbehaved the way it did when it did.[9]

My sketch of an analysis of practice beyond constraint is now just about complete, but I need to introduce one more concept, that of *accom-*

9. Remember that the aim here is to deal with practice as it happens. Of course, Morpurgo did, on occasion, construct after-the-fact explanations of why particular experimental configurations had proved unsatisfactory, but such explanations are themselves the products of the kinds of practices that are to be analyzed.

modation. Morpurgo's typical response to resistance was to seek to exploit once more the openness of modeling. In the face of resistance, he tinkered with both his material procedure and his interpretive model, revising each until he achieved what I call the *interactive stabilization* of the three elements of his practice, until, that is, he found a material procedure that translated through an interpretive model into one of his phenomenal models (always, as it turned out, the no-quark model). Whenever he reached such a situation he reported a fact. This tinkering and interactive stabilization, designed to get around resistance and to achieve a particular association of elements, is what I call accommodation. And the production of facts, I claim, is the outcome of a goal-oriented, temporally extended dialectic of resistance and accommodation of the form just described.[10] Three aspects of accommodation are worth stressing. First, accommodation, like resistance has to be seen as situated—with respect to particular resistances, goals, and fields of cultural resources. Accommodations are made in the spaces between cultural elements; like resistance and unlike constraint they are not proper to any single one. Second, again like resistance and unlike constraint, accommodations are made in real time; there is no sense in which they are already there. And third, like resistance once more, an element of contingency enters into accommodation: "it just happened" that Morpurgo's attempted accommodations usually failed but sometimes succeeded.

Modeling, association, resistance, accommodation, and contingency are my ways of thinking about practice beyond constraint.[11] My technical claim is that these concepts enable one to get past the placebo of constraint-talk and grapple with the difficulty of practice. I think this is an important advance in itself, in that one cannot claim to have a satisfactory understanding of science if one does not have a pretty well developed analysis of why doing science is hard work. I also think my way of unraveling the difficulty is important because it foregrounds two aspects of science that fade away in traditional accounts of science-as-knowledge, namely, the topology and, especially, the temporality of scientific practice. My analysis makes it possible to think of science as embedded in time in a way that constraint-talk blocks. And there is yet a third reason why I think the analysis is important. This brings me back to the epistemological argument I promised in my opening remarks. Now for objectivity, relativism, and historicism (see also Pickering 1991, 1995).

10. The general form of this practical dialectic is what I now call "the mangle of practice"; for more on the mangle, see Pickering 1991, 1993, 1995.

11. Most of these concepts seem to me to be primitive; I cannot see how to analyze them further. Modeling is different: see Pickering 1990b and Pickering and Stephanides 1992.

2. The Historicity of Knowledge

Once upon a time, we all knew that scientific knowledge was objective, and it was the job of philosophers to spell this out for us. That they did such a dismal job was what opened the way for constructivists and relativists to claim that knowledge was not objective, in the sense that it did not float free from the culture that produced it; instead knowledge had to be understood as knowledge-relative-to-a-particular-culture. And the sentiment that underlay the outburst of constraint-talk in the 1980s was, as I suggested earlier, that the constructivist/relativists had gone too far. Constraint-talk, in a confused way, was intended to reassert the objectivity of scientific knowledge, its distance from us. I am now in a position to remark that constraint-talk fails miserably in this task—Peter Galison may have refuted the bad dream of wish-relativism, but his analysis of constraint makes no distinction between the natural sciences and, say, the beliefs of the Azande—but, returning to my more agreeable mode, I want to put that aside and to ask where my analysis of the temporality of practice leaves me on the objectivist-relativist spectrum. The answer is complicated.

On the one hand, I have tried to catch the structure and regularity of practice in the concepts of modeling, association, resistance, and accommodation. And in so doing, I have, it would seem, cast in my lot with the cultural relativists. The role I give to modeling, for example, serves to tie new knowledge directly and distinctively back to the culture in and from which it is made. Like Galison, I have no way of marking a difference between us and the Azande; scientific knowledge is knowledge relative to scientific culture in just the same way as knowledge of chicken oracles is relative to Azande culture. But, on the other hand, I have not claimed that *all* of practice can be caught up in notions that point to structure and regularity. Thus I have not only admitted but proclaimed that there is an irreducible element of contingency in the emergence of resistance and the achievement of accommodation, and I could have said the same about the process of goal formation that gets the dialectic of resistance and accommodation going.[12] And this recognition would seem to put me somewhere

12. I can only go so far, for example, in explaining how and why Morpurgo set himself the goal of an empirical search for isolated quarks. I can see that he was working within a particular field of cultural resources (he was a key figure in the development of quark models of elementary particles; his colleagues were experimenters; the Millikan oil-drop experiment is well known to physicists in its performance and interpretation), and I can see that appropriately extended and combined these resources can condense into a vision of the kind of experimental program that Morpurgo indeed undertook (Pickering 1984b, 117–18 n. 32). But still I can see no way to a causal explanation of his formulation of this particular goal. See also Pickering 1987b.

on the wild side of cultural relativism. If the structure and regularity of practice point toward the relativity of knowledge to culture, then the contingency of practice points toward relativity-to-chance, otherwise known as historicity.

Now, I do indeed want to defend a historicist understanding of scientific knowledge. I do want to insist that all those "it just happened's" are constitutive of what scientists know and believe, and I do want to deny that there is any causal, mechanical linkage between present and future culture. Any culture is always on the edge of evolving in an indefinite number of disparate ways, and there is nothing already in culture or the world that selects between the possibilities. But two qualifications need to be entered. The first is that I insist that contingency is not all that there is to be found in scientific, or any other kind of, practice. There *is* structure there as well, and it is by virtue of this structure that one can make sense of the obvious fact that different cultures do sustain different practices and produce different knowledges—and that, for example, it is not about to "just happen" that we all start poisoning chickens or speaking Japanese. At this level, then, one can rescue a notion that our culture constrains us, though it is a thin one that needs to be filled out along the lines just discussed. What I advocate is thus, I would say, a culturally situated historicism.[13]

The second qualification is more important. My take on the historicity of scientific knowledge can certainly be made out to be the polar opposite of the objectivity of science that traditional studies of science have either sought to defend or, more usually, taken for granted. But that, to me, shows only that the tradition has looked for objectivity in the wrong place. If one is willing to move beyond an exclusive concern with knowledge and to displace the sense of objectivity toward practice, especially toward the difficulties of practice caught up in a concept of resistance, I can see no problem in viewing scientific knowledge as being as objective as can be. In fact, I think it is—but this is not to deny its situated historicity. Further, cultural relativism has been the traditional enemy in

13. Insistence on the intertwining of structure and contingency differentiates my analysis from others that give contingency a constitutive role in scientific practice. Thus Kuhn (1970) locates structure but not contingency in normal science, and vice versa in revolutionary science. Collins (1985) sees contingency everywhere but no structure (Pickering 1987a). Smith's (1988) analysis of literary production and consumption is, I think, quite close to my own. It is worth noting that in line with the standard lack of interest in the temporality of practice, the history, philosophy, and sociology of science literature (not just the literature that speaks of constraint) has a way of making contingency disappear completely. A generic fault of traditional approaches to science studies is, I think, that they *explain too much* (and thus achieve all sorts of weird effects in our understanding and appreciation of scientific knowledge).

history and philosophy of science by virtue of its apparent location of knowledge in society *rather than* in the world; this is the "mere construct" reading of constructivism, the wish-relativism, that constraint-talk opposes. My version of historicism cuts right across this dichotomy. Contingencies of resistance and accommodation lie precisely in the *engagement* of "nature" and "culture": the material world can only resist us by virtue of our culture and our operations on it, our formulation of ends and means; but, reciprocally, culture evolves precisely in accommodation to such resistances. Indeed, the very term "culture" loses its customary sense of difference from nature in my analysis: it does not refer to something particularly human that exists apart from the material world. The "cultural" field in which Morpurgo moved had been formed in the same kind of intimate engagement with the material world that he himself sought and achieved.

So, my analysis of practice leads me to a historicist position that escapes the standard dichotomies of nature and culture, objectivity and relativism. And this, I think, is the mark of a wider phenomenon. All sorts of tired distinctions, concepts, and conceptualizations start to crumble when one begins to explore the aspects of practice that have been traditionally neglected by historians and philosophers: its topology (on which I have said only a little here), its temporality, and its materiality (on which I have said almost nothing). Attention to just these three aspects—which bear a surprising resemblance to space, time, and matter, the holy trinity of modern physics—can, I suspect, bring new life to science studies and beyond. Along with "constraint," the concepts of "reason" and "interest" can likewise be called into question and thought through at a new level. Along with objectivity, dissolution of the nature/culture dichotomy invites a displacement of worn-out arguments on realism onto the new terrain of practice. And so on. A whole new view of knowledge and practice, society and the world is at stake. Well, I think so anyway (Pickering 1995).

Acknowledgments

This material is based in part upon work supported by the National Science Foundation, History and Philosophy of Science Program, grant no. DIR-8912095. I am grateful for interesting, rewarding, and amusing conversations about the themes of this paper with many people, starting with Alison Wylie, who commented formally on it at the Toronto meeting. Everyone loves constraint; no one agrees with me.

FOUR

EXPERIMENTING IN THE NATURAL SCIENCES: A PHILOSOPHICAL APPROACH

Hans Radder

1. INTRODUCTION

RECENT YEARS HAVE SEEN AN INCREASING WAVE of studies on experimentation in, especially, the natural sciences. By now a lot of materials and insights have become available concerning this often ignored but crucial aspect of scientific research. In this paper I will first describe a systematic, philosophical framework for analyzing the practice, process, and products of experimentation. Then I will discuss certain crucial philosophical issues on the basis of this framework. In doing so, I will make use of and evaluate a number of studies from the recent experimental wave, not only philosophical but also historiographical and sociological ones.

The present study is subtitled "A Philosophical Approach." This implies a somewhat different perspective than we find in historiographical and sociological studies, which have constituted the larger part of the experimental wave up to now. In my view, philosophy of science should first of all contribute to a clarification and a critical discussion of the position and function of science in present-day society, including the intimate connections between science and technology. This, of course, requires a thorough understanding of how science is practiced, and for this purpose detailed historical and sociological case studies are indispensable.

Yet, the philosopher's aim cannot be an exact and complete description of all phases and aspects of scientific practice. There are, I think, three good reasons for this. The first is that the one and only true account of science does not exist. Scientific practice is a rich and complex activity which can be fruitfully studied from different perspectives at once.[1] In the second place, philosophy approaches science from the angle of a number

1. Thus, my approach is not that of the Testing Theories of Scientific Change project. See Laudan, Laudan, and Donovan 1988, and compare the critical remarks in Nickles 1986.

of distinctively philosophical, theoretical issues, such as realism, rationality, and (social) critique. Issues like these are usually not central to the empirical study of science. Finally, philosophy of science has, in my view, also an important normative component. It does not only aim at understanding but also at assessing science and science-based technology. The most interesting of such normative assessments are those in which the intrinsic connections between epistemic and social aspects are explicitly taken into account. Given the subject of the present study, I can of course not deal with these metatheoretical issues in a systematic manner. Instead, at several places the theoretical and normative aspects of this philosophical account of natural-scientific experimentation will be illustrated with examples.

A central topic of the paper is the reproducibility of experiments. As we will see, reproducibility is a rich and complex notion. By differentiating between various types and ranges of reproducibility I will show that, contrary to what is claimed sometimes, reproduction does play a significant role in experimental practice. In addition, reproducibility also functions as a nonlocal norm, the application of which leads to experiments or experimental results that are stable against a number of variations in their local contexts. Therefore, experimental natural science cannot be adequately understood as a mere sum of "local cultures."[2]

A second major issue to be discussed is that of the social legitimation of experimental natural science as such. I will argue that the procedure of legitimizing through witnessing science, as it was practiced in the early days of the Royal Society, necessarily has a restricted scope and, moreover, involves fundamental ambiguities. Therefore, I turn to the more obvious legitimation of experimental science through its possible technological utilization. My conclusion is that also in this respect reproducibility, and especially the reproducibility of the material realization of experiments and the skilful work it presupposes, plays a crucial role.

The plan of the paper is as follows. In section 2 I present an analysis of experimentation in terms of the notions of material realization and theoretical description or interpretation. I distinguish three types of reproducibility, whose role in experimental practice is discussed in section 3. In section 4 the analysis of experimentation and reproducibility is compared

2. As explained above, the philosophical discussions in the present paper mainly turn on the reproducibility of experiments. In addition to this the proposed analysis of experimenting immediately suggests several further elaborations. In particular, I think of an analysis and evaluation of the so-called experimenters' regress (Radder 1992a); the relevance of experimenting for the problem of scientific realism (Radder 1988 and 1993); aspects of the relationship between science and technology other than those dealt with here (Radder 1986); the similarities and dissimilarities between experimentation and observation; and the problems concerning the nature of experimental testing.

to a recent account of the distinction between "data" and "phenomena." Next, section 5 deals with reproducibility as a nonlocal norm and investigates the stability of experiments and their results. In section 6 I enter into some aspects of the social legitimation of experimental science by means of a comparison of the notions of "witnessing" and "materially realizing" experiments and experimental technologies. Two related views on the issues in question, notably those of Albert Borgmann and Bruno Latour, are briefly discussed and evaluated in section 7. The last section concludes the paper by arguing for a philosophical and social revaluation of the role of craftwork in experimental science and technology.

2. The Realization and Description of Reproducible Experiments

In an experimental process we deal with an object to be studied and with a number of apparatus. Both object and apparatus may be of various kinds.[3] Now, the experimental process involves the *material realization* and the *theoretical description or interpretation* of a number of manipulations of, and their consequences for, the object and the apparatus, which have been brought into mutual interaction.[4] The general idea is that some information about the object can be transferred to the apparatus by means of a suitable *interaction*. That is, the interaction should produce an (ideally complete) correlation between some property of the object and some property of the apparatus. From this it follows that the theoretical descriptions of object and apparatus should also "interact": they need to have at least some area of intersection. The theory of electrons and the theories of the instrumentation used to experiment on electrons cannot be completely different (see Morrison 1990, 7–8). In the *detection* stage of the experiment the information can be obtained by measuring or observing the relevant property of the apparatus. A typical feature of the practice of experimentation is that neither object nor apparatus is "simply available." They have to be carefully *prepared* in agreement with the goal and plan of the experiment.

Consider, for example, an experiment for determining the boiling point of a particular liquid. Then, obviously, this liquid is our object under study. Our apparatus consists of a heat source, a vessel, a ther-

3. Thus, the general term "object" may stand for a thing, a process, a phenomenon, etc. Janich (1978, 9–19) discusses various kinds of apparatus and their functions.

4. The discussion in the present section is partly a summary of and partly a more precise statement and further development of the analyses presented in Radder 1988, 59–76. The idea of differentiating the theoretical description of an experiment from its material realization was stimulated by Habermas's distinction between theoretical discourse and instrumental or experimental action. See Habermas 1973 and the Postscript of Habermas 1978.

mometer, and possibly some supplementary equipment. On the basis of our knowledge of the interaction process between thermometer and liquid, we assume that our readings of the thermometer inform us about the temperature of the liquid. Part of the preparation procedure involves making sure that the liquid in question is pure. That is why it may be necessary first to clean the vessel that has to contain the liquid.

2.1. The Theoretical Description of Experiments

Let me first discuss the theoretical description. Here it is, I think, helpful to make a further distinction between the (theoretical) *result* q of the experiment and the other theoretical descriptions p which enable us to infer this result. Then, $p \rightarrow q$ is the specific overall description of the experimental process, whereas q refers to its outcome. For instance, in our example of the boiling experiment, the result q will be the claim that at a certain time the fluid in question boils at temperature y°C. The description p, a conjunction of a number of subdescriptions, is already quite complicated, even in this simple experiment. Description p involves, among other things, claims about the liquid (e.g., it is pure; its temperature distribution is homogeneous), about the apparatus (e.g., the temperature remains constant at y°C, even if we continue to supply heat), about the interaction between object and apparatus (e.g., the thermometer does not have a substantial influence on the temperature of the liquid), and about the interaction between the experimental system—that is, the sum total of object and apparatus—and its external setting (e.g., the air pressure surrounding the experimental system is constant during the measurements).[5] Generally, the theoretical description contains three main components. It refers to the preparation of object and experimental equipment; to the staging of the processes of interaction and detection; and to the screening and control of potential disturbances from the outside, that is, to the closedness of the experimental system.[6]

The following observations may further clarify the meaning of this notion of theoretical description. First, "theory" is taken here not in the sense of a systematic and comprehensive theory, but in a vaguer and more restricted sense. The prime aim of the theoretical description is not to offer a systematic explanation of the phenomenon under study but to supply a "singular causal explanation" of how the result q is produced (see Bogen and Woodward 1988, 322). Theory in this sense includes, in terms of Ian Hacking's taxonomy of the elements of laboratory experiments, background knowledge about object and apparatus, topical hypotheses, and assumptions about data processing (Hacking 1988a, 508–11). Alterna-

5. For more details of this example, see Radder 1988, 59–69.
6. For this notion of closedness, see Radder 1988, 63–69. See also Radder 1986.

tively, one may think of Andrew Pickering's notions of instrumental and phenomenal models (Pickering 1989b).

Second, the theoretical description $p \rightarrow q$ does not give us a complete account (whatever that may be) of the experimental process. However, such an account is not necessary. We need to consider only those aspects of the experimental process which are deemed *relevant* to obtaining the intended result. For instance, in the boiling experiment discussed above most of the specific characteristics of the heat source as well as a description of the gravitational field in the laboratory are irrelevant to successfully determining the boiling point of the liquid.

Third, in experimental practice the argumentation by which we infer the result q from the premises p will not always be explicit. I assume, however, that if asked, the experimenter will or should be able to come up with a plausible story about how the experimental result is produced.

2.2. The Material Realization of Experiments

Next we need an explication of the notion of the material realization of an experiment (see also Radder 1988, 69–76). The question is whether there is a way to describe and analyze the performance of an experiment other than in terms of the particular theoretical description that is in fact used to perform it. The main problem can be illustrated with this example. Suppose we want to determine experimentally the mass of an object which is at rest in relation to the measurement equipment. Two scientists each carry out such an experiment in the same way. Nevertheless, one interprets the actions performed as a measurement of the Newtonian mass, and the other as a determination of the Einsteinian mass. But both performed "the same" actions and thus—in a certain sense—the same experiment. Therefore, if we want to describe experimental action unambiguously, we have to find some sort of abstraction of these various specific theoretical interpretations. On the other hand, it is an indisputable fact that concrete experimental action is always action on the basis of certain theoretical ideas: without theoretical ideas there can be no experiments.

I propose the following maneuver to make the distinction and the interplay between the theoretical description and the actual performance of experiments more explicit and manageable. Let us suppose that in the example of the experiment to determine the boiling point of a liquid, the experimenter A orders B, a complete layperson in the field of heat theory, to actually carry out all the necessary actions. A tells B for example: "Take this thing (here) and put it in the holder (over there), in such a way that its lower part hangs in the liquid; then watch the rising silver column; when it does not rise anymore for a bit, then write down (here) which point (which number) is reached by the upper end of the column"; and B does all this *correctly according to A*.

The general supposition is that experimenter A can delegate the actual carrying out of experiments to one or more laypersons B, that is, to people who have no knowledge of the theoretical interpretations. I am referring to actions like the installation and operation of various experimental apparatus; the preparation of the necessary substances and material; and the reading of pointer positions on graduated scales, of color changes, of traces on photographic plates, etc. This all happens under A's supervision and with perhaps A's correcting instructions to guarantee that the experimental result is obtained in the right way according to the experimenter. With the help of this maneuver I can now define the concept of *the material realization* of an experiment as the whole of the experimental actions which are carried out by B in a correct way according to A and which can be described in A's instructions to B in the language in which A and B communicate with each other.[7]

This definition of the material realization of an experiment is based on two general characteristics of the human condition. The first is the possibility of a *division of labor.* Not only experimenters but all human beings are themselves part of nature. Therefore, by materially interacting with surrounding nature, all human beings are able to produce certain situations in experimental systems. In the second place, the procedure of material realization exploits the fact that, in any particular experiment, some sort of language exists in which a layperson and an experimenter can communicate with each other about certain fundamental aspects of the experimental production process in a usually reliable and successful way. I will call the language thus defined "common language."[8]

So far I have dealt with the case in which experimenter A already knows how to successfully perform the experiment. In this case the procedure of material realization introduced above boils down to a specific kind of repetition of the experiment. The prime aim of the proposed maneuver is to make visible the significance of experimental action and production. Apart from this it may also, in the form of A's *correcting* instruc-

7. Already in the 1920s and 1930s the importance of experimenting as concrete action was stressed by Hugo Dingler. However, in his wish to *ground* natural science in practical action by means of operational definitions, he understated the significance of theoretical knowledge, including its importance in carrying out experiments. See Dingler 1952 for a concise statement of his views on the matter. For a recent Dinglerian view, see Tetens 1987. (Remarkably enough, Dingler and his philosophical ideas were intimately connected with the attempts to build up a "German physics," a physics that should agree with the *Blut und Boden* doctrines of National Socialist ideology. See Richter 1980.)

8. Note that the above definition neither fixes the nature of this common language nor claims any ontological privilege for it. In practice, it may have various characteristics, as long as it enables us to materially realize experiments in the way specified above. For philosophical purposes this is a crucial point: see Radder 1988, 76, 105–6.

tions, exhibit aspects of the process through which B learns to carry out the experiment. Therefore, if B's performance succeeds, the maneuver also shows us the "learnability" of A's productive skills. But the procedure of material realization through division of labor may also be applied in the more general case in which both A and B still have to find out how to do the experiment and what experimental success means in the case in question. In this case the *inter*action between A's developing theoretical insights and B's growing practical skills will come more to the fore.[9] However, for most purposes of the present paper, a central topic of which is after all the *re*producibility of experiments, it will be sufficient to consider cases of material realization that aim to describe A's previously acquired practical abilities.

What about the role of theory in materially realizing experiments? As I have stressed, without theoretical knowledge experimenting would be impossible. The instructions and corrections of the experimenter are guided by his or her scientific insights concerning the preparation of object and equipment, the suitable staging of the interaction and detection, and the effective control of disturbing influences. Nevertheless, the resulting description of the material realization is phrased in everyday terms and thus it is, in this specific way, theory independent. As we will see further on, this procedure suggests an *empirical mechanism* by which it is possible to step from one theoretical description of an experiment to another. Put differently, it will enable us to define the reproducibility of the material realization of an experiment under a whole set of, possibly radically different, theoretical interpretations.[10]

In sum, with the term "material realization" I intend to refer to experimental action and production *either* by the experimenters themselves *or* by laypersons. But in order to make explicit these aspects of experimentation and the specific way in which they are theory independent, we need the proposed maneuver that makes use of the possibility of division of labor and of a common language as a means of communication between scientists and laypersons.

9. See the first of the two illustrations given below.

10. Note that this account does not make use of the assumption that the theoretical interpretation of experiments can be done in terms of stable, low-level theoretical "home truths" and therefore does not depend on the ever changing high-level theories (see Hacking 1983, e.g., 265). On this issue I agree with the criticism put forward by Morrison (1990, 6–14). In Radder 1988, 91–93, 144–47, I have presented counterexamples to the claim that home truths remain stable across time. On the other hand, the above analysis of experimenting seems to be more in line with an aspect of Hacking's position which is emphasized by Stump 1988, namely, that skilful experimentation is relatively autonomous because it need not be committed to one specific theoretical interpretation.

2.3. Two Illustrations

I will continue the discussion of the material realization of experiments by addressing the question of the practical feasibility of carrying out experiments in the way described by the definition of material realization. For this purpose, consider the following quotation regarding an ethnomethodological study carried out by Friedrich Schrecker in a completely different context:

> In Schrecker's study, a methodological set-up was used for the purpose of perspicuously identifying the mutual dependence of chemical reasoning and embodied action. Schrecker volunteered his services to aid a handicapped student in his laboratory work for an undergraduate chemistry course. He then received the student's permission to use his experimental work as a research topic. There resulted a division of labour and responsibility between Schrecker, who was largely ignorant of the field of chemistry—and ignorant as well of the specific lab assignments he bodily assisted—and Gordon, the chemistry student, who because of a spinal injury was paralyzed from the neck down with very limited use of his hands. Gordon depended upon Schrecker to bodily perform the work at the bench for the weekly lab experiments assigned to students in the course on "Quantitative Analysis."
>
> The isolation of Schrecker's handiwork from its theoretical basis in chemistry necessitated that Gordon and he make explicit for one another how the experiment was progressing as a witnessable production of chemistry. Gordon provided instructions for Schrecker on what to do next, while at the same time he relied upon Schrecker's developing work to show him what "next" meant in terms of where they stood in the course of the experiment's events.[11]

This case shows very directly the practical possibility and the main features of materially realizing experiments by means of a division of labor between an experimenter and a layperson. In this study there is, just as in the definition of "material realization," clearly question of a somewhat unusual maneuver; we have a layperson, Schrecker, who was "recognizably doing chemistry, whether he knew it or not" (Lynch, Livingston, and Garfinkel 1983, 228); and we also see the specific interplay between theoretical knowledge and experimental action and production.

A second practical illustration is slightly less direct but nevertheless very illuminating. It concerns Robert Boyle's so-called void-in-the-void ex-

11. This study has been reported extensively in Lynch, Livingston and Garfinkel 1983. The quotation is from p. 225.

periment carried out in the 1650s, which is summarized by Steven Shapin and Simon Schaffer as follows:

> This is what Boyle did: he took a three-foot-long glass tube, one-quarter inch in diameter, filled it with mercury, and inverted it as usual into a dish of mercury, having, as he said, taken care to remove bubbles of air from the substance. The mercury column then subsided to a height of about 29 inches above the surface of the mercury in the dish below, leaving the Torricellian space at the top. He then pasted a piece of ruled paper at the top of the tube, and, using a number of strings, lowered the apparatus into the receiver. Part of the tube extended above the aperture in the receiver's top, and Boyle carefully filled up the joints with melted diachylon. . . .
>
> Pumping now commenced. The initial suck resulted in an immediate subsidence of the mercury column; subsequent sucks caused further falls. . . . After about a quarter-hour's pumping . . . , the mercury would fall no further. (1985, 42–43)

Although in this summary no explicit mention is made of laypersons, Boyle in fact nearly never carried out the experiments himself. He had this done by all sorts of assistants, "who had skill but lacked the qualifications to make knowledge" (Shapin 1988b, 395). Apart from this, the above passage provides a beautiful illustration of how an experiment is materially realized. Shapin and Schaffer, moreover, explicitly distinguish between, on the one hand, the experimental production of "matters of fact" and, on the other, Boyle's theoretical interpretations on the basis of his ideas about the nature and properties of the air, such as its pressure and "spring": "So far, the account we have given has been restricted to what Boyle said was done and observed, without any of the *meanings* he attached to the experiment" (Shapin and Schaffer 1985, 43).

From such examples it is, I think, justified to conclude that materially realizing an experiment in the way defined above is not merely a philosophical construction but is also feasible and plays a role in actual experimental episodes.

2.4. Three Types of Reproducibility

Many philosophers assume, implicitly or explicitly, that successful experiments are or should be *reproducible*.[12] However, since "experiment" is a general term for what in fact is a rather complex practice, the precise meaning of this assumption is not very clear. To clarify the notion of reproducibility we need to address the following question: reproducibility *of what* and *by whom?* In answering this question I will first distinguish,

12. See Popper 1965, 45–46; Habermas 1978, esp. chap. 6; Bhaskar 1978, esp. chap. 2. Other terms used are "repeatability," "replicability," and "regularity."

on the basis of the above analysis of experimentation, between three types of reproducibility and discuss their mutual relations.

Reproducibility of the material realization of an experiment. This type of reproducibility obtains when the same material realization can be reproduced under different interpretations, such as $p \rightarrow q$, $p' \rightarrow q'$, $p' \rightarrow q$, or $p \rightarrow q'$. Since in this case a reproduction may be achieved on the basis of any member of a whole class of theoretical interpretations, reproducing the material realization of an experiment does not depend on (a shared belief in) any *particular* interpretation from that class.

Consider, for instance, the experimental determination of the Newtonian mass of an object that does not move relative to the measuring apparatus. Suppose that experimenter A has a setup available and has the experiment performed successively by layperson B and by experimenter A′, who, for the time being, acts as a layperson. A instructs both B and A′, in common language, to carry out the experiment. Now it may well be that the performance of the experiment requires some special skills that are, for instance, mastered by B but not yet by A′. Then of course the two descriptions of the material realizations resulting from this first run will be different, and another run will be necessary. The assumption is, however, that after some time of training the two descriptions of the material realizations will converge. Yet, having reached this phase the theoretically informed experimenter A′ may rightly claim to have measured the Einsteinian mass, in contrast to A, who claims to have determined the Newtonian mass. In this case we can say that the material realization of the experiment has been reproduced, even though its theoretical description or its result is radically at variance in the two experiments.

The possibility of reproducing material realizations is based on the fact—demonstrated in processes of division of labor—that performing an experiment may be learned by people who are ignorant of, or disagree with, the theoretical interpretation of the (results of the) actions they produce. In such processes of division of labor communication through common language plays a crucial role. The range of this type of reproducibility is greatly enhanced by the fact that experiments can be made more robust and less dependent on the skills of specific individuals, as happens, for instance, in their "demonstrative" or "showing" stage.[13] A reproducible material realization indicates that *a* stable phenomenon has been detected. It does not imply any agreement about *what* phenomenon it is. It is even

13. Since reproducing experiments is a very important way of making them "public," the fact that achieving the reproducibility of the material realization of an experiment is more difficult in an experiment's "exploratory" or "trial" stage than in its "demonstrative" or "showing" stage is of course precisely what we would expect. For these distinct stages of experimentation, see Gooding 1985 and Shapin 1988b, 399–404.

possible that some interpreters will argue that the phenomenon is an artifact, because, though it is stable, it is not to be attributed to the object under study but to certain features of the apparatus.

Reproducibility of an experiment under a fixed theoretical interpretation p → q. We might find out, for example, that in our boiling experiment the description $p \rightarrow q$ does not lead to reproducibility if q is the claim that the liquid boils at y°C. However, if we add an error interval z and change q into the proposition that the boiling point lies within the interval $(y \pm z)$°C, then the experiment may well be reproducible under this (newly) fixed theoretical description. Or, we might find that a description p that does not include the air pressure as a relevant factor does not lead to a reproducible result, whereas inclusion of the air pressure does. Thus, this type of reproducibility implies a repeatability of the experiment from the point of view of the theoretical interpretation in question. If a particular reproduction succeeds, the people involved believe that what has been reproduced is the experimental process as described by this theoretical interpretation. As we shall see further on, in section 3, reproducing an experiment under a fixed theoretical interpretation will not always imply exactly reproducing the same material procedures. Sometimes slight differences occur, in which case we have an approximate reproduction of the material realization.

Reproducibility of the result q *of an experiment.* This type of reproducibility applies when it is possible to obtain the same experimental result by means of a set of different experimental processes. Again, "sameness" and "difference" mean: sameness and difference from the point of view of the theoretical interpretation. Thus, we may have $p \rightarrow q$, $p' \rightarrow q$, or $p'' \rightarrow q$, where p, p', and p'' are (possibly radically) different descriptions that will usually (but not necessarily) describe different material procedures for realizing q. An example is a determination of the boiling point of our liquid by means of different types of thermometers, for instance, a mercury and a gas thermometer. I propose to call an experiment that reproduces q but not p a *replication.*

It follows from the above definitions that the reproducibility of an experiment under a fixed overall description and the reproducibility of the result do not mutually imply each other. Furthermore, it will be clear that when we reproduce a result, we do not necessarily reproduce the material realization of the first experiment. Though it may be that p' is simply a different interpretation of precisely the same material realization, p' may just as well describe a slightly or even completely different experimental process which can be used to determine the same result q.

Reproducing an experiment, in any of its senses, requires two things. In the first place, it requires agreement on the description of the original experiment. This description provides a first specification of the ways in which original and reproduced experiment should be similar. However, having such a description is not enough. Also required are the abilities and resources for skilfully re-producing the experimental situations described. In spite of these conditions, claims on reproducibility possess a definite surplus value. This value rests on the fact that agreement on descriptions and availability of the relevant skills and resources cannot guarantee success. Thus, obtaining reproducibility constitutes a significant experimental achievement.[14]

Given the three kinds of description of an experiment (the description of its material realization, the theoretical description of the overall experimental process, and the theoretical description of its result), the question Reproducibility of what? can now be answered as follows. What is reproduced are the experimental situations identified by any of these three kinds of description.[15] Finally, with respect to the question Reproducibility by whom? four ranges suggest themselves: reproducibility by any scientist or even by any human being, in the past, present, or future; reproducibility by contemporary scientists; reproducibility by the original experimenter; and reproducibility by the lay performers of the experiment. Since reproducibility "by anyone" is, for reasons that will become clear, *very much* in principle or counterfactual, in the discussion below I will mainly deal with the other three possibilities.

2.5. Nature and Scope of the Approach

So much for the definitions of the three notions of reproducibility. In the next section I will come back to this issue and scrutinize what role these notions play in actual experimental practice. In concluding this section I want to say something about the nature and scope of the foregoing analysis of the practice, process, and products of experimentation. As stated in the Introduction, the nature of my analysis is philosophical, which implies that it is both theoretical and normative. Here I will deal with the first aspect; a discussion of the normative aspect will be given in section 5.

Consider the notion of material realization. It is important to distinguish between the process of material realization itself and its description by means of the theoretical notion of material realization as introduced above. The claim is that material realization, the process of acting and producing, is a fundamental aspect of every experiment. The *notion* of material realization, however, is a theoretical-philosophical concept. Like

14. The above argument is discussed in detail in Radder 1993. See also Giere 1988, 108–9.

15. Occasionally, I will use the less accurate phrase "reproducibility of the theoretical description" or of the "result" as a shorthand for the reproducibility of the experimental situations to which these descriptions refer.

many other theoretical concepts, its chief function is not to describe but to explain or interpret a fundamental aspect of experimental practice. It purports to explain or interpret both the interaction between and the relative autonomy of experimental action and theoretical reasoning. Put differently, it attempts to explicate the mechanism through which theoretical interpretations are underdetermined, not by data or empirical statements, but by the process of experimental action and production. On the other hand, it will be equally clear that theoretical notions should not be mere speculations that are totally unrelated to the practice of science. Therefore, I have also pointed out that the concept of material realization can be empirically substantiated through the phenomenon of division of labor in experimental practice.

What about the scope of the above approach to experimenting? My claim is that it is generally applicable to experiments in the natural sciences.[16] Although the illustrations given so far concern small-scale experiments, this does not entail that the analysis only applies to yesterday's experiments. Contrary to popular opinion, small-scale experiments still occupy a prominent place in contemporary natural science. Think, for instance, of the recent cold fusion and high-temperature superconductivity experiments. Or consider the experiments in fluid dynamics that played such an important role in the recent development of chaos theories (Gleick 1987, 125–31, 191–211). Moreover, the analysis does not seem to be restricted to tabletop experiments either. Of course, in complex, large-scale experiments it will be difficult to carry out the analyses along the above lines in full detail. Yet, as far as I can see, there do not seem to be any fundamental reasons why the approach would not, in principle, be applicable to "big science" experimentation too.

3. Reproduction in Experimental Practice

I will now consider the three types of reproducibility and examine how far they range in actual experimental practice. Of course, there is a difference between the claim that scientific experiments are or should be reproducible and the claim that they, in fact, are or should be reproduced by scientists (see Radder 1993). I think that many philosophers, when discussing this issue, intend to refer primarily to the possibility and not to the actual occurrence of reproduction. On the other hand, without a thorough empirical investigation into the actual occurrence of reproduction, we cannot

16. Of course, for more particular purposes further specifications may be useful and clarifying. See, for instance, the taxonomy of the elements of laboratory experimentation offered in Hacking 1988a, 508–11.

decide whether the philosophical claim is merely an idle speculation or has a firm footing in experimental practice. At this point, however, we are confronted with the claims of empirical students of science. Collins, for instance, states that "replication of others' findings and results is an activity that is rarely practised" (1985, 19). And Hacking asserts that "roughly speaking, no one ever repeats an experiment" (1983, 231).

I do not think that these statements, formulated in this manner, are strictly speaking false. My main objection, though, is that they are inadequate because they cover only some of the issues surrounding experimental reproduction. By distinguishing between reproduction "of what" and "by whom" we obtain a richer notion of experimental reproduction, which enables us to draw a more finely grained map of its place in experimental practice. As I will show now, this more differentiated approach makes it possible to uphold the claim that various types of reproducibility do in fact play a significant role in experimental practice.

3.1. Reproducing the Material Realization

Let me start with the first type of reproduction: the reproduction of the material realization of an experiment being compatible with a whole set of theoretical interpretations. This type concerns reproducing an experiment as defined by the actions of laypersons or, as the case may be, of the experimenter. I will first give an example and then point out the problems one may come across in trying to reproduce a material realization.

Consider, for instance, Boyle's experiments with the air pump. As we have seen, Boyle almost never performed the physical manipulations himself. This was done by various sorts of assistants, being for the most part laypersons. According to Shapin, "they made the machines work, but they could not make knowledge" (1988b, 395). Therefore, every successful repetition of an air-pump experiment performed in this manner—for instance, a successful trial at Boyle's home and a showing in the rooms of the Royal Society—reproduces its material realization. The material realization remains the same, even if the theoretical interpretations differ. If, for example, we place a burning candle in the receiver of the air pump and notice that the candle goes out after some time of pumping, we have a stable material realization, whether we interpret the ceasing of the flame as due to the vanishing air, as Boyle did, or as due to the violent winds produced by the pumping, as Hobbes did (see Shapin and Schaffer 1985, 122).

In general, this type of reproducibility applies to the demonstrative stage of experiments and a fortiori to the large number of more or less standardized experiments, such as the measuring of a boiling point. As we have seen, reproducing an experiment under a fixed theoretical interpretation implies the (approximate) reproduction of its material realization. Consequently, the examples of the former type of reproduction given below are also examples of the latter type.

Nothing I have said so far should be taken to imply that this type of reproduction can be achieved easily, let alone in an algorithmic manner. In many cases craftsmanship is required to reproduce material realizations.[17] If this skilfulness is absent and if it cannot be learned by laypersons or other experimenters, attempts at reproducing the material realization will fail. As Shapin remarks about Boyle: "Time after time in Boyle's texts, technicians appear as sources of trouble. They are . . . responsible for pumps exploding, materials being impure, glasses not being ground correctly, machines lacking the required integrity" (1988b, 395).

In general, there seem to be three kinds of trouble with respect to reproducing a material realization. First, the required craftwork might be so extraordinary that only one or a few people are able to master it. In such a case successful intervention in nature would require a very individual "feeling."[18] In the second place, it may be impossible for all experienced performers, including the original experimenter, to reproduce the required material actions and processes. Reproducing the material realization will fail if one cannot correctly repeat the preparation of object and equipment, the staging of their interaction, and the creation and maintenance of the external conditions required for the closedness of the experimental system. Third, we also need social control of the people involved in the experiment. We must be able to supervise and discipline the experimenters or laypersons in such a way that they will exercise sufficient care in performing their tasks (see Rouse 1987, esp. 220–26). And we must control any other people who might come into contact with the experimental setting, in order to prevent disturbing influences from the outside. That is, we must also create and maintain the social conditions for the closedness of the experimental system (see the analysis and examples in Radder 1986).

Notwithstanding these potential problems, the claim that many, if not most, material realizations in the natural sciences can in fact be reproduced by the original experimenters, contemporary scientists, and lay assistants does seem to be plausible. Moreover, in the case of more or less

17. Cf. Collins's distinction between algorithmic and enculturational models of experimentation (Collins 1975, 206–8). See also Ravetz 1973, chap. 3.

18. Here lies an interesting parallel to the individual bodily "sensibility" of human beings and the problems this poses for an exclusively natural-scientific approach to medicine and health care. See Kunneman and Hullegie 1989 for an illuminating account of these problems.

standardized experiments, the possibilities of reproduction will stretch considerably beyond the period of their original performance.

3.2. Replication of the Experimental Result

Next there is what I have called the replication of the theoretical result *q* of an experiment. In most cases this concerns experiments carried out by different experimenters, in which it is possible to determine *q* by means of a number of different methods. In these cases, therefore, we do not have a reproduction of the same material realization. The scientists' reasons for preferring replication to exact reproduction are recorded by Collins:

> Some said that an exact copy could gain them no prestige. If it confirmed the first researcher's findings, it would do nothing for *them,* but would win the Nobel prize for *him,* while on the other hand, if it disconfirmed the results there would be nothing positive to show for their work. But, if their apparatus was better in some way, in the case of positive results, they would be ahead of the field, and in the case of negative findings they could be seen as being better experimenters than the first researcher. (1975, 210)

In particular, there is a drive to take advantage of local opportunities. As one of Collins's interviewees said: "So what you copy is the part that you are not an expert on and the part you are an expert on you build something you think is better" (1975, 211; see also Rouse 1987, 86–92).

The last quotation shows nicely what may go wrong in attempts at replication. First, building an exact copy of part of the setup may fail. After all, for most replications the original experimenter will not be on the spot to instruct, supervise, or correct the replicators. Furthermore, published accounts of the original experiments are apt to be schematic and incomplete, so that problems can be expected to arise. Second, a general consensus about what constitutes good equipment for measuring *q* may (still) be lacking. A fortiori, this implies that there will be no agreement among experimenters on whether or not some part of the equipment is better for measuring the experimental result. So here too problems for replication may arise. Some well-known problems with replication documented in the literature concern attempts at detecting gravity radiation (Collins 1975), experimental tests of local hidden variable theories (Harvey 1981), and efforts at replicating a number of Boyle's air-pump results (Shapin and Schaffer 1985, chap. 6).

Yet even radical relativists concede that in scientific practice sooner or later these controversies fade away and a more or less stable consensus on procedures of replication develops. A paradigm case is the experimental replication of tests of Avogadro's hypothesis, which says that different substances of 1 gram mole all contain the same number of molecules. In

1913, the French physicist Jean Perrin listed a large number of different experimental methods for checking Avogadro's hypothesis.[19] By that time it had been tested by a set of quite different experimental procedures: for example, in Brownian motion, alpha decay, X-ray diffraction, blackbody radiation, and electrochemical processes. By decreasing the chance of systematic errors and by increasing the systematic nature of Avogadro's hypothesis, these independent replications contributed much to its plausibility. So here is a further reason for preferring replication to strict reproduction. If successful, it not only adds to the credibility of the experimenters in question but also enhances the systematic character and hence the plausibility of the experimental result.

3.3. Reproductions under a Fixed Theoretical Description

In the case of reproducing an experiment under a fixed theoretical description we are, in contrast to replication, concerned with a strict repetition of the experiment from the point of view of the theoretical characterization. It seems to me that this is the primary sense of reproducibility most philosophers have in mind. Thus, in Dingler's view it is only the realization of the "completely unambiguous" theoretical concepts of what he calls the "ideal sciences" (arithmetics, chronometrics, geometrics, and mechanics) that makes possible and guarantees the unambiguous reproduction of experiments (Dingler 1952, 21). Karl Popper states that "the scientifically significant *physical effect* may be defined as that which can be regularly reproduced by anyone who carries out the appropriate experiment in the way prescribed" (1965, 45). And Habermas claims that repeating a (successful) experiment under exactly the same conditions must lead to the same effect (1978, 127).

Note that in the context of my analysis of experimentation it is the theoretical characterization which decides what is meant by "exactly the same." This implies that the actual material circumstances are allowed to vary in a series of reproductions *as long as* these variations are deemed irrelevant and, thus, do not show up in the theoretical description. Below we will meet with examples of such slight variations in the material realization.

Let me first consider attempts by different experimenters to exactly reproduce an experiment. Such attempts at what has also been called "isomorphic reproduction" (Collins 1985, 170–71) are rather exceptional, for reasons explained in the discussion of replication. Yet they do occur.

19. See Salmon 1984, 213–17, and Cartwright 1983, 82–85, for brief accounts of some of these replications.

Colins classifies two experiments from his case studies, one concerning gravity radiation and one concerning psychokinesis, as "fairly isomorphic" (1985, 170). In general, when reproductions of this type, performed by different scientists, occur, much is at stake in confirming or disconfirming the claimed results; in other words, the claims of the original experiment are both very important and very controversial.

Second, in experimental practice many reproductions of the type under discussion are carried out by the original experimenters. They do so in order to make sure that their experiment really works. For instance, Shapin and Schaffer report about Boyle's void-in-the-void experiment, discussed in section 2: "The experiment was quickly repeated in the presence of witnesses, and the same result was obtained" (1985, 43).[20]

As a second example consider the experiments by the Italian physicist Giacomo Morpurgo in the 1960s and 1970s. These experiments were aimed at the detection of free, fractional electric charges, that is, free quarks (see Pickering 1989b). At one stage of a series of experimental runs stability broke down: Morpurgo started to find both integral and fractional charges. After some time, however, he discovered how to restore an interactive stability between his material procedures and a new theoretical interpretation of the working of his apparatus. As Pickering reports: "He increased the separation between the metal plates . . . and found that he could obtain consistent measurements of only integral values in this way" (1989b, 288). Thus, the latter measurements constituted a series of reproductions under a (newly) fixed theoretical interpretation.

Third, and most important, this type of reproduction by the same experimenter also plays a role in most experiments in a slightly different way. Consider, again, Boyle's void-in-the-void experiment. Boyle repeated this experiment with a smaller receiver but still observed exactly the same fall of the mercury column. He concluded that the size of the receiver was not a relevant factor in his theoretical interpretation (Shapin and Schaffer 1985, 43).[21] We can render this procedure more generally as follows: any experimenter who repeats an experiment under varied (material) conditions and finds that these conditions do not make a difference will conclude that they are irrelevant to the theoretical interpretation of the experimental process. Therefore, in this way he or she will have reproduced

20. I will discuss the role of witnesses in section 6.

21. Thus, this experiment exemplifies the fact that reproducing an experiment under a fixed theoretical description is compatible with slightly different material procedures (a glass globe of a somewhat different size had to be mounted upon the pumping device). That is, in this case a reproduction of the experiment under a fixed theoretical interpretation implies an *approximate* reproduction of its material realization.

the experiment from the point of view of this theoretical interpretation. Since such a procedure is very common, this type of reproducibility is prominently present in experimental practice (see Collins 1985, 67–68, for another example).

A fourth way of reproducing an experiment under a fixed theoretical interpretation also occurs frequently. This is reproducing an experiment as part of another experiment. It happens whenever established experimental procedures or stable, experimentally produced phenomena and entities are employed as parts of a larger experiment that is performed for the first time. Cases like these concern the reproduction of experiments that are more or less standardized or are otherwise deemed uncontroversial. There is agreement about some theoretical description and about how to correctly perform the experiments on the basis of this description. Examples of these cases abound. They range from measuring time by means of a clock as part of an experiment to spraying artificially produced positrons into an experiment for detecting free quarks.[22]

Of course, in all these cases success is not guaranteed, for reasons that will be clear from the preceding discussion: agreement about the theoretical description may break down and/or problems may arise in trying to (approximately) reproduce the material realization. Yet, I think that the previous analysis shows that reproducing an experiment under a fixed theoretical interpretation does occur frequently in scientific practice, even if attributing reproducibility to a particular experiment is fallible in principle. Moreover, this type of reproduction is not at all trivial or insignificant. In most cases, it requires hard and ingenious theoretical and experimental work to achieve.

4. DATA VERSUS PHENOMENA?

For the purpose of further clarifying the preceding account of experimentation and reproducibility, it is instructive to briefly examine its implications for the distinction between experimental data and phenomena. The importance of this distinction has recently been emphasized in an illuminating paper by James Bogen and James Woodward. Their main philosophical conclusion concerns the relation between explanation and observation. They claim that systematic scientific theories do not purport to explain the observable data but only the nonobservable phenomena:

22. Cf. Hacking 1983, 22–24. The reproductions of a TEA laser, discussed in Collins 1985, chap. 3, are also of this kind; the scientists in question reproduced this device in order to use it in other experimental work. Calibration techniques which, if successful, show the ability of parts of the experimental setup to reproduce known results provide another illustration; see Franklin 1986, 175 ff., and Hones 1990.

> Data, which play the role of evidence for the existence of phenomena, for the most part can be straightforwardly observed. However, data typically cannot be predicted or systematically explained by theory. By contrast, well-developed scientific theories do predict and explain facts about phenomena. Phenomena are detected through the use of data, but in most cases are not observable in any interesting sense of that term. (Bogen and Woodward 1988, 305–6)

Bubble-chamber photographs and thermometer readings are examples of data. The existence of phenomena, such as weak neutral currents or the melting point of lead, is inferred from the data in combination with other (e.g., statistical) premises.

I will not discuss Bogen and Woodward's main philosophical claim but instead focus on the rather sharp contrast they make between data and phenomena (Bogen and Woodward 1988, 305–22). Data, they claim, are *observable;* they are *created* by experimenters; and they are highly dependent on the *multicausal* and *local* experimental contexts in which they have been produced. Phenomena, in contrast, are inferred and are not observable; in general, they are real and not created; their occurrence results from the interplay of a relatively small number of causal factors; and they can be detected in a variety of experimental contexts.

Now, how does this distinction between (experimental) data and phenomena relate to the analysis presented in the present paper? Suppose we have a theoretical description of the form $p \rightarrow q$. Then it seems to me that the experimental result q will, in many cases, provide a description of what Bogen and Woodward denote as phenomena, while what they call data will be included in the composite theoretical premise p. After all, the experimental result, just as the claimed existence of a phenomenon, is the outcome of a process of inference from the theoretical description p. However, if this interpretation of Bogen and Woodward's notions of data and phenomena holds good, the above analysis of experimentation implies that their sharp contrast between data and phenomena cannot be upheld.

In the first place, it is not the case that the data included in the theoretical description p of an experiment are all or even for the most part observable in Bogen and Woodward's sense, in which what is observed "should figure somehow in retinal interactions" (1988, 347). For instance, even when we want to detect such a simple phenomenon as the boiling point of a liquid, we have to make use of, among other things, the data that "the liquid is pure" and "the air pressure remains constant during the measurements" (see Radder 1988, 60–69). In this case these crucial data are by no means straightforwardly observable in the required sense. Since it is not difficult to find many more counterexamples, especially in more advanced experiments, the claim of Bogen and Woodward is simply not plausible as it stands.[23] From my point of view it would rather be the material realization of the experiment that is "observable." However, its description in common language can definitely not be used to infer the experimental result. For this purpose we do need a *theoretical* description including theoretically interpreted data.

Second, *if* data are "created" in the laboratory,[24] a fortiori the same must apply to phenomena. After all, the specific realization of a phenomenon (e.g., the boiling point of a liquid) depends crucially on the realization of the created situations described by the data (e.g., that the liquid is pure). Bogen and Woodward's claim to the contrary seems to rest on a confusion of epistemological and ontological matters. They are right in maintaining that many phenomena are replicable; that is, they can be shown to occur in a variety of experimental situations. In these cases we have good epistemological reasons to conceptualize the phenomenon or experimental result "as such" and ask for an explanation of it in abstraction from the various ways in which it has been experimentally produced. For instance, if we are successful in experimentally replicating a fixed boiling point for our liquid, we may plausibly proceed and look for a systematic explanation of it in terms of a molecular theory of phase transitions. Hacking's view of the creation of phenomena, however, is ontological. He claims that a phenomenon, such as the Hall effect, does "not exist outside certain kinds of apparatus" (Hacking 1983, 226). This claim cannot be refuted by pointing out that when a phenomenon is replicable, the phrase "certain kinds" may be broadened into "a variety of kinds," since any particular realization of the phenomenon remains dependent upon *some* specific set of laboratory conditions.

Third, Bogen and Woodward claim that not only phenomena (1988, 317) but also data should be reproducible: "Data must also be such that it is relatively easy to identify, classify, measure, aggregate and analyze in ways that are reliable and reproducible by others" (320). But, how can these requirements be met if data are at the same time "idiosyncratic to particular experimental contexts, and typically cannot occur outside of those contexts"? (317). Thus, there is a tension in Bogen and Woodward's view between the reproducibility of the data and the reproducibility of the phenomena. This tension is due to the claimed locality and idiosyncracy of data as contrasted to the nonlocality and stability of phenomena. In my view both data and phenomena may be reproducible and may therefore

23. Note that they could retain their philosophical claim if they weaken it somewhat to the effect that "systematic scientific theories aim at explaining nonobservable phenomena and not data, whether the latter are observable or not."

24. For a discussion of whether or not this claim is plausible, see Radder 1993.

be nonlocal and stable in certain respects. Yet, as we will see in the next section, the form this nonlocality and stability takes may be different in the two cases as a consequence of differences in type of reproducibility.

From the point of view of the present paper we can, I think, conclude that Bogen and Woodward have the right intuition that there is a substantial difference between reproducibility under a fixed theoretical description $p \rightarrow q$ (where p includes the data) and the replicability of the phenomenon or experimental result described by q. The questions Reproducibility of what and by whom? may get different answers in the two cases. As a consequence these types of reproducibility play different roles in scientific research. Nevertheless, it will also be clear that these differences do not entail the sharp contrast between data and phenomena that has been claimed to exist by Bogen and Woodward.

5. Normativity, Stability, and Nonlocality

So far I have argued for the claim that many experiments in the practice of the natural sciences are reproducible in one or more of its senses. At the same time a number of qualifications were added to this claim. First, I have pointed out the types of problems and failures that may and do turn up in actual attempts at reproducing experiments. Next, the essential fallibility of any particular attribution of reproducibility was noted. It also appeared that the question Reproducible by whom? may be answered differently, depending on the relevant experiment and on the type of reproducibility. Finally, I emphasized the fact that reproducibility does not result from a passive registering of regularities, but that it demands theoretical and practical work in order to get it.

5.1. Normativity

In addition to the factual occurrence of reproductions, we may ask whether or not reproducibility, in one or more of its forms, functions as a *norm* of experimental practice. I think it does (see also Hones 1990). But, of course, this does not entail that every experiment should in fact be repeated by different scientists. In many cases its reproducibility is taken for granted. Yet, as we have seen, aspects of the norm of reproducibility have been incorporated into experimental practice through procedures such as the division of labor between laypersons and experimenters and the testing for (ir)relevant factors. Moreover, the operation of this norm becomes explicit when the experiments are ambiguous, controversial, or very significant. Boyle, for example, thought the replication of his air-pump experiment by other researchers very desirable, and he even expressed despair when many attempts turned out to be failures (see Shapin and

Schaffer 1985, 59–60). Faraday, in his experiments on magnetic rotation in the early 1820s, designed a more robust "pocket" rotation device and even sent it to some of his fellow scientists in order to facilitate the reproduction of his experiments by them (see Gooding 1985, 120–22). And in the case of Joseph Weber's gravity-wave experiments, the significance the involved scientists attached to replication was very obvious (see Collins 1975).

The argument so far is, I think, sufficient to refute the claim made by a number of authors that norms (or rules, standards, criteria, guidelines) that exceed the context of their local use do not play any role in scientific practice (see, e.g., Rouse 1987, 93, 119–25; Woolgar 1988, 45–48; and cf. Latour and Woolgar 1979, 24). These authors reason as follows: applying a norm requires locally situated judgments; hence, norms do not and cannot determine practice; therefore, nonlocal rules or norms do not play any guiding role in science; instead, they merely form post hoc rationalizations of already established practice. However, although the first two statements are correct (or even trivial), the third does not at all follow. The claim that a rule should *compel* particular courses of action in order to have any practical impact at all is mistaken. The much more plausible alternative is, of course, that nonlocal rules may be practically effective *along with* all kinds of local factors.[25] As we have seen, the requirement of reproducibility was obviously consequential in such diverse contexts as that of Boyle, Faraday, and the gravity-wave experimentalists, even if it did not determine the particular courses of action.

Next, following upon the empirical issues of the occurrence of reproductions in scientific practice and of the actual functioning of the norm of reproducibility among scientists, there is the normative question of whether or not experiments *should* be reproducible. That is, can reproducibility be taken as a necessary condition of epistemic success and do independent reasons for advocating such a norm exist? Again, the differentiation between the three types of reproducibility is required to obtain a sensible answer.

It seems to me that the reproducibility of the material realization ought to be generally satisfied. In other words, I consider the norm of the reproducibility of material realizations a constitutive norm of (good) experimental natural science. My main argument for this is that, without such a material stability, performing experiments and, very important, us-

25. On this specific issue I agree with Collins, who states that "there must be something more to a rule than its specifiability" (1985, 14). Cf. also the discussion of "values, norms and practice" by David Henderson (1990, 128–33). For instance: "Much of what we learn is socially taught; yet once learned, it informs and constrains social practice. Thus, practice and value should not be set apart as largely distinct matters" (128).

ing them in technologies would be unreliable and dangerous.[26] With respect to the other two types of reproducibility the situation is less clear. In general, reproducibility under a fixed theoretical description will facilitate communication among scientists, but it will also block creative reinterpretations of the experiment in question. Therefore, it does not seem advisable to put this type of reproducibility as a general constraint on experimenting. To a lesser extent the same applies to the case of replication. As we have seen, successful replication by means of a number of different experiments will establish the experimental result more firmly and will increase the systematic character of our scientific knowledge. Yet, in this case, too, completely different interpretations of this result can and should not be excluded. It would be inappropriate, therefore, to argue for replicability as a general norm of experimentation in the natural sciences. In sum, the norms of reproducibility under a fixed theoretical interpretation and of replicability of an experimental result are best regarded as regulative, rather than constitutive, norms.[27]

5.2. Stability

By reproducing an experiment or experimental result it becomes more stable, that is, less sensitive to locally varying factors. But, as we have seen, reproducibility is a complex concept. Therefore, the same holds good for stability: distinguishing types of reproducibility leads to different forms of stability.

A reproducible material realization requires that laypersons or other experimenters who possess or are able to acquire the necessary skills and who are willing or can be made to do the required craftwork can successfully carry out the experiment. Therefore, that an experiment is reproducible in this sense implies that these craft skills are learnable and, thus, that the experiment is more stable against, or less sensitive to, local variations in individual skilfulness. Moreover, in this case there is a very important second form of stability, since local controversies concerning the theoretical interpretation of the experiment in question do not matter as long as they allow for reproducing its material realization.

Another form of stabilization works by means of replications. If an experiment can be replicated in a number of very different situations, the

26. Please note that I do not say that a reproducible material realization is *sufficient* for accepting and realizing the experiment (or technology) in question. Other (social, ethical) considerations are also important in this respect. See Radder 1986, 674–79.

27. Constitutive norms or rules (e.g., of a game such as chess) are definitory: if we do not follow them, we simply are not playing the game (correctly). Regulative rules are like the strategic or tactical rules of a game: on the average it helps to follow them, but they cannot guarantee success in a particular case.

plausibility of its result will increase accordingly. The result will become less sensitive to criticisms of any particular experiment. That is to say, the more systematic the position of this result within a body of knowledge, the more stable it is. Apart from this, many replications aim at *increasing* the stability of an experimental result, in the sense that the effect is amplified, the noise reduced, or the result made less sensitive to potential disturbances. Above we met with an example of this in the case of Faraday's more robust magnetic rotation device. The same process of stabilization through replication occurred in the case of Boyle's air pump: during the 1670s air pumps became routinely applicable and commercially available experimental devices (see Shapin and Schaffer 1985, 274–76).

Finally, consider the reproduction of an experiment under a particular theoretical description. An example is a pneumatic experiment performed by Boyle first at home and later in the rooms of the Royal Society. Such a repetition stabilizes the experiment under the description in question in the sense that it turns out to be practicable at different times and places. Put differently, even when the experiment involves specific skills and a specific interpretation, it still shows a nontrivial insensitivity to variations in space and time.

We may conclude from the discussion in the present section that both the norms of reproducibility and the stable *aspects* of the experimental knowledge produced by their application can be seen to be specific, nonlocal patterns of scientific practice.[28] This is not to say that the knowledge obtained has been decontextualized or universalized, but rather that it does, or has been made to, apply to a more or less large number of local situations that may vary in certain respects. In *these* respects the experimental knowledge has been delocalized.[29]

6. Experimental Science and Its Social Legitimation

How do experiments that have been tried out successfully by individual experimenters acquire a more widespread public recognition? This question has been investigated by some historians of experimentation recently. David Gooding, in his studies of the work of Faraday and his contemporaries, points to the significance of public demonstrations of experiments

28. This conclusion agrees with Peter Galison's account of experimentation, especially with his views on experimental constraints and experimental stability (see Galison, 1987, chap. 5). For more general discussions of the significance of nonlocal patterns in scientific practice, see Radder 1992b, 150–55; 1993.

29. In this sense I disagree with Joseph Rouse's claim that even where stabilization, or, to use his term, standardization, obtains, experimental knowledge remains *essentially* local knowledge. See Rouse 1987, 111–19.

with the aim of making particular phenomena and processes seem evident and natural. By being *witnessed* by lay audiences, experimental results would become recognized as part of a collective body of knowledge (Gooding 1985; 1989, 191, 202–3). At a more fundamental level there is the question, not so much of the public recognition of *specific* experimental results, but of the experimental method as such. It is this question of how experimental natural science is socially legitimated which is at issue here.

For the days of the early Royal Society this issue has been discussed in detail by Shapin and Schaffer (1985, 55–60; Shapin 1988b). They emphasize the central role of witnesses in generally establishing the authority of the experimental way of producing knowledge. These witnesses testify to the fact that the experiments in question have really been performed and that the knowledge produced is reliable. For this purpose witnesses should be credible, which was taken to mean that they should originate from the same (higher) classes of society as the natural philosophers themselves did. In this way the new experimental philosophy opposed all at once the secret experiments of the alchemists, the thought experiments of the rationalist scientists, and the antiexperimentalism of the school philosophers.

Yet, although the early British experimentalists were successful in establishing the experimental "form of life" as a social phenomenon, it will be clear that this procedure for legitimizing experimental science has a rather restricted scope. In fact, even in the beginnings of the Royal Society the procedure involved two ambiguities. These concern who should witness and what should be witnessed. Should witnesses be laypersons who have no real connection to experimental philosophy or should they "have knowledge of the things they deliver," as Boyle once put it (Shapin and Schaffer 1985, 59)? And should they testify only to the existence of "matters of fact" or also to the truth of experimental hypotheses? In practice all different possibilities occurred. It will be clear, though, that this way of legitimizing experimental science leads to a dilemma. On the one hand, when witnesses are knowledgeable and are required to validate experimental hypotheses, they are likely to be prejudiced in favor of one claim or the other. Consequently, their role as independent and credible testifiers may be questioned. As Peter Dear (1985, 156) has argued, this problem was to some extent recognized by the members of the Royal Society. On the other hand, when witnesses are genuine laypersons required to establish matters of fact, their testimony cannot be of any help in settling controversies about different theoretical interpretations of the same matter of fact, such as the debate between Boyle and Hobbes on the void-in-the-void experiment. What is more, in the overwhelming majority of modern experiments the theoretical interpretations will simply be incomprehensible to such laypersons.

In conclusion, the procedures for validating experimental knowledge employed by the Royal Society involve an inherent and irresolvable tension. Moreover, it will be clear that modern experimental science is no longer socially legitimated by means of the method of direct witnessing. The modern laboratory is much more a private place than its seventeenth-century precursor (see Shapin 1988b, 404). Nowadays, debates on the interpretation of scientific experiments take place exclusively among the scientists themselves. This is, of course, not to say that researchers may not make use of all kinds of resources from the society at large. But it means that, in general, people in society do not themselves actively *participate* in the technical-scientific debates.[30]

6.1. Experimental Science and Technology

Given this state of affairs the question arises as to what form of social legitimation has replaced the early procedures of the Royal Society. Why is society willing to provide ever larger means for an experimental science which has become ever less intelligible to the general public? To be sure, one cannot expect this question to have one and only one answer. Different times, different places, and different sciences may have produced different legitimations.

Yet, concerning experimental natural science as such, I think two legitimations play a fairly prominent and general role. The first is the claim that science is valuable because it delivers the truth about nature or, at least, promises to eventually give a true account of nature. The second major social legitimation is framed in the claim that experimental science is practically useful, that its results can often be fruitfully incorporated into all kinds of technological projects. Gooding, for instance, mentions the fact that, already in the 1820s, British experimenters on electromagnetism were eager to stress the significance of their experiments for navigation and the discovery of ores (1989, 202–3). Actually, in present-day society the "technological" legitimation seems to be the most influential, and I will restrict myself to it in the following. As a matter of terminology I will speak of *experimental technology* in cases where materially realized scientific experiments are used as parts of technological systems.[31]

So far this answer to our question seems rather evident, but it is also

30. Shapin stresses the notion of participation in the context of the Royal Society: "The Royal Society expected those in attendance to validate experimental knowledge as participants, by giving witness to matters of fact, rather than play the role of passive spectators to the doings of others" (1988b, 390).

31. The point is that experimental natural science often is *one* of the factors contributing to a successful technology. It is by no means implied that the process of transforming scientific experiments into technological artifacts is a simple or unidirectional matter of "application." See Latour 1983 and Radder 1986.

rather uninformative. The more difficult task is to specify the notion of social legitimation through the technological potential of science. My suggestion is that an important part of the social legitimation of experimental natural science rests on the possibility of actually achieving reproducible material realizations of experiments within the context of experimental technologies. That is, it is due to the action and production aspects of experimenting and to the availability of *some* theoretical description on the basis of which we can obtain reproductions. What is not required is a general agreement on specific theoretical interpretations. Consider, for instance, the manned U.S. space missions to the moon in the 1960s. In this case, as long as the experimental technologies continued to work in the sense of remaining materially realizable in a stable and reproducible manner, the average politician and taxpayer did not at all bother about controversies concerning the correct theoretical interpretation of some feature of the project. For example, a debate on the question of whether the trajectories of the spaceships should be calculated by means of classical mechanics or by means of the special theory of relativity does not seem to be relevant to the social legitimation of this experimental technology.[32] As we will see in the next section, this view has some significant consequences with respect to the role of experimental science and technology in present-day society.

7. Devices and Black Boxes

On the basis of the above analyses we can also provide a general explanation of the *possibility* of the worldwide use of experimental science in technological projects. Given the complexity and the spatiotemporal variability of theoretical interpretations, the relative autonomy of the material realization with respect to particular theoretical descriptions is a crucial and fortunate feature both for producers and for consumers of experimental technology. On the one hand, if it were the case that first a general, scientific consensus must arise about all the details of a particular account of how an experiment works, then in view of the present state of natural

32. Please note that the above considerations refer to de facto functioning legitimations and aim at elucidating these somewhat further. I myself do not think that these legitimations are satisfactory. In fact, I even think that the entire enterprise of justifying experimental natural science "as such" or "as a whole" (in contrast to specific developments, results, or uses) is mistaken. A systematic argument for this is that, although experimental science does exhibit nonlocal patterns (see section 5), it is equally the case that its concrete development, results, and uses depend upon local characteristics. As a consequence there cannot be a general justification of science as such. See also Radder 1988, 118–19, for my objections to Habermas's notion of a "technical interest" as being generally constitutive and justificative of the natural sciences.

science its technological adoption would be very difficult. On the other hand, the large-scale use of experimental technology would become virtually impossible if every user should first acquire and accept the knowledge about even one of its theoretical interpretations. Thus, it is this relative autonomy of reproducible material realizations which makes it possible to produce and use experimental technologies on a large scale, even in the face of theoretical controversy and theoretical ignorance.

This explanation of the possibility of experimental technology can be fruitfully used in commenting upon some other recent views on the issues under discussion, notably those of Borgmann and Latour. Borgmann, in his book *Technology and the Character of Contemporary Life,* on the one hand fully accepts what he calls the "apodeictic" results of natural science and he interprets these results in a strongly realist way (see Borgmann 1984, 15–31). On the other hand, he is very critical of the form modern, science-based technology has taken in our present-day society. He characterizes this form as a "device paradigm" (1984, esp. chap. 9). A device, such as a central heating system, is a technological artifact that consists of machinery and that procures a commodity. One of the central characteristics of modern technological devices is that their machinery is hidden from, and their operation unfamiliar to, its users. This admits of a certain variability of the machinery as long as the device continues to provide the same commodity. The core of Borgmann's criticism of modern technology, then, is that this paradigmatic device character disengages and alienates people from their "ontic roots." According to him, it makes them incapable of a truly human interaction both with their natural and with their social environment.

If we leave aside Borgmann's inadequate views on science, we see that the analysis of experimental science along the above lines goes some way in explaining the device character of a technology that makes use of an immense body of specialized, fragmented, and ever changing experimental knowledge and artifacts. As we have noted, demanding an intensive engagement of every user with this body of knowledge would make experimental technology practically impossible. For large-scale experimental science and technology a division of theoretical and material labor is a crucial requisite. As a consequence, for the average consumer modern experimental technologies will necessarily have the form of devices or, to use a related term, black boxes.

Next, this relation between the notions of "device" and "black box" enables me to make some comments on Latour's views on the development of, to use his term, techno-science (see Latour 1987). He interprets this development as essentially a struggle for power within networks of heterogeneous, antagonistic "actants." In this process various powerful actors, or Machiavellian "Princes," strive for control and stability of their networks. The more black boxes a Prince is able to uphold, the more stable his power. According to Latour, it is a defining characteristic of a techno-scientific black box that well-established scientific facts and unproblematic technological artifacts are intrinsically connected in it (Latour 1987, 131). Put in my terms, Latour's view implies that a stable, experimental technology requires both a general, be it an implicit, agreement on its theoretical description and a reproducibility of its material realization.

On the basis of the present analysis such a view appears not to be fully adequate. It is true that theoretical controversies that question the reproducibility of the material realization—and thus the working—of an experimental technology will affect its stability. In contrast, all theoretical interpretations that are compatible with the same, reproducible, material realization are in this respect[33] equivalent, so that a controversy about these interpretations will not directly touch the stability of the techno-scientific black box. In other words, in many cases experimental technologies are black boxes in the original sense of the word: they are systems for which only input and output matter, *independent* of whether the account of their internal workings is scientifically well-established or controversial.[34] The important implication of this is that it enables us to analyze and evaluate the worldwide pervasion of materially realized experimental technologies without having to assume the existence of Latourian super-Princes, who make all the elements of the technological black boxes "act as one" (Latour 1987, 131). The imperialistic expansion of technological devices is compatible with a considerable amount of pluralism at the level of their theoretical interpretations.

8. Conclusion: Revaluing the Invisible Actors

In concluding the paper, I want to return one more time to the early days of the Royal Society. My view so far has been that the current social legitimation of modern experimental science cannot be based on the witnessing of credible lay participants. Instead, it rests (at least partially) on the potentialities of materially realizing experiments in a stable and reproducible way. One of the main goals of this paper has been to make explicit

33. Of course, there might be other differences between these "materially equivalent" interpretations which may make some difference with respect to the stability of the relevant experimental technologies.

34. Latour (1987) notes this meaning of "black box" on pp. 2–3, but further on in his book he changes it in favor of the definition given above.

these crucial action and production aspects of experimenting by means of the notion of material realization and to point out its actuality in processes of division of labor.

In the context of the Royal Society this means revaluing the importance for legitimating experimental science of the "invisible actors," such as Boyle's many laborers, operators, assistants, and chemical servants (see Shapin 1988b, 395–96; 1989). This implies a shift to an even stronger form of participation of laypersons: from participation through witnessing to participation through performing the experiments and producing the experimental systems. In this way the ambivalence of the Royal Society procedures for legitimizing knowledge noted in section 6 has been resolved. In the struggle to establish particular theoretical interpretations, only knowledgeable scientists participate. The legitimation of experimental knowledge claims, however, does not presuppose the truth of a specific interpretation. Instead, it derives from the possibility of the reproducible material realization of the experiments by "lowly folk," whose performance is stable just because it "stands above" any of the particular, competing interpretations (see Dear 1985, 156; see also Radder 1988, 74–75).

More in general, the argument of this paper entails *both* a critique of a theory-biased philosophy of science (which has its roots in a much broader, cultural bias toward intellectual work) *and* a revaluation of the mostly invisible but crucially important craftwork in the experimental natural sciences, as it is performed by the experimenters themselves or by what I have called laypersons.

Acknowledgments

I would like to thank the following persons. Martha Féher, at the Technological University of Budapest, posed the stimulating question of the relationship between materially realizing and witnessing science. Brian Baigrie and other participants in the Table Top Experiments conference in Toronto, March 1990, made a number of valuable comments on an earlier draft. I have also profited from a fruitful discussion with my colleagues in our research group on Knowledge and Normativity at the Vrije Universiteit in Amsterdam.

FIVE

SCIENTIFIC PRACTICE: THE VIEW FROM THE TABLETOP

Brian S. Baigrie

1. INTRODUCTION

ONE OF THOMAS KUHN'S EARLIEST ESSAYS, "The Essential Tension" (1959), anticipated what is perhaps the critical philosophical issue in the burgeoning sociological literature on science: the tension between innovation and tradition in scientific practice. The reward system of science is geared toward innovation, but producing desirable results often implicates scientists in the constraints and traditions of their community(s). Kuhn's influential *The Structure of Scientific Revolutions* (1962) attempted to relieve this tension by locating innovation and tradition at different *temporal* stages, the former during "extraordinary" periods rife with institutional chaos, and the latter during comparatively tranquil periods of "normal science" (see Nickles 1988). Though the Hobbesian device of normal science proved useful as an idealization for highlighting the skills reflected in even the most mundane scientific procedures, scholars protested that the *stability* of much scientific achievement must be understood in an institutional context where opportunism and its associated bickering are standard fare—where Hobbesian elements stand shoulder to shoulder with Machiavellian ones.

The *homogeneity* of Kuhn's conception of normal science—the fact that shared methods and procedures are all that stands between scientists and the unproductive wrangling of sociologists and philosophers—testifies that the author of *Structure* sided with tradition against innovation and, in so doing, with a view of scientists as members of a homogeneous, unitary community. Three of the papers in this volume—by Galison, Pickering, and Radder—signal a rejection of Kuhn's resolution to the tension between tradition and innovation, as well as the growing conviction that, in order to come to grips with the success of science, it must be seen in terms of the interaction between *heterogeneous cultural practices* that collaborate so as to produce new knowledge: "Normal science" is not a single set of shared practices and techniques but is a number of heterogeneous activities that work together to push scientific culture in new and sometimes unexpected directions.

Though these scholars are unanimous in resisting the homogeneity of Kuhn's paradigms, the confrontation between Galison and Pickering (with which I will be principally concerned here) testifies that they are at loggerheads about how experimental practice is to be described.[1] The view that I will advance in this paper is that so long as historical inquiry is restricted to Kuhn's favored examples (i.e., to the leviathans associated with the names of Newton, Darwin, and Einstein), the emphasis placed by Kuhn on the Hobbesian constraints on scientific practice may seem reasonable. However, when we turn to science on the tabletop—to the science of Boyle, Hertz, and Morpurgo (no paradigm names here)—we need a new vocabulary to enable us to speak about science as heterogeneous cultural practices that change over time, in consequence of the activities of scientists with widely differing interests and competencies. Indeed, when viewed from the tabletop, the elements of scientific culture are more readily seen as proper to the field of human consciousness, as resources to be used in one's ongoing work, than as abstract social practices that constrain the behavior of scientists.

The view that I will develop therefore sides with Pickering in holding that this new post-Kuhnian vocabulary will be one of resourcelike cultural elements that are stamped with the aims and goals of scientists. Of course, the tabletop is an unlikely place to glean the whole story about science. If we want to extend our intellectual horizons to address the larger institutional framework of science, with its complex ties to industry and the universities, we will need to adopt a different social perspective and perhaps a different vocabulary. It is here that the sociologism advocated by Galison may come to the fore, though I will leave this other story for another time.

2. The Babel of Practice

In a recent paper published in *Isis*, a widely read journal in the history of science, Jan Golinski praises the "pioneering" influence that the work of sociologists, such as Collins (1985), Pickering (1984b), and Pinch (1986a), has exerted on Gooding (1985), Galison (1987), Shapin and Schaffer (1985), and other historians of science; historians and sociolo-

1. My presentation of the views of Galison, Pickering, and Radder is based on their original papers as presented at the conference on Table Top Experiments held at the Institute for History and Philosophy of Science and Technology, University of Toronto. Though the papers published here are essentially the same, the authors have made some alterations that do affect my argument in places.

gists are portrayed as mutually collaborating to produce what Golinski calls a "theory of practice."[2] If Golinski is right, this will certainly be cause for celebration. By the same token, however, his suggestion that our understanding of the development of scientific ideas and institutions has been insensitive to the activities of scientists will come as something of a surprise to many scholars.

At the dawn of the modern period, Juan Luis Vives, a tutor at the English court and a scholar of great erudition, declared in his *De Tradendi Disciplinis* (1531) that scholars "must not be ashamed to enter into the workshops and into the factories, asking questions of artisans and trying to become cognizant of the details of their work" (cited by Rossi 1970, 6). The French potter Bernard Palissy subsequently gave more vivid expression to what in some respects appears to be a blueprint for many contemporary ethnomethodological strategies pursued by the likes of Latour and Woolgar (1979). In 1580 he declared that

> through practice I prove that the theories of many philosophers, even the most ancient and famous ones, are erroneous in many points. Anyone can ascertain this for himself in two hours merely by taking the trouble to visit my workshop. Marvelous things can be seen here (demonstrated and proved in my writings and arranged in an orderly manner with texts at the bottom so that the visitor may be his own instructor). I assure you, dear reader, that you will learn more about natural history from the facts contained in this book than you would learn in fifty years devoted to the study of the theories of the ancient philosophers. (Cited by Rossi 1970, 2)

Descartes, for many sociologists the veritable devil in modern philosophy, himself declared in his *Principia Philosophia* (1644, pt. 4, art. 203) that those artisans and engineers well versed in building mechanical contrivances are in the best position to discern what arrangements of insensible particles are at work in producing natural phenomena.[3] Granted

2. Philosophers of science (with the exception of Ian Hacking) are depicted by Golinski as defenders of orthodoxy, as holding that science is "no more complex in principle than the knowledge of tables and chairs" (1990, 501). It is true that some philosophers place scientific knowledge on a par with our beliefs about mundane objects, but Golinski draws the wrong conclusion. What "some philosophers" believe, rather, is that our knowledge of tables and chairs presents a very complex problem. Though Golinski is confident that "it is the complexity (the use of instruments, for example) that makes scientific knowledge much more problematic [than everyday knowledge]" (501), his conception of the cultural factors involved in constructing even mundane artifacts seems to me to be too narrow, particularly in view of our lack of anything like a reliable "meter" for comparing the complexity of diverse cultural practices.

3. André Baillet, Descartes's biographer, reports that as a member of the research team assembled by Queen Christina of Sweden, Descartes devised the idea of a school of arts and

that Descartes conceived of God as an "artificer," it is hardly shocking that he regarded human hands as indispensable aids in cognition. A generation later, another so-called rationalist, Gottfried Leibniz, remarked in a little known paper on the importance of the labor of artisans that "we need a real theater of human life, taken from the practice of men, very different than that which learned men have handed down to us, in which, great as it is, there is only that which turns out to be useful for the production of harangues and sermons" (Gerhardt 1875–90, 7:181).[4]

The passages courtesy of Vives, Palissy, Descartes, and Leibniz do not amount to an early statement that we can understand science by watching what scientists do. There is an asymmetry between *truth about nature* as revealed in workshop practice and *truth about science* as revealed in scientific practice. In the former case, the practices of artisans and engineers are consulted so as to know nature, whereas in the latter case the practices of experimenters and technicians are studied so as to know science itself. The difference roughly is between seeing practice as a means to an end (i.e., knowledge of nature) and seeing science as an object of study in its own right (e.g., the sociology of science). Though these authors did not advance a theory about the activities of artisans, their writings testify to a growing conviction that knowledge of the devices fashioned by the divine artificer is enhanced by the collaborative effort of many hands engaged in

---

trades to foster collaboration between natural philosophers and artisans. In Baillet's words: "Ses conseils alloient à faire bâtir dans le collége Royal & dans d'autres lieux qu'on auroit consacrez au Public diverses grandes salles pour les artisans; à destiner chaque salle pour chaque corps de mêtier; à joindre à chaque salle un cabinet rempli de tous les instrumens méchaniques nécessaires ou utiles aux arts qu'on y devoit enseigner; à faire des sonds sussisans, non seulement pour seurnir aux dépenses que demanderoient les expériences, mais encore pour entretenir des maîtres ou Professeurs, dont le nombre auroit été égal à celuny des arts qu'on y auroit enseignez. Ces Professeurs devoient être habiles en Mathématiques & en Physique, afin de pouvoir répondre à toutes les questions des Artisans, leur rendre raison de toutes choses, & leur donner du jour pour faire de nouvelles découvertes dans les arts" (1691, 2:434). Clearly, Descartes would have resisted the suggestion that experiment can proceed in the absence of theory, but the ties that he attempted to forge between natural philosophy and the mechanical arts, and the admission that it is the artisans who ultimately are responsible for making new discoveries, testify that even Descartes, caricatured by Voltaire and subsequent generations of Newton's legions as attempting to manufacture the cosmos with the sole device of a disembodied *res cognitans,* was alive to the importance of the practices of artisans and engineers for scientific understanding.

4. Leibniz's paper bears the title *Discours touchant la méthode de la certitude et l'art d'inventer pour finir les disputes et pour faire en peu de temps de grands progrés.* The text reads: "il nous faut un veritable Theatre de la vie humaine tiré de la practique des hommes bien different de celuy que quelques sçavans hommes nous ont laissé, dans lequel tout grand qu'il est, il n'y a gueres que ce qui peut servir à des harangues et à des sermons." Of course, Descartes and Leibniz had Scholastics in mind when they chastised those who regarded practical activities as insignificant.

workshop labor. These authors would have disagreed about whether or not these practical activities were the whole of *sciencia*. Some believed that the meditations of the cloistered academician could still be of service. Descartes, for instance, held that the better part of science is derived from first principles that are plain to the attentive mind; though practical activities were an invaluable aid in cognition, the better part of science was produced by clear thinking.

This issue is complicated by the fact that during the seventeenth century the very concept of science underwent a series of pronounced shifts as the technical arts and natural philosophy continually bounced off one another. What we can extract from these passages, however, is that many writers in the early modern period regarded scientific behavior in practical terms—not as laboratory skills (laboratories were still in the future)—but in terms of the practical know-how involved in fiddling with artificially prepared substances and mechanical devices. My point, then, is that some scholars have long regarded science as a practical activity, at least since the dawn of the early modern period. What is new and invigorating about the recent literature on science, therefore, cannot be that science is suddenly being regarded in terms of practical activities. It is rather that these practical activities are now viewed (in theoretical terms) as an object of study.

Even this transition from thinking about science in terms of practical activities to thinking about scientific activities as an object of study is not as recent as one might suppose. The introduction to Theodore Brown's dissertation (1968), for instance, proclaims:

> I have come to believe that *scientific ideas cannot be separated from the activity that produces them* . . . [and this study] treats physiology as a dynamic enterprise, not as a static aggregate of ideas, and it considers the influence of the mechanical philosophy in special sets of circumstances which . . . affect the timing, scope and direction of that enterprise. ([1968] 1981, vii, emphasis added)

Kuhn is acknowledged as "the principal intellectual father of this dissertation" (viii). In his engaging 1977 study of the development of Galileo's military compass, Stillman Drake remarked that "few readers today will care to retrace Galileo's actual work on the first universal mechanical calculating instrument. Most historians prefer to search for profound philosophical tenets supposedly embedded in Galileo's thought, instead of following its formation in his mere practical activities" (1977, 50). These sentiments recently have been echoed by another student of Kuhn, who writes that

> living sciences cannot be corralled with verbal generalizations and definitions. Attempting to capture a vibrant science in a precise, logical structure

> produces much the same kind of information about it that dissection of corpses does about animal behavior; *many things that depend upon activity are lost.* When such a thing eventually happens, as it often does, the science is either very dead, of interest only to a few people at the margins of the active discipline, or else it has been transformed into a technical artefact and relocated to an engineering department. (Buchwald 1992, 1–2, emphasis added)

Experimental practice has traditionally taken a backseat to theory and formalism in the history and philosophy of science (in part because of the deeply embedded supposition that the mathematical structure of Newton's *Principia* [1687] was responsible for its unprecedented success, as well as making it a standard to which all science should be held accountable).[5] However, with the possible exception of the most unrepentant a priorist philosopher, it appears that hardly anyone is still infused with a Scholastic disdain for practical activities. A more likely explanation for the discrepancy between the ubiquitous references to scientific practice in the literature of the last quarter century and the insistence that scholarship has suffered under the weight of a deeply embedded prejudice against practical activity is that scholars today widely disagree as to what counts as scientific practice. The philosopher may discuss the pros and cons of assorted methodological strategies for certifying belief; the historian may analyze a series of mathematical formulae from a manuscript to enrich our comprehension of the process of discovery; the sociologist may itemize the ways that cognitive and material resources are mobilized to render inscriptions authoritative. Each presumes to speak about scientific practice. Hence, the Babel of scientific practice.

Each of these approaches to scientific practice will be sampled in the remaining sections of this paper. What will emerge is the consensus that science is best seen in terms of heterogeneous activities. Even though Golinski may be right in a general sense to portray the recent literature as keying on practice, when we get down to cases the suggestion that scholars are "converging on a theory of practice" will be seen to underestimate the range of opinion among scholars. Our specimen philosopher, historian, and sociologist may agree that heterogeneity holds the key to a new post-Kuhnian vocabulary for speaking about the locality and temporality

5. The disappearance of experiment in the history of science is often taken by philosophers and sociologists as a mere prejudice against manual labor (see Galison 1987, ix; Hacking 1983, 149–51; Gooding, Pinch, and Schaffer 1989, xiii). While admitting that there is some truth to this claim, it seems to me to overlook the tenacity of the Enlightenment belief that Newton gave to science a form that it was bound to keep and a style of mathematical reasoning that would enable subsequent scientists to emulate his achievements—a belief that renders experiment inferior to theory.

of scientific practice, but their consensus belies widely different beliefs about how this heterogeneity is to be understood. As the many occurrences of the word "culture" in their works indicate, Pickering and Galison are engaged in a philosophical struggle; at stake are two deeply opposed conceptions of scientific practice. Both take it as a given that an account of scientific practice is eo ipso an account of the culture of science, but *they disagree about how practice is to be described.*

3. Durkheim, Galison, and a Sociologism of Constraints and Traditions

Galison is candid about his debt to the work of Emile Durkheim, though he gives the reader no more than a thumbnail sketch of the renowned sociologist's central theses. A few words of clarification are in order. Durkheim identified the properties of community with the *constraints embodied in historical types of social organization.* Two types of social organization are postulated in Durkheim's classic *The Division of Labor in Society* ([1893] 1984): the mechanical and the organic.[6] Durkheim contended that the mechanical society has existed throughout much of recorded history. In the mechanical society, social and moral homogeneity are the rule. All questions of individual thought and conduct are determined by the will of the small community, property is communal, and religion is indistinguishable from cult and ritual. Intimate ties of kinship and the sacred give unity and stability to the whole. The organic community, on the other hand, is a newcomer made possible by the rise of industrial technology and its associated division of labor; with the unshackling of the individual from the collective past, social order comes here to consist of free individuals pursuing different functions but united by their complementary roles. Gradually, the individual becomes disengaged from the traditional bonds of kinship, localism, and the collective conscience. Heterogeneity and individualism replace homogeneity and communalism.

A critical issue for Durkheim concerned the relationship between these two kinds of communities, whether the organic represents a continuation or a replacement of the mechanical society. Nineteenth-century progressivists, under the banner of idealist philosophies that emphasized the capacity of individuals to choose their social relations, staunchly defended a replacement thesis. Durkheim, however, held fast to the thesis

6. Many writers now refer to the Middle Ages as a period of organic communities in order to contrast it with industrialized and highly bureaucratized modern society. The mechanical/organic pairing is employed by Durkheim, however, to register markedly different kinds of social solidarity—the cogs of a machine having no real independence, unlike the members of a species.

that the institutional stability of organic society must be rooted in a continuation with the mechanical society; indeed, *The Rules of Sociological Method* ([1895] 1964) transmutes the properties of mechanical solidarity into the properties of all social facts whatsoever. However human behavior is studied, in Durkheim's view *constraint and tradition are the only units of explanation proper to sociology*. For an approach to be a genuinely social one, Durkheim insisted that sociology must have a reality distinctly its own: social facts (*faits sociaux*) cannot be reduced to individual, psychological, or biological data[7] but must be studied as things, as realities external to the individual. Failing this, the only possible subject of observation would be mental states and individual behaviors (that fall under the rubric of psychology) and whatever social and political consequences these psychological phenomena are deemed to possess.[8]

Peter Galison here advocates an account of scientific practice that is invigorated by Durkheim's sociologism, notably by Durkheim's contention that constraint and tradition are the principal units of sociological analysis.[9] Galison characterizes constraints as "*endpoints of inquiry, the boundaries beyond which inquirers find it unreasonable to pass . . .* bounds on the field of the *reasonable and persuasive*" (Galison draft, 2, emphasis added). As conceived by Galison, constraints regulate "allowable actions," not "mechanically" as determining a specific performance, but "organically" as delimiting a range of activities (see Galison

7. See Durkheim 1973: "Social facts exist *sui generis;* they have their own nature. There truly exists a social realm, as distinct from the psychic realm as the latter is from the biological realm and as this last, in its turn, is from the mineral realm" (16). The method articulated in *The Rules of Sociological Method* is a development of ideas that first appeared in *The Division of Labor in Society* ([1893] 1984) and that were subsequently applied in an explicit manner in *Suicide* ([1897] 1930) and in *The Elementary Forms of Religious Life* ([1912] 1976). These works share a common vision of the social.

8. Durkheim is often portrayed as an opponent of individualism sui generis, but what he resisted, rather, was a brand of individualism that he identified with the utilitarian egoism of Herbert Spencer (see Schoenfeld and Mestrovic 1991).

9. I use the expression "invigorated" with caution. Galison suggests that the notion of constraint serves as "a demarcation criterion" for Durkheim. In support of this suggestion, he draws our attention to a passage where Durkheim remarks that the "domain of sociology" comprises social facts. Durkheim's remark, however, is not meant as a general thesis about sociological practice; making it one is analogous to extracting a view of dynamical explanations from an ontology of attractive and repulsive forces. What would get overlooked is an account of what we are to do with these forces. Galison has conflated the value-free, analytical sociology that we often encounter in undergraduate sociology texts with Durkheim's sociologism, which has for its object morality—notably, principles of justice and freedom. When we examine Durkheim's actual practices, it becomes apparent that he regarded sociology as a means toward ends that he called moral ends; i.e., his work is infused with the conviction that the sociological mission is to see what can be done to improve the world (see Schoenfeld and Mestrovic 1991, 84).

1987, 257). Constraints, in Galison's view, are *abstract social practices* that explain whatever stability science possesses: they are the source of theoretical acumen and technical expertise in science, and they individuate fields of inquiry and, ultimately, close off debate. Constraints are the very substance of that which is cultural about science.

Galison's contribution to this volume regards virtually any scientific practice that persists through a period of time, however short ("constraints have histories"), as synonymous with a constraint. His *How Experiments End* is more circumspect. Here three kinds of abstract social practices are distinguished on the basis of their duration: long-, medium-, and short-term constraints. Specimen long-term constraints are theoretical beliefs about the discreteness (or continuity) of matter and methodological convictions respecting the unity of science. Such beliefs have great tenacity and are widely accepted by scientists. Some readers may discern in Galison's discussion of long-term constraints a residue of R. G. Collingwood's notion of "absolute presuppositions," or those convictions that are fundamental for each historical epoch. Collingwood did not regard these absolute presuppositions as "constraints" on practice, but Karl Popper and Imre Lakatos advocated a view of scientific practice where certain absolute presuppositions, now understood as unrefutable (though criticizable) assumptions, constrain scientific behavior. Moreover, these thinkers regarded these programmatic elements as constitutive of scientific rationality—in Galison's terms, as bounds on the reasonable and the persuasive. Galison should be bracketed with these thinkers, at least so far as these long-term constraints are concerned.

The suggestion that middle- and short-term constraints are operative as well sets Galison's view apart from the programmatic account of science that flourished during the 1970s. Middle-term constraints are attached to specific institutions and to individual programs of research. Specimen examples are Millikan's conviction that gamma rays could be linked with the formation of more organized matter in space, Einstein's goal of explaining the zero-point energy, and the "technological presuppositions" that are built into experimental devices. Galison notes that many different phenomenological and instrumental models shape the design and interpretation of experiments and, accordingly, that a number of these models may be compatible with each research program. In order to demarcate discrete research units and to explain how the scientist is led from grander metaphysical and programmatic schemes to quantities that can be precisely measured, Galison posits short-term experimental constraints as well. Examples are the specific quantitative models that helped Millikan make sense of data.

Galison's argument may be seen as an attempt to extend the conceptual elements of Kuhn's *Structure* that nevertheless represents an original and important view. The sovereignty of Kuhn's paradigms is reflected, not only in the fact that scientists adopt them uncritically and without reservation, in the way that a speaker adopts a native tongue (see Hattiangadi 1989, 193), but, more fundamentally, in Kuhn's conviction that the sovereignty of paradigms distinguishes scientific behavior from the unproductive wrangling of schools of philosophy. Without a common set of techniques and procedures—without a unitary, homogeneous community—scientists will waste their time and effort criticizing one another. Instead of a science grounded in technical achievement, we would have philosophy or (something very much like it) sociology. Retaining a weakened version of Durkheim's view that tradition and constraint are constitutive of cultural phenomena (i.e., that the satisfaction of constraints underlies all scientific activity),[10] Galison weakens this Hobbesian aspect of Kuhn's paradigms. In place of a catholic paradigm, Galison substitutes many "island empires," each of which is constituted by a different cluster of constraints. Although Galison insists that scientific groups sometimes overlap like so many subcultures, he strongly emphasizes that the constraints upon these cultures may mark them off as different communities. Where Kuhn (1970) weds technology, experiment, and theory into a homogeneous whole, Galison provides a framework for thinking about the social dimension of scientific practice in terms of "often quite disparate subcultures [that] follow their own rhythms, and with their own conflicting views on the appropriate laws, objects, and demonstrations" (Galison draft, 4).

Though Galison remarks on a strong tendency toward centralization, he contends that the constraints on scientific practice carve out three principal kinds of communities: experimental, instrumental, and theoretical.[11] Sometimes these communities completely overlap (e.g., instrument makers and experimenters in the cosmic-ray physics of the 1930s), but by and large they occupy different cultural spaces. These communities are broken down into smaller subcommunities, each conditioned by the same long-

10. Durkheim maintained that some social facts are restricted to local communities, whereas others are sufficiently general to be constitutive of all social life. Classic examples of the latter are the institutions of logic and arithmetic, which are bounded only by our human physiology and perhaps by the structure of the world (see Durkheim [1912] 1976). Though rooted in group perception, we extend these primitive classifications to all things in the natural and social environment. For Durkheim, these institutions act as a general social environment that enables different communities to take their place in the greater social order. Many of the functions that Galison attributes to trading languages are accounted for by Durkheim with these general social facts. See the thorough discussion by Hund (1990, 212–13).

11. Galison sometimes (draft, 7) speaks as though experimentalists and theoreticians are the two principal communities, but in other places he adds communities of engineers and technicians (2), and even computer programmers (17), to this list.

term constraints but differentiated by middle-term constraints. Some of these subcommunities may study electronic logic boards, whereas others may work on magnets and spark chambers. Finally, there are groups within these subcommunities, exploring the viability of different models and running autonomous subexperiments.[12]

What is especially striking about Galison's view is the description of the laboratory as a trading zone: "the site of an immensely complex set of trading relationships, and it is here in the everyday exchange between tabletop and blackboard that contact languages play an ever-present role in crossing the boundaries thought forbidden by theorists of incommensurability, and in linking these subcultures to the culture around them" (Galison draft, 26). Communication between these Hobbesian fiefdoms occurs through a trading language—a pidgin of sorts. Though each of these fiefdoms is a unitary, homogeneous community, the culture of science is seen to consist of a medley of heterogeneous subcultures, each with its own peculiar syntax and semantics that only partially overlap those of the trading languages (Galison draft, 23). Local syntax and semantics are structured primarily by local constraints, but trading languages (e.g., Monte Carlo programs and computer simulations) can act as conduits between different subcultures.[13]

It seems clear that Galison's frame of reference is the institution of science writ large: "the history of large-scale experiments cannot be written as if the experiment issued from a single mind. We are faced with a new kind of historical phenomenon that must be accorded the multiple structures of a true and heterogeneous community" (Galison 1987, 274–75). It seems fair to hold that the large-scale experiments, which are Galison's specialty, extend beyond any individual scientist or single community of scientists. If we are to speak, as does Galison, about the many communities that fall under the umbrella of a large-scale experiment in one breath, on the supposition that there is a single unifying thread to their many activities, then it seems that we must invoke the institutional

12. Though this seems to be a typical scenario, Galison notes that at times scientists "may well jettison the more abstract, high-level commitments while maintaining the lower-level ones" (1987, 255).

13. Galison's provocative discussion of trading languages must of course include paper communication, since results are ultimately formally frozen in journals. Theoreticians need not speak directly to experimentalists, but both need to speak the trading language of the journals (and granting agencies) simply because the form of these journals, the redactional process, the standards proclaimed, and those adhered to by referees are held stable in the face of changes in the scientific community, which, in turn, gives scientists a sense of the stability of the entire enterprise. The trading language of the journals may change, of course, but in general it is not in synchrony with changes in substantive practice. See Baigrie and Hattiangadi 1992.

framework in order to tell any story at all. It is the very scope of our institutional framework, however, and not anything proper to the activities of scientists, that removes them from the domain of consciousness and gives them an abstractness that they need not possess when viewed from the tabletop. Whatever rigidity these activities seem to possess may reflect the complexity of this institutional framework, and not the cultural elements themselves.

4. Weber, Pickering, and the Making of Culture

Despite Andy Pickering's professional training as a sociologist at Edinburgh, he dismisses constraint talk as symptomatic of a weary view of science as knowledge, promulgated at a time "when it was permitted to talk about scientific knowledge as though practice were irrelevant" (Pickering draft, 2).[14] He protests that constraint-talk is "anti-constructivist, anti-relativist" and "gets one nowhere in understanding what the difficulties of scientific practice are" (2).[15]

Pickering's *Constructing Quarks* (1984b) was assailed by Peter Galison (1987) because of its failure to address elements of tradition that are, he maintained, beyond the control of scientists. Recently, Pickering has written a series of articles, including the one in this volume, that develop some of the themes of his earlier work so as to disarm Galison's criticism. These articles (Pickering 1989b, 1991) capitalize on a single historical episode, namely, the attempt of the Italian physicist Giacomo Morpurgo, between 1965 and 1980, to detect experimentally the particles called quarks.

In 1964, Murray Gell-Mann and George Zweig, working indepen-

14. There is some truth to this allegation. During the late seventies, philosophers of science did indeed debate the issue of constraints in attempting to develop a theory of scientific problems. *The discussion proceeded almost exclusively in the orbit of scientific rationality.* For example, in his "Scientific Problems and Constraints," Tom Nickles (1978) formulated a question similar to Galison's, namely, "what types of constraints are there, and how do the various types function to determine [scientific] problems and their solutions?" (134). For Nickles, constraints are conditions that are imposed on solutions to various kinds of problems; indeed, he submits that "scientific problems, whether empirical or conceptual, cannot even be formulated, cannot exist, in the absence of rational constraints on their solution" (1980, 288). Moreover, Nickles supposes that constraints can be methodological (such as simplicity), metaphysical, logical, and empirical in character. The only condition is that they serve as bounds on what is acceptable in science. Nickles even stresses the flexibility of constraints, so as to allow for deep change; in fact, it is central to his view that all constraints—even logical consistency—have been violated many times. Galison, however, emphasizes the temporal properties of constraints, whereas Nickles does not.

15. This passage advances three related criticisms. I will focus on Pickering's complaint that constraint-talk does not help us to understand the difficulty of science.

dently of one another at Caltech, suggested that all particles that experience the strong nuclear force (hadrons) were composed of three fundamental particles (and their three antiparticles). The distinguishing feature of quarks, as they were called by Gell-Mann, was their electric charge. All previously discovered particles had integral charges, whereas quarks had nonintegral charges ($\pm 1/3$ or $\pm 2/3e$, where e denotes the charge of an electron). This model accounted beautifully for the properties of the hadrons that were known in the 1960s; for example, since nine different pairs consisting of quark and antiquark could be formed from the three quarks and three antiquarks, it explained why only nine mesons of spin zero had been found at energies below 1500 MeV. Bolstered by the explanatory potential of quarks, Gell-Mann suggested that experimenters might search for entities with fractional electric charge.

Pickering interprets Morpurgo's social situation as an opportunity structure, not only because he was one of a great number of physicists who accepted the challenge of experimentally detecting third-integral charges,[16] but also because he attempted to modify Robert Millikan's oil-drop experiment in order to optimize his chances of success (Pickering 1990b).[17] According to Pickering, Morpurgo started with an interpretation of how his apparatus should work that was explicitly modeled on Millikan's practices; the model of Millikan's oil-drop device (see fig. 5.1) functioned as a "resource," or, on Pickering's reading of Kuhn, as an "exemplar."[18]

Though Morpurgo's instrumental model was tailored to his goal of detecting quarks, it failed in practice because he did not realize that his samples might exchange charges with the suspending liquid. In response to this "resistance," as Pickering calls such a situation, Morpurgo designed a magnetic suspension system, which, in turn, engendered its own

16. Pickering (1990b, 236 n. 20) credits Knorr-Cetina (1981) for recognizing "the opportunistic quality of the dynamics of scientific practice," but it is difficult to ignore the specter of Feyerabend.

17. In 1909, R. A. Millikan carried out a brilliant experiment that produced the first accurate measurement of the electron charge. Minute droplets of oil were charged by means of X-ray irradiation and observed with a microscope as they floated between two parallel charged plates (see fig. 5.1). The charges of the droplets were calculated by measuring the speed of their fall under different electric field strengths. Millikan's measurements proved to be integral multiples of a fundamental charge (approx. $1.6 \times 10-19$), which was reckoned to be that of a single electron.

18. It could be suggested that the many experiments that scientists have performed to detect quarks are all variants on Millikan's oil-drop experiment. I will not pursue this line of argument here, but if this claim is sound, perhaps Pickering overestimates the options available to Morpurgo. It does not follow that his instrumental model acted as an autonomous constraint on his activities, however, since it could also be seen as a "resistance" in Pickering's sense of the word.

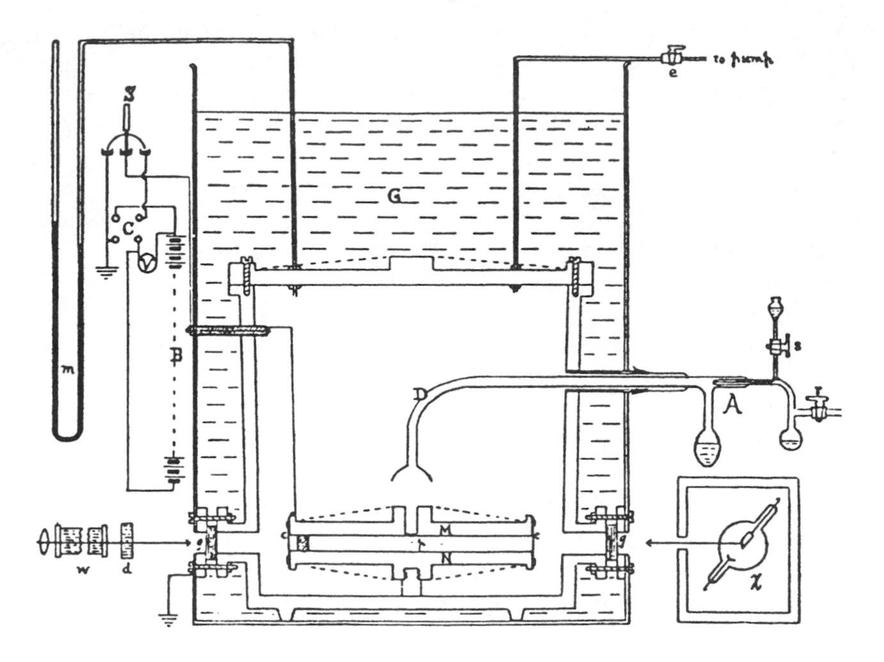

Fig. 5.1. Morpurgo's apparatus (Millikan 1917, 232)

resistance, or, in Pickering's words, "interference of particular open-ended extensions of bits of culture within the context of [his] goal-oriented practice" (draft, 6). Finally, Morpurgo suspended particles of graphite between two metal plates and obtained measurable responses.

In Pickering's portrait of experimental practice, scientists strive "to associate three disparate cultural elements" or "plastic resources" with the objective of achieving an "interactive stabilization," or a three-way coherence, in which facts are sustained.[19] These resources are (1) *a material procedure,* which includes such activities as setting up an apparatus

19. The moment of fact production is characterized by Pickering (1989b, 277) as a three-way *coherence* between plastic resources, but the expression "collaboration" is arguably more appropriate to describe the production of facts. Coherence has many meanings, but primarily it is a judgment that a given belief or practice is justified to the extent that it is a member of a coherent set of beliefs or practices; in short, it is a matter of fit and comfort—a static measure of the relationship between fairly rigid entities like mundane beliefs. What gets left out of account, then, is that these resources collaborate in such a manner as to produce something which is itself not a member of the supposed coherent set. It seems to me that what Pickering is after here is a notion which registers the *active participation* of the elements of the resource base in the manufacture of something new—namely, the fact carried by them.

and attending to its operation; (2) *an instrumental model,* which embodies the experimenter's beliefs about how the apparatus works; and (3) *a phenomenal model,* which expresses a view about whatever aspect of the phenomenal world is under investigation (Pickering 1989b, 276–77).

While conceding that Galison is right to regard scientific practice as movement in a complex cultural space, Pickering objects that constraint-talk is invalidated by the supposition that bits and pieces of culture exist independently of those who fashion them. Culture should be seen, rather, as "a space in which all sorts of moves can be made, all sorts of configurations constructed" (draft, 3), whereas constraint-talk "encourages us to think that this property (that limits practice) is already and enduringly there in the elements in question" (3). Galison's approach accordingly invites a Durkheimian sociology of intractable pressures, and it is to be rejected on this account. Pickering concedes that the phenomenological model invoked by Morpurgo is a cultural element that appears to possess a Durkheimian rigidity (since Morpurgo was steadfast in his conviction that the only relevant finds would be integral or third-integral charges). He insists, however, that this restriction was considerably less limiting than it should have been on Galison's account because it *changed with time* in a radical and discontinuous manner. Whatever resistance or "mismatch" (1984a, 114) the experimenter encounters in attempting to get cultural elements to cohere is not located within discrete bits of temporally static culture, as Galison (1987) maintains, but notably between dynamic elements now seen as resources instead of constraints. Accordingly, where Galison speaks about enduring constraints that are "already there," Pickering stresses the temporal emergent structure of practice:

> [R]esistance is the emergence of obstacles on the path to some goal. And the response to resistance is the search for accommodation: the revision of open-ended modeling sequences, the exploration of new directions for cultural extension, with, as one possible upshot, closure, meaning the successful achievement of the desired association of transformed cultural elements—a machine that works or . . . a new instrument, a new interpretation of that instrument, and a fact, a new piece of knowledge. (1991, 412; see also Pickering 1990b, 217)

A distinctive feature of Pickering's pragmatic realism is the role it grants to the aims and values of scientists. Morpurgo, for example, "had a goal—the search for fractional (third-integral) charges in his laboratory—and he set out to design an apparatus appropriate to the achievement of that goal" (Pickering 1989b, 283). By a "goal," Pickering means "an image of a future state of culture, a state in which some piece of work has been done which has not, in fact, yet been done, an image of something which does not exist" (1990b, 217). "The key idea here," Pickering asserts, "is that of an interdependence between obstacles—resistances—and goals: the former can only exist in relation to the latter, and are discovered in practice" (1990b, 235 n. 13). This interdependence enables Pickering to speak to the difficulty the experimenter invariably encounters in the attempt to get the experiment to work—difficulty which is manifested in feelings of frustration and thwarted desires—but it does not concede anything to constraints.

When we focus on the role that he grants to the aims of individual scientists, we can see that Pickering is abandoning the idea that culture is an element distinct from the individual, a view that has been a major stream in sociological theory from Durkheim and Merton to Galison, in favor of an approach that is decidedly more *egoistic* in that it sees culture as shaped by the *activities* of individual scientists. Though he is less forthcoming than Galison about what particular bits of culture he is seeking to extend, it seems that Pickering's pragmatic realism bears the signature of Max Weber (1864–1920), the German sociologist.[20] The importance of

20. Variations on the views of Durkheim and Weber are not the only ones in the current sociological literature on science. Setting aside the influence of political theories on the work of contemporary sociologists, such as the theory of the organic state (associated with Kant, the idealism of Hegel, the historical methods of Savigny in the field of jurisprudence, and the school of Schmoller in the domain of economics) and the theory of the utilitarian state (which can be traced through the works of Rousseau, Diderot, d'Alembert, Bentham, and Mill), three conceptions of the social come readily to mind. (1) Under the sway of Locke's New Way of Ideas, which restricts even the most abstract and sublime ideas to sensible particulars, many scholars seek to reduce social phenomena to human psychological properties (i.e., behaviors or practices) that an observer, such as a historian or a sociologist, can observe and record. (2) Another approach is "disciplinary eclecticism" (an expression suggested to me by Pickering in private communication), which is growing ever more popular in the literature. Disciplinary eclecticism expresses the social loosely in terms of the discourses of the various social sciences. Scientists are seen to be primarily producers of hard copy. The only kinds of scientific practices that are regarded as epistemologically significant involve the production of linguistic artifacts (see Fuller 1989, 4). This view is congenial to the treatment of laboratories as rhetorical devices, but it seems to leave no room for the study of the activities (e.g., building and monitoring an experimental device) that many scholars now regard as critical. Another difficulty with this approach is that (to borrow from Weber) the generality of the term "social" rests on nothing but its ambiguity. "It provides when taken in its 'general' meaning, no specific *point of view,* from which the *significance* of given elements of culture can be analyzed" (Weber 1949, 68). Latour and Woolgar (1979, 281) echo Weber's assertion in conceding that "by demonstrating its pervasive applicability, the social study of science has rendered 'social' devoid of any meaning." (3) An approach that bears family resemblances to disciplinary eclecticism is to be found in the work of Latour, Woolgar, Gilbert and Mulkay (1984), and Bazerman (1988). These scholars also assume that the difference between words and things is functional and not ontological, but they maintain in addition that there is nothing privileged about existing disciplines; indeed, we need to get past the idea that it makes sense to think of the psychological, the social, the natural, etc., as corresponding to essentially distinct realms in which thought can be situated (psychology, sociology,

Weber's sociologism for Pickering is that it furnishes an account of community in terms of the *values* that individuals impose on bits and pieces of nature. Weber asserts that

> the significance of cultural events presupposes a value-orientation towards these events. . . . Empirical reality becomes "culture" to us because and insofar as we relate it to value ideas. It includes those segments and only those segments of reality which have become significant to us because of this value-relevance. (1949, 76)

Weber's point seems to be that knowledge of cultural events is inconceivable except on the basis of the significance that concrete manifestations of reality have for us in certain individual concrete situations. Expressing a similar sentiment, Pickering contends that the phenomenal model "endows experimental findings with meaning and significance" (1989b, 277). Accordingly, if we are to speak of practice as practice of scientific culture, then we must think of this practice in terms of the meaning and significance that scientists place on the world. Scientists are cultural beings, endowed with a capacity to take a deliberate attitude toward the world and to lend it significance. It is this capacity to which our accounts of scientific practice must be held accountable.

5. Scientific Practice and the Making of History

Pickering's contention that such stability as is to be found in science is engendered by the collaboration between (three) plastic resources has received some support in the literature (see esp. Hacking 1988a but also Baigrie 1991).[21] However, in a reply to a paper by Hacking presented at a symposium on experiment, Galison resisted Hacking's contention that the stability of laboratory science reflects a collaboration between various plastic resources. Galison objects that

> stability, as I see it, must come from more than the convergence of experiment and theory. . . . The stability of the scientific enterprise rests (in this

natural science, etc.). My 1994 article brackets Latour with disciplinary eclecticism, but Pickering has convincingly argued (private communication) that these modernist distinctions are just what Latour is attempting to dismantle. Evidence for Pickering's contention is to be found in the fact that the word "social" is dropped from the subtitle of the second edition of *Laboratory Life* (1989).

21. There are apparent differences between the views of Hacking and Pickering. For instance, it is central to Pickering's view that resistance is encountered within the laboratory in consequence of the attempt to stabilize material and conceptual resources. Hacking's view seems to be that, within the context of the laboratory, these elements are constantly manipulated, and that it is outside the laboratory that the facts carried by them become passive (see Hacking 1988a).

> scheme) not on the total unification of science based on experimental or theoretical reductionism, but on the contingent fact that (1) there are traditions within experiment, theory, and instrumentation; (2) the dislocations within these "subcultures" of physics are not all synchronous; and (3) there are only piece-wise connections between the different strata, not a total convergence or reduction. (1988c, 526)

Galison insists that "constraints other than the convergence of experiment and theory are at work (e.g., instrumental tests of superconductivity and theoretical constraints based on renormalizibility and symmetries)," in particular, elements of experimental practice (e.g., knowledge of the instrument, hypotheses that bind theory to experiment, high-level theory) that are in place prior to experimentation (1988c, 525). Pickering and Hacking explicitly reject this thesis (though Ackermann [1985] endorses the thesis that some elements of experiment are passive, namely, the devices that generate data that are then interpreted in light of theory).

Both Pickering and Galison contend that their respective approaches yield novel and important perspectives for coming to grips with the activities of scientists. Though these approaches furnish us with important ways of conceiving science in cultural terms, neither of them seems to address the sorts of hands-on activities that are the bailiwick of the history of science. Galison provides us with interesting ideas concerning trading languages, whereas Pickering reminds us that science is shaped by the goals of social actors. Neither author, however, has much to say about the activities of scientists. This omission is grounds for concern.

Kuhn's (1970) description of science in terms of skills of a particularly local and routinized character has now become a byword for scholars everywhere. Three decades have passed but as the endless platitudes about the importance of practice suggest, the history of science still seems to privilege theory over practice. Pickering's appealing account of Morpurgo's experimental practices is a case in point, since it affords us a highly *schematic* picture of working scientists. If we restrict our attention to his account of Morpurgo's apparatus, for instance, we encounter an essentially schematic theoretical representation with just enough detail to persuade us of the plasticity of Morpurgo's material and cognitive resources. *It is in fact less a full-fledged representation of practice than a theoretical reconstruction as to why Morpurgo reported a result as a fact at a certain time.* Galison's (draft, 13) complaint that Pickering gives sustenance to the belief that theory dominates all aspects of physics is not altogether unfounded. Practice is ordinarily identified with routinized behavior of some sort. For Pickering, however, the purely localized goals and dispositions of scientists are primary; Galison rightly remarks that they are held by Pickering to shape two elements of the scientist's area of cognitive and material resources. But it is not immediately apparent what any of this has to do with practice in the usual sense of the term.[22]

I've remarked on the tendency of scholars to regard the disappearance of experiment in the history of science as little more than a prejudice against manual labor. Even if there is some truth to this view, it certainly does not explain why our histories seem unduly prejudiced in favor of theoretical claims even when we are fully cognizant of the importance of practice. Only Robert Ackermann (1988, 330), it seems, has advanced a positive thesis as to why this should be the case, namely, that the history of science is an instrument that picks up theory in science but that can only give us a very indirect and schematic characterization of practice. Ackermann has coined a metaphor to depict the historian's situation: "written history interacts with the past like an x-ray that reveals the bones of theory, while barely indicating the flesh of experimentation" (1988, 332). If the historian is insensitive to the activities of scientists, the written product will not give us an account of the things that depend on those activities; but a historian sensitive to practice will be compelled to sift out properties that are essential from those that are dispensable, in order to give *any* kind of representation at all. This sifting of properties gives the impression that scientific practice has a structure or rigidity that, in Pickering's view, it does not possess. Ackermann contends that

> history and theory share a temporality that is different from the temporality of experimental activity. History presents either experimental data, or an account of how experiments work, thus bringing experiment into a statemental (theoretical) form that is commensurate with theoretical statements. The writing of history does not conspire against experiment; its most honest deployment simply doesn't have the resources to do experiment justice. (1985, 63)

If we think of the history of science itself as a set of cultural practices that are detected with certain instruments, then Ackermann's claim is that the signals from history shift and change according to the particular situation of the historian. These signals need to be made to settle down (i.e., they have to be simplified and reduced) in a way that gives a foothold to historical narrative or, if the case dictates, to philosophical analysis. The historians' instruments are refined until they yield consistent data within reliable parameters that, in turn, *are shaped by the goals of the historian* or (as Pickering might reflexively assert) by a model of the historical phenomenon under investigation. When this process is successful, the signals

22. Galison's account of scientific practice is no less schematic. Consider his remarks about trading languages. He does not tell us how these languages arise, what sustains them, and how scientists actually use them for the purposes of communication.

are reconstituted as constraints or as resources, depending on the historians' craft and agenda.

We can consult scientific treatises and articles, laboratory notebooks, tables of data, etc., in order to reconstruct some scientific achievement, but *there is no simple historical analogue to practice.* Hackmann (1989, 33–34) has recommended that historians of science study and even learn to operate scientific instruments, as though a hands-on approach would make theory any the less necessary for historical reconstruction. If we look at a single attempt to reconstruct a comparatively straightforward scientific instrument, such as Stillman Drake's reconstruction of Galileo's steps in developing the military compass, the necessity of reference to published and unpublished writings seems to be indisputable.[23] If Drake is right, in order to say anything useful about Galileo's practice we need to consult his writings, which, in turn, puts us in the science-as-knowledge orbit. Scientific illustrations are useful, of course, but critical details about the working of the instrument are not communicable by pictures. Indeed, in the course of time, it becomes increasingly difficult to recover the identity of the exact kind of instruments used in the past, as well as the local modifications and techniques used to facilitate their production of relevant data. We need ideas about how they were employed, which, in turn, means that the so-called "textual deposits left by experiment" (Gooding, Pinch, and Schaffer 1989, 5) are not merely a feature of experimental life; they are also indispensable tools by which the historian seeks to reconstruct laboratory practice.[24]

6. Radder and the Remaking of Knowledge

Even if we grant Ackermann's thesis that our knowledge of scientific practice is mediated by historical representation, we can still speak coherently about the activities of scientists *as mediated by these histories.* Pickering may be optimistic in suggesting that his view "aim[s] to get to grips with practice as it happens" (draft, 7 n. 5), but we can nevertheless seek to represent the temporality of scientific practice with the devices at our dis-

23. Similar reconstructions of practice have been advanced by others. Ryan Tweeney's (1985) reconstruction of Faraday's work on induction and Holmes's (1985) work with Lavoisier's notebooks illustrate how the study of texts can reveal the subtleties of scientific practice.

24. I will not discuss here in detail the kinds of devices employed in the sociological literature. At least two kinds of devices seem indispensable to social history: organizational and representational devices. An organizational device is something that is conducive to storytelling—e.g., a chronology that typically is based on the goals that the hero seeks to realize and that, in turn, elicits the reader's sympathies. A representational device is any narrative feature that helps to persuade the reader of the narrative's truth. The assertion that a study is based on an examination of scientific practice is a typical representational device.

posal. The heart of Pickering's account of scientific practice seems to be the suggestion that scientists constantly *rework* previous results and techniques so as to *concretize aims* that are ethereal compared to weightier experimental procedures. Pickering identifies this concretization of aims—however brief in duration—with the emergence of substantive laboratory procedures, ones that can be performed routinely and that possess a *reliability* that is denied to mere goals and aspirations.

Though Pickering gives us a theoretical perspective for representing the activities of scientists, he does not appear to take his own view to heart. In his considered judgment, scientists constantly attempt to refashion extant cultural resources. His own sociological practices give us a different impression. Facts get constructed. A resistance emerges relative to a particular configuration of cultural elements. A scientist avoids this resistance, thereby achieving a new stabilization or encountering a new resistance. For practical purposes, this is the end of Pickering's story. What, then, of the general constructivist thesis that cultural resources constantly get reworked and redistributed?

If we consider the production of what seems in retrospect to be a fairly nonproblematic fact, say, the elliptical planetary orbit, we can see that it was reworked many times. The elliptical hypothesis was first secured by Johannes Kepler on quasi-dynamical/quasi-Aristotelian grounds. It was then reworked by Newton in his *De Motu Corporum in Gyrum,* a small treatise that served as the blueprint for book 1 of the *Principia.* A scholium to Proposition 3 of *De Motu* affirms Newton's conviction that the planetary orbits might be exact ellipses, with the sun fixed at one focus. This scholium is deleted in the *Principia.* Newton proceeds instead to consider in Propositions 12 and 13 bodies moving in hyperbolas and parabolas, finding that the force directed to the focus varies as the square of the distance in both cases. The reason for this change is that, during the composition of the *Principia,* Newton became convinced that there is no instance of undisturbed motion (a view that is encapsulated in his third law of motion). Kepler's ellipse hypothesis therefore could at best be true of planetary bodies that do not interact with the central body. Consequently, Newton reworked Kepler's ellipse in terms of a model in which the force of attraction between sun and planet is mutual and opposite. The ellipse hypothesis is shown to be to a near approximation true because the action of the planet on the sun is minimal; each of the two bodies will move in an elliptical orbit with respect to the center of mass of the system. Finally, Newton reworked the ellipse so as to account for mutual interplanetary forces. However, as Newton remarks in book 3, Proposition 13, of the *Principia* (1934, 421), "the actions of the planets one upon another are so very small, that they may be neglected." The only sensible interactions occur between Jupiter and Saturn at their conjunction and between Earth and the Moon, and these were quite small but the limits of observation in Newton's time.[25] Once we possess a reliable way of calculating for deviations, the ellipse could be safely regarded as a fact of planetary motion (see Baigrie 1991).

In 1847, William Whewell characterized Kepler's laws as "the facts on which mathematicians had to found their mathematical theories" (1857, 47–48). Kepler's ellipse hypothesis has been reworked many times, but at some point between 1687 and 1847 it was transmuted from a lively argument for a modified Copernican theory into a comparatively unimaginative and rigid fact (see Baigrie 1991) that effectively concealed its social character (though not its history). The laws that we associate with Kepler, and that appear prominently in astronomical treatises, are not the same ones that Kepler discovered; it is the practice of teaching and the names that give the impression that the old science of Kepler is preserved intact by the science of Newton.

Sometimes the reworked fact is identical to its precursor; sometimes it is transformed almost beyond recognition. Sometimes the reworking capitalizes on extant material and cognitive resources; sometimes it does not. Sometimes this process appears to delocalize previous knowledge (i.e., by relocating it somewhere else); sometimes it does not.[26] In some cases, this reworking of facts appears to reach a terminus—they ossify into rigid entities that are relegated to engineering faculties. For any proponent of a social approach, this phenomenon is cause for concern, since it suggests that there is an important (but not very well understood) connection between the reworking of facts and the seeming objectivity of scientific products.

Hans Radder's contribution to this volume capitalizes on this connection, though he takes the reworking of facts more narrowly as the activity of reproducing or replicating them. Radder regards this phenomenon in terms of what Allan Franklin (1989) aptly calls *the epistemology of experiment,* namely, as grounds for investigating the existence of abstract standards in the scientific community that supposedly demarcate genuine experimental results from spurious ones. Though Radder's conviction that scientific practice constitutes grounds for epistemological deliberation is not universally shared, his paper (steeped in the conviction that scientific practice should be seen in terms of heterogeneous elements) iden-

25. Newton appears to have underestimated this value, since the error for Saturn can run up to a degree and for Jupiter to about 20′.

26. Joseph Rouse (1987, 125) contends that knowledge is initially localized and that, in extending it from the home laboratory, scientists transfer a package *comprising the experimental phenomena and the techniques through which they were produced* (see Golinski 1990).

tifies something that many scholars would readily classify as a scientific practice.

The practices on which Radder seeks to found an epistemology of experiment are those that seem to involve the reproducibility of experiments. He identifies three questions that we can ask about reproducibility.

1. Are experiments in fact reproducible?

2. In what *sense* are experiments reproducible? Successive data runs using the same apparatus with an eye to reducing error constitute reproductions in a trivial sense. Sometimes the original apparatus is not available, however, and even when it is available, subsequent workers often will employ it in a different way. These elementary considerations testify that reproducibility is not a simple matter at all. In order to clarify matters, Radder distinguishes three different ways in which we can speak about this activity: (*a*) reproducibility of the performance, or "material realization," of the experiment; (*b*) reproducibility of the experiment under a fixed theoretical interpretation (where "theory" is understood as a nonsystematic, easily revisable set of beliefs about the apparatus and the phenomena under investigation); and (*c*) reproducibility of the result of an experiment. Reproducibility in this third sense is distinguished from the other two senses by use of the term "replication." When we take Radder's own classification of the elements of experimental practice into account, we are warranted in speaking about the stability of the theoretical result, the stability of the theoretical description, and the stability of the material realization of the experiment.

3. Should experiments be reproducible? Radder maintains that "reproducible experiments are, for the time being, regarded as *stable elements of natural science*" (draft, 15, emphasis added). Reproducibility, in other words, constitutes a de facto philosophical criterion for the stability of these elements.

The normative question that Radder raises is whether scientists are warranted in taking reproducibility as a criterion of stabilization. He contends that reproducible experiments transcend the local context in which they were produced for the first time, but it is not clear how we are to interpret this assertion. We can take this passage as an account of what scientists think about reproducibility, in which case we could respond that scientists simply are mistaken in supposing that the stabilization of an experiment for a community of practitioners warrants the supposition that it is thereby "objectified" or certified in some philosophical sense, as Radder's assertion implies. On the other hand, we can take this assertion as Radder's account of the significance of reproducibility, namely, that the element itself is stabilized in some epistemological sense, irrespective of what scientists think about it. Only with this latter interpretation could we address the normative issue, unless, of course, we were to take scientific opinion as the grounds for normative views about knowledge. Irrespective of how this assertion is interpreted, Radder offers us an explanation (or at least part of one) for the persistence of certain elements of experimental science, whether it is only the way in which laboratory samples are prepared or something more. His account is that a stable feature of experimental science is one that scientists can (or like to think that they can) treat as an observer or as a witness, and not as a participant.

This reading would explain Radder's worries about "for whom" these elements are reproducible. If reproducibility is restricted to a group of skilled practitioners, then it might be suggested that the element has not thoroughly transcended the context in which it was first stabilized (if we extend the context to include scientifically skilled workers). It would be stabilized only for those with the requisite skills, and so scientists would not be warranted in supposing that the result could now be regarded from the observer's perspective. It would also explain Radder's supposition that "a reproducible material realization implies that all laypersons who possess or are able to acquire the necessary skills, or who are willing or can be made to do the required work [!], can successfully carry out the experiment" (draft, 15). On the face of it, it is not clear what the term "layperson" means in this passage. One would suppose that once a person possesses the specialized skills, then he or she is no longer a layperson or at least not a layperson in any ordinary sense of the term. Perhaps Radder's point is that a skilled layperson can be taught in a cookbook-like way to do the work, whereas a skilled practitioner needs to know lots about how devices work. Many of us, for instance, are skilled laypersons with respect to fixing computer problems, but we are not thereby skilled technicians, since we may not be able to easily probe the deeper sources of computer glitches.[27] We may not even be able to get the machine to work, let alone produce knowledge with it. Though the distinction between a skilled layperson and a skilled technician may be meaningful from the point of view of amateurs who tinker with mechanical devices, it does not translate very well into experimental practice. If anything, Radder's suggestion would seem to *underestimate both the skill involved in engineering even simple experiments and also the degree of autonomy that this element of experimental science enjoys.*[28] (I'll have more to say about Radder's failure to consider the actual position of the layperson in the next section.)

An important implication of Radder's paper is that some elements of

27. This analogy was suggested to me by Jed Buchwald.

28. Radder considers himself a follower of Habermas and his theory of communication, which underwrites much of Radder's conception of idealized reproducibility.

experimental science may be quite stable even when there is intense controversy about the other elements; scientists may be able to reproduce a specific performance even while they disagree about its theoretical description. Conversely, scientists may agree about the theoretical description but disagree about the material realization of an experiment. By classifying the different ways that scientists are able to reproduce tangible results, Radder's analysis yields a framework for addressing Kuhn's essential tension between tradition and innovation. Radder's systematic approach to the question of reproducibility in terms of *the activities that are associated with the closure of debate* gives substance to the principle, advocated by Pickering and Galison, that scientific culture consists of heterogeneous elements.

Radder's philosophical approach, however, is undermined by his general thesis that a result that has been detached from the circumstances of its first realization is better, epistemologically speaking, than one that is tied to a particular apparatus or theory.[29] Though it seems eminently reasonable to hold that two different experimental procedures make a hypothesis more plausible than two repetitions of the same experiment, it is the tacit supposition that *the aim of experiment is to test hypotheses* that motivate this judgment. If a result is tied to a particular theoretical outlook, this is taken by Radder as evidence that it is liable to be undermined (i.e., that it is defeasible). For a proponent of this result, Radder's conviction is that justification can be found by detaching it from these initial conditions. The difficulty with this thesis, however, is that it only makes sense with respect to those experiments (e.g., Morpurgo's oil-drop experiment) where the theory is still in question. Though this class of experiments traditionally has received the lion's share of the press, in consequence of the traditional view of experiment as a way of putting nature to the test, it is statistically only a small part of experimental practice. Most experiments measure quantities (see Franklin 1986, 104). A convincing case can be made that even when the quantities measured by scientists are determined in whole by theory, these measures do not test theory or otherwise give it added epistemic currency. This raises a difficulty for Radder's view about the relationship between measurement experiments and theory: if experimental reproductions rarely are epistemologically significant, what role do they play in science?

One persuasive answer to this question has been given by Kuhn. His *Structure* (1970, 25) itemizes three foci for "factual scientific investiga-

29. See Franklin 1989 for a discussion of a number of "epistemological strategies" designed to establish the credentials of an experimental result. Like Radder, Franklin regards the warrant for these strategies to be found in scientific rationality.

tion." It is the third class of experiments that concerns me here. Kuhn writes that

> this class proves to be the most important of all. . . . In the more mathematical sciences, some of the experiments aimed at articulation are directed to the determination of physical constants. Newton's work, for example, indicated that the force between two unit masses at unit distance would be the same for all types of matter at all positions in the universe. But his own problems could be solved without even estimating the size of this attraction, the universal gravitational constant; and no one else devised apparatus able to determine it for a century after the *Principia* appeared. Nor was Cavendish's famous determination in the 1790's the last. Because of its central position in physical theory, improved values of the gravitational constant have been the object of repeated efforts ever since by a number of outstanding experimentalists. Other examples of the same sort of continuing work would include the determinations of the astronomical unit, Avogadro's number, Joule's coefficient, the electronic charge, and so on. Few of these elaborate efforts would have been conceived and none would have been carried out without a paradigm theory to define the problem and to guarantee the existence of a stable solution. (1970, 27–28)

The chief lesson to be extracted from this passage is that many measurements are carried out to determine constants that occupy a central place in the paradigm but that represent nothing more than its ongoing articulation. If Kuhn is right about this, then it is really the paradigm that guides experimental practice here and not a supervenient epistemological interest. There is no story to be told about reproductions over and above the ongoing articulation of Newton's paradigm.

It is with Kuhn's suggestion in mind that I now turn to Henry Cavendish and his attempt to measure the gravitational constant.[30] The history of science furnishes no better instance of a tabletop experiment than Cavendish's gravitation device. With an apparatus no larger than the outhouse in the garden where he conducted his experiment, Cavendish was able to weigh the earth and so produce the first experimental measure of the gravitational constant. The example is a staple in the literature. Kuhn

30. Cavendish is remembered as an assiduous experimental scientist, celebrated as much for his experimental demonstration of the compound structure of water (and the ensuing priority dispute with Watt and Lavoisier) as for weighing the earth. Among Cavendish's 20 (or so) published articles, only a single study of electricity, published in 1771, can be described as predominantly theoretical; the remainder detail careful and deliberate experimental studies. Cavendish liked to work alone, with apparatus that he could run and monitor by himself. He would work with any device that required his special blend of patience and precision, but thermometers were a particular favorite.

uses it to illustrate one focus of experimental articulation. Hacking (1983) draws on it to illustrate some aspects of measurement. Besides, few experiments have been repeated as many times.

7. Reworking $d = 5.48$

During the eighteenth and nineteenth centuries, several experiments were proposed in order to determine the gravitational constant G (if indeed it was a constant).[31] To ascertain G one must measure the gravitational force F_g between two known masses situated a measured distance apart and then compute the value of the constant by means of the rearranged form of Newton's celebrated law of gravitation:

$$G = F_g \frac{R^2}{m_1 m_2}. \tag{1}$$

Two experimental strategies were pursued by scientists. The first exploited an apparatus readily arranged in a pure state of nature. By measuring the deviation by a hemispherical mountain on the plumb line of a quadrant at its side, a mountain (or some measurable part of the earth's surface layers) was used as a benchmark to ascertain the earth's pull. The second strategy attempted to measure the mutual attraction of two artificially prepared masses. Both experimental strategies were first proposed by Newton (and then set aside) in his *Treatise of the System of the World.* Had he employed the figure of 40,000,000 feet as the earth's diameter, rather than the 20,000,000 stipulated in the *Principia* (book 3, §22), Newton might not have concluded that the gravitational force is too small to measure in the case of terrestrial bodies.

Experiments involving natural elements were conducted by Pierre Bouguer (1740), Nevil Maskelyne (1775), Baron Zach (1814), Francesco Carlini (1824), G. B. Airy (1854), Von Sterneck (1883), E. D. Preston (1893), and others. These experiments were carried out on the chance that they would yield, not only a better measure of the total mass of the earth, but also the proportional density of the matter near the surface compared with the mean density of the whole earth. Though these experiments are interesting in their own right, it is impossible to ascertain the mean density by the attraction of natural masses with any accuracy, because of irregularities in the earth's strata. Experiments involving prepared masses in the laboratory proved to be more satisfactory. The experiment that we readily associate with the measurement of the gravitational constant was carried

31. See J. H. Poynting 1894 for a thorough discussion of attempts to determine the density of the earth.

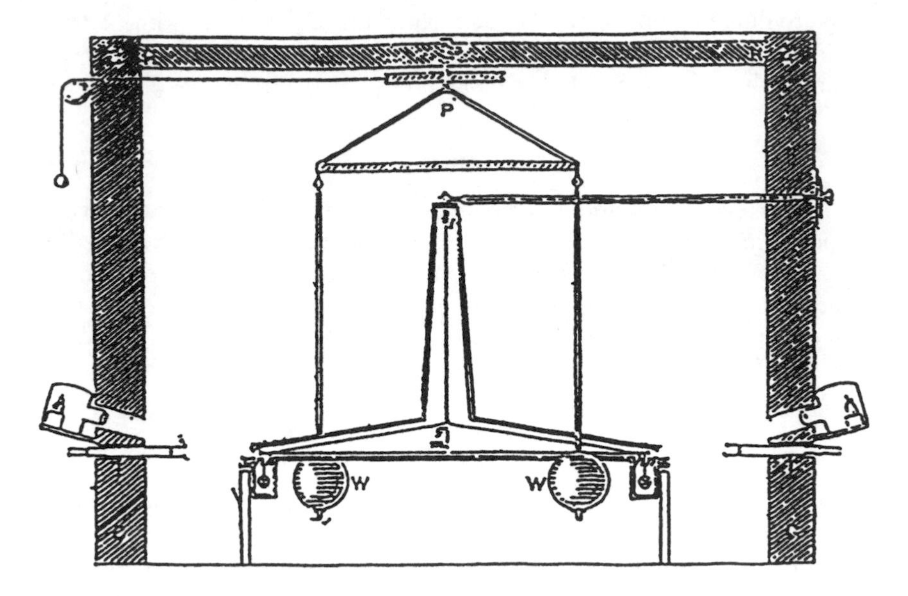

Fig. 5.2. Cavendish's apparatus (Cavendish 1798, 526)

out by Henry Cavendish in 1798, as reported to the Royal Society on 21 June of that year.[32] Cavendish (see Fig. 5.2) employed a delicate torsion balance, consisting of a wooden arm (6 feet long), supported in a horizontal position at its center by a torsion wire 40 inches long, with leaden balls about 2 inches in diameter suspended at each extremity. The success of the experiment relies on the fact that a small force is sufficient to twist a delicate wire about the axis of its length, and that the value of the twist yields a direct measure of the force.

The device was set on a stand, and the entire apparatus enclosed in a wooden case in order to prevent agitation of the air and other factors from affecting the behavior of the balls.[33] Only two telescopes and two lamps were inserted in the device. Resolving that the apparatus should be placed in a room which should constantly remain shut, Cavendish set up the entire device in an outhouse in the garden of his home at Chapham Common. Gravitational forces were brought into play on the two suspended masses when two 12-inch lead balls were placed outside the case contain-

32. Cavendish's report was published as "Experiments to Determine the Density of the Earth" in *Philosophical Transactions* 88 (1798): 469–526.

33. Cavendish reports that variations in heat and cold can undermine the experiment: "if one side of the case is warmer than the other, the air in contact with it will be rarefied, and, in consequence, will ascend, while that on the other side will descend, and produce a current which will draw the arm sensibly aside" (1798, 471).

ing the small lead balls, which therefore twisted the wooden arm, causing it to oscillate because of the torsion wire. The torsion constant of the wire was found from the moment of inertia. If T is the period of the torsion bar, I the moment of inertia, and kq the restoring torque corresponding to an angular deflection q from the equilibrium position, then

$$T = 2\pi\sqrt{\frac{I}{\kappa}} \text{ or } \kappa = \frac{4\pi^2 I}{T^2}. \tag{2}$$

Substituting $2md^2$ for I, we have

$$\kappa = \frac{8\pi^2 md^2}{T^2}. \tag{3}$$

The center of swing for small oscillations was taken as the equilibrium position. The shift in this position, $\Delta\theta$, was observed when the pair of 12-inch spheres were moved from a position where their attractions for the pair of lead balls gave additive clockwise torque to a position where they gave counterclockwise torque. Since the mean distance d between centers in each position was about 9 inches,

$$\kappa\Delta\theta = 2G\frac{Mm}{d^2}a, \tag{4}$$

where a is the distance from the axis to the centers of the small spheres. After making several corrections for oversimplifications in this equation, Cavendish[34] found that

$$G = 6.754 \times 10^{-11} \text{ Nm}^2 \text{ kg}^{-2}. \tag{5}$$

Cavendish's activities support Pickering's view of scientists as seeking to extend cultural resources. Histories often credit Cavendish with the device that "weighed the earth," though it was John Michell, professor of geology at Cambridge and subsequently rector of Thornhill, who both conceived the method of determining the earth's density and built the device after his own design.[35] Michell died before he could experiment

34. Cavendish ascertained that our globe is five and one-half times heavier than the same bulk of water, a figure which more or less bears out Newton's assertion (book 3, Proposition 10, Theorem 10) that "the quantity of the whole matter of the earth may be five or six times greater [*quasi quintuplo vel sextuplo major*] than if it consisted of all water" (1934, 418). Since the gravitational constant G and the mean density of the earth are practically the same, knowledge of the dimensions of the earth and the acceleration of gravity at its center enable one to determine G if one knows D, or to determine D if one knows G.

35. Michell is remembered as well for his defense of Boscovich point-particles against Newton's atoms, as well as a treatise on magnets published in 1750. For more on Michell, see Wilson 1851.

with his apparatus, and it was sent along to Cavendish by their mutual friend, Francis John Wollaston. Michell's apparatus served as a model that Cavendish was compelled to modify when he encountered resistances in the course of observing vibrations of the torsion bar. When the masses of the large balls were left in one position for a time, the attracted balls crept in one direction, then in another, as though the attraction were varying in direction. This result conflicted with his phenomenological model, according to which the only force between the spheres would be an attractive one. Cavendish found that the anomalous effect was due to convection currents in the case containing the tension rod that were produced by temperature inequalities; when a large sphere was heated, the ball near it tended to approach, and when it was cooled, the ball tended to recede. By placing a casing around the torsion balance itself, he was able to obliterate this effect and so produce the anticipated results.

In the previous section, I took Radder to task for underestimating the skill involved in realizing even the simplest experiments—that the skilled layperson may not be able to grasp the deeper workings of laboratory devices. Part of the explanation for Radder's confidence in the skilled layperson is that he is after a criterion of the identity of concrete laboratory skills, one that would support the contention that the same experiment has been performed at a later date. Radder admits that experimental activities always involve "theoretical ideas" (draft, 4), but it is central to his view that the concrete action involved in experimentation can be mentally dissected from the theoretical dimensions of laboratory practice. Radder's criterion stipulates that the material realization is a concrete action that makes visible (in a cookbook-like way) the instructions of an experimental scientist (draft, 5).

In order to fully isolate this concrete action, Radder (draft, 4) places theories about the phenomenon under investigation and theories about the workings of instruments under the same heading; both are subsumed under the rubric of theoretical description (see table 5.1). Though they are sensitive to the interaction between these elements of scientific practice, Pickering and Galison regard them as distinct elements of scientific practice. They are treated by Pickering as distinct cultural elements that stabi-

Table 5.1. Elements of Scientific Practice

Galison	Pickering	Radder
Experimental tradition	Material procedure	Material realization
Instrumental tradition	Instrumental model	Theoretical realization
Theoretical tradition	Phenomenal model	

lize so as to produce knowledge, and by Galison as different traditions that partially overlap one another. Perhaps more than anything else, it is his failure to see that instruments involve a distinct element that undermines Radder's philosophical analysis. I've already mentioned two unwelcome consequences. One is that Radder underestimates the skill involved in setting up and running an experiment; that is, addressing the deeper sources of experimental glitches demands a technical and theoretical expertise that is often quite distinct from that involved in constructing phenomenological models. The second consequence is that Radder underestimates the autonomy that instruments and theories about their workings have in laboratory science. This autonomy is often a factor in experimental practice. Indeed, many prima facie reproductions of experiments have little to do with validating knowledge claims. What fuels reproductions in many cases, rather, is the desire to improve laboratory devices and instruments.

This point becomes clear if we follow the Cavendish experiment through some of the reproductions that were carried out during the eighteenth century. Although Cavendish's result was distilled from 17 distinct experiments, his work was criticized by Francis Baily for the paucity of experiments. In 1838, F. Reich reproduced Cavendish's experiment 57 times with a smaller apparatus, obtaining the figure 5.44 for the mean density. Three years later, Francis Baily carried out some 1,300 experiments with the enthusiastic support of the Royal Astronomical Society and financial backing courtesy of the British government, in order to detect a possible anomaly in the motions of the torsion rod. An apparatus of the same dimensions as Michell's was employed, but Baily varied the composition of the attracted balls (he used lead, zinc, glass, ivory, platinum, and even the torsion wire alone without balls). Dissatisfied with the results, Baily modified the apparatus by covering the torsion box with a gilt case to prevent thermal radiation produced by heat from affecting the balls. Many hundreds of experiments later, he reported the density as 5.6674. During the next 60 years, the experiment was repeated many times, by Boys (5.5270), by Braun (5.52725), and the prize-winning effort of Poynting (5.4934). These and other endeavors involved considerable adjustments in the structure and composition of the torsion system, size of the attracting balls, the time of variation, etc.

It seems intuitively clear that these experiments were reproductions of Cavendish's original activity. Recall that Radder gives us three ways of thinking about this activity: as reproducibility of (*a*) the performance, (*b*) the experiment under a fixed theoretical interpretation, and (*c*) the experimental result. Measurements of the gravitational constant were subsequently carried out with very different apparatus. In 1887, A. Wilsing employed a vertical double pendulum instead of Michell's horizontal torsion system, and three years later J. Joly performed some experiments that involved suspending a simple pendulum in a vacuum. A different experiment involving a balance to determine the decrease in weight with increase in height from the earth's surface had been suggested by Robert Hooke and Henry Power, but it was J. Poynting (1891) and F. Richarz working with O. Krigar-Menzel (1898) who first recognized that the common balance could be exploited for gravitational work. Each of these successive variations did not even approximately reproduce the material realization of the original experiment. Since the experimental setup in these later activities was not identical to the original, (*a*) is ruled out as a candidate, and (*c*) is ruled out for a similar reason. Though we now judge that these activities led to no significant correction to Cavendish's original result, the determinations of Baily and others were not identical to Cavendish's result. It would appear that these subsequent activities *improved* Cavendish's measure, but this suggestion does not square with any of Radder's three options.

The candidate—if any—that seems best suited to redeterminations of *G* is (*b*). The Cavendish experiment was essentially a measurement carried out under a fixed theoretical interpretation. The problem with (*b*) is that it leads us to suppose that this activity was designed to make the value of *G* more confidence inspiring by detaching it from Cavendish's particular apparatus. Though scientists displayed great ingenuity in varying what is basically a simple and elegant device, it does not follow that the motivation was to validate Cavendish's result by showing that it could be detached from its initial material realization. In fact, instantiating the result under a variety of conditions would seem to have rendered additional tests unnecessary. Nevertheless, these scientists behaved as though the returns of information were not diminishing with each new variation of Cavendish's original activity.

It seems very obvious that scientists were attempting to improve the accuracy of an important numerical value; refinements to the apparatus and experiments with different substances were clearly motivated in part to render this value more precise (see Hacking 1983, 231). Kuhn is right to suggest that Newtonian theory provided the rationale for this activity, but there is no basis to his suggestion that this activity signified continuing work on Newton's theory of gravitation. On this point, Hacking (1983, 239) reminds us that the value of *G* is not part of this theory, a point that Kuhn appears to grant when he asserts that Newton's "problems could be solved without even estimating the size of this attraction." It seems to me that it is Kuhn's supposition that scientific practice is a homogeneous activity that is at issue here. Experiment, in his view, is an activity that is geared to articulate a paradigm. This unitary conception leaves no room for other theoretical and instrumental concerns that are not directly related to the law of gravitation. For example, improvements to the apparatus seem to have been motivated as well by attempts to further investigate what Thomson and Tait (1888) called "residual phenomena," or effects that remain unexplained at the conclusion of an experiment. A classic example is the smell produced by the operation of electrical devices. Though widely recognized during the early period of experimenting as a residual phenomenon, it was subsequently explained in 1840 by Christian Schönbein as the formation of ozone. Turning to the Cavendish experiment, redeterminations of *G* were designed to examine residual phenomena that are peculiar to this kind of experiment (e.g., convection currents carried by different kinds of materials, the mechanical structure of torsion devices, the laws peculiar to their operation). The history of the Cavendish experiment is also the story of the perfection of a technology. With a reliable measure of the gravitational constant in hand, experimenters were in a perfect position to vary their apparatus as much as possible in order to investigate these residual factors, to produce a less noisy experiment, and to carry out better measurements.

My point here is that activities that look to be mere reproductions of earlier experiments often bear little or no relation to them. The devices and phenomenal models may appear to be more or less the same, but *the only way to make sense of these activities is to recognize that the aims of science have shifted in order to address these residual phenomena.* If the Cavendish experiment had been performed circa 1690 with balls made out of a variety of substances, and the same value for *G* had been obtained for each experimental run, then this activity could have supported claims to universal gravitation. However, the law of gravitation would have still been a matter of dispute. In this event, this experiment would have been on a par with recent experimental work on free quarks. Though this situation is purely hypothetical, it underscores the underlying weakness of Radder's philosophy of experiment: its bias toward the validation of knowledge claims. What his approach does not address is that the only way that we can distinguish experiments that are geared toward numerical and instrumental improvement from experiments that are striving to validate knowledge claims is if we take into account the aims and goals of scientists. If this is so, however, Radder's belief that an element of experimental practice (reproductions) can be mined for epistemological standards that are detached from the purely localized goals and aspirations of scientists appears to lack substance.

8. Conclusion

Despite the differences between Pickering, Galison, and Radder, they are unanimous in rejecting the suggestion that science is a unitary, monolithic, homogeneous culture. For Pickering, scientific culture is composed of heterogeneous elements that collaborate in various ways to produce stable products that move outside the culture in which they originated. Galison, on the other hand, portrays scientific culture in terms of many heterogeneous subcultures that collaborate through the conduit of trading languages. The debate between Pickering and Galison is about many things, of course, but when we focus on their respective views of scientific practice, it seems to me that it is about whether scientific culture is best described in terms of heterogeneous *cultural elements* or heterogeneous *subcultures*. Both advocate heterogeneity as the key to post-Kuhnian science studies, but they disagree about how it is best described.

The difference between "cultural elements" and "subcultures" is subtle but important. Constraint-talk furnishes us with a cultural perspective that tends to leave the activities of individual scientists out of account and to obscure how constraints are realized in discrete human activities; that is, Galison's brand of sociologism is to a great degree saddled with the historic difficulties that beset Durkheim and Kuhn. If our object of study is big experimental science, those difficulties will perhaps be tolerable simply because our framework here will be a comparatively stable institution, even in the face of quite sweeping changes in concrete laboratory settings. Galison's perspective is that of comparatively self-contained subcultures that exchange information through an institutional framework afforded by trading languages, which, in turn, gives to each of these communities a great sense of the stability of the entire enterprise.

Pickering's pragmatic realism, on the other hand, is beset by those difficulties that we have come to identify with idealism in all its guises. Galison, for example, insists that Pickering's view collapses into a kind of wish-relativism in which anything is permitted, provided only that leaders in the relevant community agree that it is so. Resistances may emerge between elements of culture, but this does not satisfy deeply embedded realist convictions that the world—even the social world—contains objects that resist our best attempts to change them. If our point of view is the larger institutional framework of science and its many ties to industry and the university, then Galison's objection will seem plausible. But if our perspective is the tabletop, then these cultural elements will be seen to lie in the field of human consciousness, that is, as resources. Though Pickering holds that the phenomena are manufactured by scientists, it is not part of his view to suppose that they do it in circumstances that are entirely of their own choosing. What scientists do not manufacture are the resistances that emerge in consequence of their attempt to stabilize their conceptual and material resources.

We should be wary of Jan Golinski's suggestion that constraint- and resource-talk are interchangeable ways of speaking about scientific practice, as though scientific practice were a homogeneous thing ready for subsumption under a general theory of the social. Indeed, the suggestion that these discourses are *general philosophical frameworks* for coming to grips with the activities of scientists inadvertently reinforces the very unitary conception of scientific practice that many scholars are striving to dismantle. If anything, these discourses are suited to different kinds of cultural perspectives, and consequently, we can make clear-cut choices about their strengths and limitations only so long as *we are clear about our object of study.* As we continue to work out a new vocabulary for speaking about science as heterogeneous cultural practices, we need to be reminded from time to time of Weber's claim that *culture is always culture from a particular point of view.*[36] In their supposition that there is a privileged way of speaking about scientific practice—whether it be from the view of the tabletop, where experimenters seem to fiddle with active cultural elements, or from the view of a comparatively rigid institutional framework, where scientific behavior seems to be inscribed by abstract social practices—Galison and Pickering often lose sight of the fact that if the ubiquitous expression "culture" is to mean anything in the burgeoning sociological literature on science, it must adopt a specific point of view.

For example, in recounting attempts to weigh the earth, I have adopted the point of view of the tabletop. This perspective, in my opinion, furnishes a better representation of the activities involved in refining Michell's apparatus, as well as providing some general insights about the activity of replicating experiments. In particular, it highlights the attempts of experimenters to extend bits and pieces of culture to improve our knowledge. The suggestion that scientists seek not merely to stabilize results *but to improve them* does not square with Galison's realism of intractable pressures or with the epistemic constraints that Radder places on attempts at reproduction.

However, it is not part of my view to suppose that this is the whole story; indeed, it is not part of my view that the activities of scientists are sufficiently homogeneous to lend themselves to a whole story. I suppose that with respect to the history of this experiment, we could adopt the frame of reference of the British government or the Royal Astronomical Society and tell a different story, one that displays the elements of scientific

36. An option is to follow in the footsteps of Latour and Woolgar (1979) and hold that the "social" in the sociology of science is an ironic expression (see Baigrie 1994). Pickering's reluctance to address philosophical issues raised by the conception of culture that underwrites his pragmatic realism would suggest that he endorses a similar view.

culture as abstract cultural practices that impinge on the practices of the likes of Cavendish.

Acknowledgments
I want to express my gratitude to Andy Pickering and to Jed Buchwald for comments that greatly improved this paper.

Six

Following scientists through society? Yes, but at arm's length!

Yves Gingras

All science would be superfluous if the outward appearance and the essence of things directly coincided.

Karl Marx, Capital, *volume 3*

Over the last 15 years, sociologists of science have produced a great number of fascinating, fine-grained analyses of scientific practice, emphasizing the variability of methods and practices of scientists and engineers as well as the contingency of the results or conclusions they obtain. The main agenda of this research program was to counter the then dominant conception that scientific knowledge was simply the end product of the application of logic, observation, and experimentation, the results of which were somehow an adequate representation of the real world existing outside and independently of the scientist. The microanalysis of scientific experiments and controversies has shown that these practices are in fact quite complex and should not be taken to constitute an *unproblematic* bedrock on which science is once and for all securely founded. It is a great merit of the constructivist approach to have shown philosophers that simple models based on an abstract "rationality of science" must be abandoned in favor of much more contextualized and dynamic models of scientific practice.

That being said, it is important to note that one of the striking features of the constructivist[1] literature, particularly in recent years, has been the proliferation of code words or buzzwords and "principles" supposedly necessary to understand scientific practice: "seamless web," "heterogeneous engineering," "actor-network," "actant," "black box," and, to mention an example that has probably not yet reached the English-speaking

1. Though one should distinguish between many species of "relativists" and "constructivists," the term "constructivist" will be retained here for the sake of simplicity. It applies to all the literature discussed in this paper.

world, "*investissement de forme.*" As for the "principles" invoked, they are considered self-evident, and mentioning that they have been *transgressed* in an analysis now seems sufficient to justify discarding its results without further comment. The best-known example is certainly the "principle of symmetry." Whereas it has been widely discussed and is very useful as a *heuristic* device in sociological practice (though its *epistemological* status is controversial), "extensions" of it (such as the symmetry between animate and inanimate or between nature and society) appear without much discussion or justification, like the curious idea that "the explanation has to be at least as rich as the content" that is to be explained (Latour 1988a, 258). Frankly, I just do not see any reason to limit a priori the kind of explanation to be offered and would not be averse to accepting a "simple" explanation for a "complex" phenomenon if it were convincing. The case of chaos theory is a perfect example of the possibility of explaining the complex behavior of systems by using simple dynamical equations (see, e.g., May 1976). Latour's statement is even more problematic if we keep in mind that the categories of "rich" and "complex" and "simple" are not given or self-evident but, if I may say so, "socially constructed." Finally, other kinds of statements in the literature appear as principles but are actually more like incantations—for example, the habit of stating that "the technical and the social cannot be distinguished." These statements are often found in the introductions and conclusions of papers that describe case studies making many such supposedly impossible distinctions.

The problem about all this is not that we do not need specific concepts to study science and technology from a sociological point of view. It is rather that the dynamic of exchange in this field has been such that there is now a cacophony of discourse and ideas that makes it difficult to pinpoint, understand, and evaluate the various views put forward by different "schools." In their criticisms of Latour and Callon, Collins and Yearly also note this problem of interpretation. They even confess: "Over the years [they] have found difficulty in taking seriously the more flamboyant statements of the Actant Network School at face value but, *fearing to appear foolish,* [they] have kept quiet" (1992, 370, emphasis added).

This confusion partly results from the tendency to juxtapose many of these buzzwords, principles, and incantations without regard to their consistency. One of the objects of this paper is to show that consistency and clarity have not always been characteristics of the pronunciamentos found in the papers which use the "actor-network" or "heterogeneous-engineering" language. To do so, I will look at the way these notions are presented, defended, and used in the narratives of empirical case studies. After this critical part, which takes up the first five sections of the paper, I will propose a sociological model of the practice of science that is reflexive, is consistent with our present knowledge of the microdynamics of research, and takes into account structural constraints on scientific practice that are invisible at the microlevel of analysis. Only in this way can *invariants* be recognized in the infinite variations observed in the many microhistories of controversies and of scientific practice.

1. Heterogeneous or Homogeneous?

Nowadays, no "serious" paper in sociology of science can begin without *stating* (not *arguing*) that "technical, social, economic, etc., factors are inextricably bound together." The most developed positions in this regard are no doubt the so-called actor-network and heterogeneous-engineering approaches, which try to take seriously the impossibility of discriminating between the kinds of objects and factors that enter into the dynamic of scientific activity. For Callon and Latour, for example, not only can we not distinguish the "social" from the "technical" but even the distinction between animate and inanimate actors or objects is arbitrary—a turn to hylozoism brilliantly analyzed by Simon Schaffer (1991).

The first point to make concerning the proclaimed "indistinctness" of the entities usually referred to as "science" and "technology" is that if it is to be taken seriously, then one can hardly see how one could even talk about science and technology. It should go without saying that it is impossible to write or even think without making distinctions and that even if in the "real world" everything were in everything, any analyst (as the term suggests) would have to make distinctions to describe, analyze, or comment on a situation. This is why vocabularies devised to try to convey the idea that "everything is in everything" engender some confusion and, as we shall see, do not really reflect the content of the case studies they preface (or follow).

Take the word "heterogeneous." In conventional dictionaries it refers to objects that are different, distinct, and separated. A heterogeneous mixture of liquids is one in which the two liquids do not mix together—oil and water, for example. So it is more than curious to see those who want to convey the idea that social, technical, animate, and inanimate cannot be distinguished a priori since they form an "organic whole" (Callon 1987, 84) use the term "heterogeneous engineering," which suggests exactly the contrary. Rather, one would have guessed that they would have come up with, say, something like "homogeneous engineering." The use of "heterogeneous engineering" can only mean that all the factors involved are *distinct* and brought together in a whole that may *then* become a "seamless web." On the Cartesian scale of "clear and distinct ideas" Thomas P. Hughes (1983) is more consistent than Law or Callon for he explicitly distinguishes between animate, inanimate, political parties, engineering companies, and so on, and shows how all these components have been *linked and transformed into* a "seamless web" in which it has become difficult to disconnect the social from the technical. In other words, for Hughes, the seamless web is the *result* of the process of constructing systems.

Since one simply could not write without making distinctions, it is not surprising to see that in all the narratives constructed to understand any scientific or technological development, authors constantly distinguish between factors that are not supposed to be distinct. Since Law and Callon have been the most forceful to voice this "indistinctness" between various categories (social, technical, political, and now economical), it is fair to look at their joint paper devoted to the analysis of a British military aircraft project to see how they manage to hold to their "principles."

As usual, the introductory manifesto notes that "any attempt to separate the social from the non-social not only breaks the methodological principle of following the technologists. It is also, quite simply, impossible because the social runs throughout the technical and thus cannot be separated from it" (Law and Callon 1988, 285). The first problem here is that we are never told if the impossibility of separating these factors comes from the fact that *in practice* (at the ontological level) they are linked or mixed together and/or that these *analytical tools* are not useful in understanding this reality (the epistemological level). One can frequently *distinguish* factors that cannot always be *separated.* The second problem with this position is that the authors in fact describe the TRS 2 project and explain that its future "depended on *two factors.* On the one hand, it was important to *demonstrate the technical competence* of the project and the best way to do that was to have a *successful maiden flight.* . . . On the other hand, the *outcome of the general election was also vital.* Conservative success would assure the future of the project. Labour victory would call it into question" (293, emphasis added). For an "average" political scientist or sociologist this analysis would pose no problem, for it is quite traditional. These potential (and ideal) readers would also see nothing unusual in the fact that the experimental flight "was highly successful, the aircraft handled well and there was no hint of the destructive resonance that had plagued the engine" (293). They could also easily understand that opposing political parties had different priorities so that once elected, the Labour Party, "beset by *economic* problems, quickly ordered a detailed scrutiny of the various military aircraft projects" (293, emphasis added). They would find the detailed description of the events and negotiations that led to the cancellation of the program to be very interesting: the Treasury Board was against the project, the minister of defence was in favor of an American F-111, and so on. In short, they would recognize in the narrative the usual objects and actors making up society (Labour Party, Treasury Board, cabinet ministers, etc.) and conclude that there was nothing unusual or startling in all that. And even "traditional" sociologists or political scientists would accept their conclusion that "the project went through *different phases,* some of which were *more technical in character,* while others were *more political*" (295, emphasis added). Finally, they would also probably be convinced that, given all the factors involved, "the development of the project was contingent." At least I was and cannot see why the authors suggest in their introduction that their analysis will "lead to a conclusion that is *counter-intuitive for many sociologists*" (284, emphasis added). Again, beyond the self-proclaimed "radical" or "counter-intuitive" character of the actor-network approach, the narratives produced are fairly traditional.

To an outsider to the ongoing debates among sociologists of science, the traditional character of these descriptions and narrations (and other case studies would show the same structure) is plain, and the extracts quoted above are sufficient to show that the authors constantly make clear-cut distinctions between many kinds of objects and actors in order to make sense of the "world" they analyze.[2] Thus, not only do they locate two "political parties" in their story but they also isolate "two factors" important for the future of the project, one being "a general election," an object quite familiar to political scientists and even sociologists. Since this choice of factors was made among what is a priori a great number of possible elements that should not be distinguished, the authors did not apply their "methodological adage" (Law and Callon 1988, 285) very well and even forgot to remember Callon's order that none of the ingredients "can be placed in a hierarchy or distinguished according to its nature" (Callon 1987, 86). Moreover, in the *temporal succession* of events, where discontinuity is admittedly more problematic than between kinds of actors or objects, they even distinguished between "more technical" and "more political" phases.

In his paper on the Portuguese maritime expansion, John Law also shows his capacity to constantly distinguish (at least analytically) the social from the technical: talking about the set of relatively stable associations between the cannon, the ship, the master, etc., Law writes that "some of [the] hostile forces were physical (the oceans), while others were social (the Muslims)" (Law 1987a, 247). Note also that physical reality

2. Of course, one is always open to the charge of quoting out of context—the usual and easy argument. To this argument I can only respond by suggesting that the reader read slowly and carefully the papers referred to, applying to them the method of highlighting the traditional distinctions that are constantly made in the course of the narrative, as well as the words used to denote causality or to identify the subject of action; all this without paying attention to or being impressed by the introductory manifestos, which try to orient the reader by already imposing on the text an interpretation that is in fact inconsistent with its grammatical and semantic content.

("oceans") is introduced as an *explanatory* factor on the same level as social factors in the good old realist tradition, which has never been shy of combining social and technical factors to explain a given situation.

So, if there is anything "counter-intuitive" in their paper it is probably the fact that the reader is uneasy in the face of statements of principles that are contradicted in the main body of the text. How are we to explain the insistence in some quarters to repeat constantly, in introductions and conclusions, statements of principles that are not applied in the main body of the text? Either the authors mean what they say or they don't. Let us suggest some possible interpretations.

2. Do They Mean What They Say?

A charitable interpretation of this apparent contradiction is to suppose that they do not really mean what they say and that the contradiction between the content of the analysis and the introductory statements can be attributed to a lack of clarity: by saying that one cannot distinguish between all the elements intervening in the dynamics of science or technology, they only mean that science and technology are not done in a vacuum but with many other "tools" and that one can never say in advance how the actors will combine these different (thus "heterogeneous") elements; they just want us to see that there are complex relations between objects and actors and that neither technological determinism nor social determinism can explain the development of science and technology. To take the analogy of a cake, it is clear that once it is baked the chocolate cake is perfectly homogeneous, and a child may be convinced that it is not composed of several "heterogeneous" ingredients mixed in definite proportions.[3] But the cook knows that despite appearances it is indeed the case and could show it by a chemical analysis of the cake or, more simply, by making another one in front of the child and showing how mixing the ingredients in the right manner and in the right proportions (and, of course, with all the tacit knowledge of the true chef!) gives this homogeneous appearance.

The lack of precision of many statements concerning the role played by different factors in the analysis of science and technology also comes from a shift in the meaning of the terms and expressions used. When Law writes, for example, in the conclusion of his paper on Portuguese expansion that his approach "makes use of a vocabulary that does not distinguish among the social, the scientific, the technological, the economic and the political, and makes no a priori assumption that one of these carries

3. Thanks to Philippe Faucher for the analogy of the cake, which I developed according to my own taste. . . .

greater explanatory weight than all the others" (1987a, 252), we are not told whether the lack of distinctions refers to ontological, epistemological, or methodological levels. Does he simply mean that the *list* of possible explanatory (or at least intervening) factors is open or that trying to distinguish between these factors is not legitimate and, consequently, that no such list, however long, exists? Whereas the second part of the sentence suggests the first interpretation, the first part suggests the second. What is clear, as we have shown above, is that despite affirmation to the contrary the empirical description does in fact make the distinctions between all the enumerated factors. And if they were not distinguished, how could one talk about the *relative* explanatory weight of factors? Only if one first distinguishes them can talk about relative contribution make any sense. Shifts in meaning of the terms used or in the levels of analysis are frequent in heterogeneous-engineering and actor-network language and probably explain the difficulty of knowing exactly the sense of the interpretations offered, as Collins and Yearly (1992) have noticed. Olga Amsterdamska (1990) has also shown how deftly Latour constantly moves back and forth between semiotic and commonsense definitions of terms, thus making it impossible to attribute any definite position to the author. When Callon writes that all factors "are inextricably bound up in an organic whole" (1987, 84), he cites as confirming evidence Pinch and Bijker (1984), who, according to him, "also show the *impossibility* of separating the *definition* of *technical* problems from the *socioeconomic context* to which the investigators associate them" (Callon 1987, 102, emphasis added). Here Callon conflates two different problems: the *analytical* distinctions between the *factors* involved and the role played by these factors in the *definition* of problems. Saying that all *factors* form an organic whole amounts to denying the possibility of distinguishing them, whereas saying that the *definition* by social agents of technical problems is contingent on social situations is not equivalent to saying that the social is not distinguished from the technical. Quite the contrary: only by first distinguishing the "socioeconomic context" can one say that this context affects the way the technical problem is defined.

This tendency to confuse *relations* with *identities* is frequent. To give a last example: in an otherwise excellent analysis of the complex series of negotiations which led to the definition of the technical characteristics of guided missiles, Mackenzie and Spinardi succumb, in their conclusion, to the usual incantation about the indistinctness of, in their case, "politics" and "technology." They say: "So if we start our analysis from 'technology' we are led towards 'politics'; if we start from 'politics' we are led towards 'technology.' In this maze we indeed begin to find how difficult it is to distinguish the two" (Mackenzie and Spinardi 1988, 611). My uneasiness about that kind of argument can be made clear by the following transposition. Lieutenant Columbo—of my preferred television series—is on the trail of a murderer that he cannot localize and, after some research, summarizes his findings: "So, if we start our analysis from Paris, we are led toward London; if we start from London, we are led toward Paris. In this maze I indeed begin to find how difficult it is to distinguish the two." In fact, this "maze" is only the result of confusing a relation with an identity: why not conclude, as Columbo actually would, that "Paris" is not "London" and that what his research suggests is simply that there is a relation between "Paris" and "London" and that his job is to find the exact nature of this relation. In fact, what Mackenzie and Spinardi's analysis shows is simply that "a technological enterprise is *simultaneously* a social, an economic and a political enterprise" (Mackenzie 1987, 198, emphasis added). The fact that it is through their interaction that the factors take the shapes they have does not lead to the negation of their distinctness. Quite the contrary, it is only by starting from the distinctions of factors that one can talk about their simultaneous presence and interaction.

Thus, to summarize our charitable interpretation, if we limit ourselves to the core of most case studies, we can conclude that their authors do not really mean that "politics," "economy," and "science" *are* the same thing or are not distinguishable but simply argue that there is no definite hierarchy among the various factors involved, that this hierarchy and the nature of the relationships between elements change from one situation to another, and finally that no predefined and fixed number of factors can be identified and applied for all cases. It is a worthwhile plea against social reductionism. This seems to be the position of John Law when he writes that "sociologists of science tend to limit both the type and number of explanatory social factors" (Law 1987a, 229). The message that is transmitted in the introductory and concluding statements of the studies analyzed here thus seems to be that analytical distinctions should not be reified into natural kinds and that we should keep in mind that the Labour Party and the Treasury Board, for example, did not exist ten thousand years ago and are the result of a contingent historical process. This I think is perfectly right and important to remember as long as one adds that for a *given* time and place, however, there is a definite distribution of objects, institutions, and actors and that it is from this distribution that the analysis proceeds to show how these entities *are* interwoven into something new that will then become the "given" for a future situation. If this is what is meant by the "radical" pronunciamentos in the introductory sections of so many papers, then there is nothing really revolutionary here, for this understanding corresponds to the practice of the "average" sociologist.[4]

4. Since some colleagues thought I was suggesting that the "average" sociologist is necessarily right, I should say here that reference to this figure is ironic: it is used only to point

And when we look at the way these notions are used by other sociologists, we see that this more "conservative" interpretation is frequent (see, for examples, Sorensen and Levold 1992; Misa 1992; Westrum 1989).

If one takes seriously the temporal aspects of change, it is clear that Latour's argument that "since the settlement of a controversy is the *cause* of Society's stability, we cannot use Society to explain how and why a controversy has been settled" (Latour 1987, 144, emphasis added) is empty, for it is clear that a *previous* state of the distribution of actors and institutions can be used in the explanation. In addition to leaving out the temporal dimension of change, Latour speaks as if it was usual to use Society with a capital *S* as an explanatory resource, whereas in fact most sociologists refer to concrete groups and institutions, which is a completely different matter; if society cannot be the subject of action, particular institutions and actors can.

3. Back to Realism!

Under the guise of extending the list of explanatory factors, Law—as well as Callon and Latour—reintroduces reference to nature by insisting that physical phenomena must be taken into account to fully explain technological change. I have nothing against this form of realism as long as it is acknowledged as such and not presented under the guise of actor-network language. Though this return to realism has already been well analyzed by others (Amsterdamska 1990; Collins and Yearly 1992; Shrum 1988), it is worth noting that using metaphors of "associations," "strength," and "force" *rhetorically* unifies nature and society but hardly hides the realist position that lies behind these terms.

John Law writes that "to try to reduce an explanation of the Portuguese system to a limited number of social categories would fail to explain the specificity of the *volta,* the *caravel* or the *Regimento.* Portuguese views of the sun and the adverse winds are needed to make the explanation work" (1987b, 131). Since all the categories named in these two sentences refer to commonsense physical reality, these two sentences suggest that it is not so difficult, after all, to make a priori distinctions between the social and the technical (thus contradicting the conclusion of his paper that his approach "makes use of a vocabulary that *does not distinguish* among the social, the scientific, the technological, the economic and the political"; (emphasis added), and I think it would be difficult to disagree with these statements. Quite the contrary, a "traditional" sociologist would find

---

out that the self-proclaimed "novelty" of some of the analyses of science and technology is far from borne out by the results presented, which, notwithstanding the pronunciamentos, are fairly "traditional."

rather curious that someone could have *ever* tried to explain the Portuguese expansion (or the flight to the moon for that matter) without talking about boats and sails (or computers and rockets) or at least taking them for granted in the explanation! Again, a "traditional" sociologist may with good reason be considered too conservative, but it is clear that he or she will not see anything revolutionary in Law's text on the Portuguese expansion. He or she may just become a little upset by the grandiose style. As Collins and Yearly (1992) have already observed about Callon and Latour, this position is a return (others would say "retreat") to a "traditional" realist position. After all, taking into account the form of the sail and the direction of the wind and water currents has always been the way traditional historians of science and technology have explained Columbus's and other great voyages on the sea. In Maurice Daumas's encyclopedic *History of Invention* (surely not an example of avant-garde research) one can read: "In the Mediterranean, ships with triangular lateen sails suspended from oblique yardarms had long been in existence; the use of this sail plan *permitted* beating to windward. . . . Off the coast of Africa, south of the equator, the ships had both wind and current against them. This fact *explains* the new use of the caravels" (1969, 362–63, emphasis added). One can surely rewrite all this in more trendy language and say that the lateen sail and caravel help the explorer to "ally himself with" or to "enroll hostile forces," but one can hardly miss the fact that behind this varnish the basic realist explanation is left unchanged. Contrary to Collins and Yearly (1992), I do not object to this kind of realism, but I do find it a bit irritating to see it presented as "new" or "counter-intuitive." It may well be that a strategy of reversal, of negating a relativist position (seen as having become standard) to present a "new" position, has simply had the unintended effect of falling back on the realist position that was first negated by the relativists: a double negation is an affirmation and by declaring himself against "social reductionism," John Law was bound to reintroduce "winds," "oceans," and other "material" objects in his analysis so that the final result looks very much like old wine in a new bottle.

4. The Unconscious Search for a "Total" Sociology

On second thought, the problem with the charitable reading of actor-network and heterogeneous sociologists is that it may not sound "radical" enough. I have observed a frequent use of this term in papers presenting "new" views on the dynamics of science and technology. As far as I can see, the mere use of the term "radical" is supposed to be in itself a proof that the visions so characterized are somehow inherently superior or preferable to other ones characterized (implicitly or explicitly) as "traditional" without further argumentation. This frequent reference to supposedly "radical" interpretations seems to confirm Bourdieu in his suggestion that for intellectuals "extremes are always more chic" (Wacquant 1989, 25).

Since the above interpretation of the heterogeneous-engineering discourse is probably too traditional and not innovative enough, we are left with the possibility that they really mean that it is impossible to distinguish the social from the technical (recall Callon's "organic whole"). The apparent inconsistencies discussed above would then have their origin in the "traditional" disciplinary language, which tend to reify concepts into things and thus do not provide adequate tools to analyze the specificity of scientific practice. This lack of appropriate language would explain the contradiction between the content of the narratives and the methodological position defended, and this state of affairs would only be solved by finding a new language transcending the canonical disciplines. This seems to be the road suggested by Pickering when he writes:

> My suspicion is that scientific practice has its own unity and integrity that cuts very deeply across present disciplinary boundaries. . . . And thus the deployment of existing disciplinary concepts and categories is liable to a serious misunderstanding of what science is like. These concepts and categories have typically been formulated and refined with an eye to the delineation of autonomous disciplinary subject matters, and the danger of insensitivity to the nature of phenomena at disciplinary boundaries is evident. I do not know whether it is inevitable, but the upshot of disciplinary analyses of science has typically been the construction of disciplinary master-narratives in which a schema drawn from a single discipline constitutes an explanatory backbone around which all else revolves. (1990a, 710)

If it is the case that our problem is one of finding a "nondisciplinary" language, then we will have to wait for a new dictionary (and a new ontology?) before arguing this point further.[5] In the meantime, one may nonetheless ask how it will be possible in this "great whole" to talk about science, technoscience, or even technoeconomics (why not scientifico-techno-economics?) or anything similar. Grouping science and technology together still leaves too much else besides that should be taken into account. I am not sure that simply adding "network" after a usual word (like "money-network" or "text-network" in Callon 1991) really illuminates anything. What strikes me about this insistence in finding new words is that it suggests that those who use "traditional" disciplinary categories cannot but succumb to their reification and are bound to forget that they are simply analytical distinctions which are limited and will never catch

5. For a recent attempt to construct a new vocabulary by which to analyze society, see Boltanski and Thévenot 1987. Callon's own dictionary can be found in Callon 1991.

all the aspects of the phenomena. After all, it is not only science that can be considered to have "its own unity and integrity": a chocolate cake also has unity and integrity, and we can nonetheless analyze it to know its composition.

An example of the sterility of discussions on the "essence" of words is provided by Pickering's contribution to the present volume where he attacks my cherished word "constraint" (Gingras and Schweber 1986; Gingras and Gagnon 1988) for being supposedly "asymmetric" and wants to replace it by "resistance." First, contrary to what Pickering asserts without explanation, there is nothing *inherently* asymmetric and anticonstructivist in the word "constraint," for it can be used to refer to social, cognitive, and material constraints. And constructivism should not be conflated with relativism, for one can have the first without the second. The most important semantic difference, it seems to me, between "constraint" and "resistance" is that the latter has a more active connotation, which is an advantage, whereas the former suggests a structural aspect absent from "resistance," which is also an advantage, for it conveys the idea that not all trajectories are equally probable and not only because of external resistance but also because of internal limitations like the habitus of agents and the hierarchical distribution of institutions in the scientific field. This lack of any structural aspect to Pickering's "dialectic of resistance and accommodation" is evident in his narrative of Morpurgo, who is described as working alone in his laboratory confronted with and confronting his apparatus. As an antidote to this linear view, one should go back to Jean Piaget, who (already in 1950) had a similar but much more structural model of the dialectic of "assimilation" and "accommodation" (Piaget 1950, 1985). Pickering also opposes "resources" and "constraints," even though it is easy to understand that there is nothing absolute in constraints, for what is a constraint for a given actor in a given situation is a resource for another actor in the same situation (Gingras and Trépanier 1993). This being said, once we have accepted, with Pickering, that we live "in the material world," there should be no objection to the use of both terms according to circumstances and context (Pickering 1989b).

More important than the obsession with words that are rarely definite concepts and the search for a nondisciplinary narrative is the not always conscious dream of a *total* history of scientific activity (in the sense that Fernand Braudel [1972], for example, did not write a "social" or "economic" history but a "total history" of the Mediterranean Sea). The road toward this "total history" however is not in the direction of the negation of distinctions but of their integration. Contrary to Pickering, Callon, and Law, I do not think it will be possible to solve the problem of the "wholeness" or "integrity" of science or technology (or the chocolate cake) by simply inventing new words to make an artificial unity around terms like "entity," "force," "strength," and so on. Though it will not sound "radical" enough—and I apologize for that—I think it is more "realistic" to try to do a sociology of science and technology on the basis that it is not only possible but necessary to distinguish, and hence define (however loosely), the tools and concepts used to offer explanations of a given problem or situation. In other words, the discipline constructs its object and its tools. It often happens that in the course of research one has to redefine the object to enlarge its scope and to add tools constructed by neighboring fields (economics, sociology of work, anthropology, etc). In fact, what the "social construction" current has done is exactly that: import concepts that have been in use for a long time in anthropology, sociology of work, or ethnomethodology to enable a fresh look at scientific practice. This practice has opened the way to a "social" analysis of science and technology that now tends toward a "total history," trying to take into account aspects that were left out by previous studies. Over the last 15 years the tendency of microanalysis of science has thus been toward an integration of economic and political factors that were left out of the analysis in the first wave of ethnographic studies, rather than toward the dissolution of these macrosocial categories. This however supposes that one still wants to do sociology, a question that we now have to address.

5. Choosing Your Discipline: Engineering or Sociology?

Despairing of sociology, or simply making another attempt at the strategy of negating what is perceived as a dominant tradition,[6] Callon and Latour

6. After having been among the first to talk about the "social" construction of science in their now "classic" book *Laboratory Life: The Social Construction of Scientific Facts,* published in 1979, B. Latour and S. Woolgar have deleted the word "social" from the title of the Princeton edition published in 1986; the new subtitle now being *The Construction of Scientific Facts.* Latour's last moves—for those who like to follow them—are the replacement of "reflexivity" by "infra-reflexivity" (Latour 1988b) and of "postmodernism" (itself a "reaction" to "modernism") by "amodernism." This reminds me of Claude Lévi-Strauss's description of the habit he learned at the Sorbonne for solving any problem "grave or trivial" by using a "method that never varies": "You first establish the traditional 'two views' of the question. You then put forward a commonsense justification of the one, only to refute it by the other. Finally you send them both packing by the use of a third traditional interpretation, in which both the others are shown equally unsatisfactory. Certain verbal manoeuvres enable you, that is, to line up the traditional 'antitheses' as complementary aspects of a single reality: form and substance . . . appearance and reality . . . and so on. Before long the exercise becomes the merest verbalizing, reflection gives place to a kind of superior punning, and the "accomplished philosopher" may be recognized by the ingenuity with which he makes ever-

want to convince us that engineers are better sociologists than "professional" sociologists. They suggest that "social sciences [should] in some way or another make use of the *astonishing faculty engineers possess* for conceiving and testing sociological analysis at the same time as they develop their technical devices" (Callon 1987, 99, emphasis added). They think that "scientists and engineers . . . are much more original, daring and 'progressive' social philosophers and social theorists than most social scientists" (Callon and Latour 1992, 351). Such remarks certainly make sense as pep talks for would-be engineers in a school of mines but hardly contribute to a sensible discussion about the kinds of tools and concepts needed in sociology of science.

To say that "engineer-sociologists" "develop explicit sociological theories" (Callon 1987, 98) in order to create new technology and that they should become "the model to which the sociologist turns for inspiration" (Callon 1987, 99) constitutes another example of the typical strategy of inversion, which, like that which brought Law back to a classic realist analysis of technology, simply leads to the position expressed, for example, by American engineers during the "technocratic movement" of the inter-war years. As Akin (1977) has shown, these engineers were consciously trying to transform society with their technical knowledge. Moreover, the suggestion that engineers are sociologists plays on the different senses given to "sociology" by shifting from the actors' categories reinterpreted in terms of sociological theory and *then* saying that this theory is explicitly developed by them.[7] It would also be easy to write the same kind

bolder play with assonance, ambiguity, and the use of those words which sound alike and yet bear quite different meanings" (1967, 54). Though the Sorbonne may be the best at inculcating this habitus, it is certainly not impossible to acquire it in other French institutions as well.

7. Callon's paper comparing the sociologies of Touraine and Bourdieu with the activities of the engineers of Renault and EDF is entirely based on a metaphorical use of the term "sociology" and the *attribution* of an explicit sociological discourse to the engineers, which is then compared with what the chosen sociologists *would have said* in the situation. After a totally artificial reconstruction of Bourdieu and Touraine, Callon concludes that in "his [Bourdieu's] explanation of car users' preferences he [Bourdieu as re-created by Callon] omits most of the elements that make up and influence these preferences. . . . Although Bourdieu happens to be right and Touraine wrong, this is quite by chance. . . . the discovery of a cheap catalyst as a substitute for platinum might have proved Bourdieu wrong and rehabilitated Touraine's sociological theory after all" (Callon 1987, 97). This analysis is quite surrealist: (1) Bourdieu's "omissions" are in fact Callon's, for he is the one who constructed this "Bourdieusian explanation"; (2) the realist argument that the discovery of new catalysts would have changed the situation is trivially true and is just plain commonsense realism; it amounts to saying that the landing of a flying saucer or, the explosion of an atomic bomb would affect sociological explanations. What would one think if instead of analyzing papers written by Callon I would write what Callon would have written and then show that "his" analysis is inadequate?

of metaphorical rhetoric about politicians, prime ministers, or presidents and say that they are "better" political scientists than professional ones. Who denies that politicians do experimental economy, political science, or sociology? They vote on laws instituting medieval taxes, eliminating unions, and so on. To observe that some categories "escape completely from the vocabulary of sociology or economics" and that these vocabularies cannot "describe the relationships between fuel cells and the electric motor" goes without saying for anyone who believes in the existence of material reality. Who doubts that the relationship between fuel cells and electric motors cannot be described in "terms other than those of electric currents and electromagnetic forces" (Callon 1987, 95)? Again, insisting that traditional sociological categories leave out many concepts—which is true by definition—leads to the absurd position that if it were not legitimate to stop the analysis at some point, one would be obliged to reconstruct society (or, more exactly, "the great whole") from scratch every time one writes a paper on any subject. Even those interested in a "total sociology" (or a "total technology") could not take all these factors *explicitly* into account in their analysis, for it would be endless and hardly readable.

An argument frequently presented to show the impossibility of distinguishing the social from the technical is to say that the actors' actions "cut across" these very distinctions and ignore them. This raises the question, never really addressed by the proponents of heterogeneous engineering, of the distinction between the categories of the actors and those of the analyst. The fact that in their discourse actors use categories that are different from those used by sociologists is perfectly normal, but this fact does not show that the latter categories are not adequate to their purpose. After all, the role of the sociologist is to analyze actors' discourses and practices, and this can never be done by simply *repeating* the actors' points of view. The fact that what is labeled "technical" by one actor is labeled "political" by another actor certainly raises the question of why they do apply these different classifications. But that actors do not agree on these categories does not mean that the categories used by the analyst cannot explain why they disagree; and the existence of conflicting categories among actors does not mean that distinctions are not made but simply that they are made differently. So I agree with Mackenzie and Spinardi (1988, 612) that an actor's distinctions between the "technical" and the "political" cannot be adopted by the analyst, as long as one adds that it is not because the analyst does not make these distinctions but because the analyst makes them differently so as not to take sides with one group of actors against the other and in order to understand the reasons for (or behind) their disagreements. But whether he or she uses actors' categories or some other analytical categories, the sociologist of science always makes distinctions in the "great whole of society" in order to formulate an analysis. Once it is admitted that no one seriously pretends that these distinctions refer to a Platonic world of pure and unalterable categories, there is no reason to impute to authors who use a given set of analytical categories simplistic ideas about "political ontology" or similar anathemas, except as a strategy to avoid answering criticisms.[8]

The syncretic portrayal of sociologists offered by Callon, Latour, and Law suggests that they are more interested in engineering than in sociology. It is perfectly legitimate to prefer engineers to sociologists, but they should then realize that in so doing they have in fact "followed" engineers and scientists to such a point that they have "joined" them and have become their "representatives" and spokespersons instead of "following" them by staying at arm's length in order to look at them and observe them. No surprise then that Collins and Yearly (1992) perceive their work as traditional and in the "old" realist mode of describing science and technology, while Callon and Latour perceive theirs as trying to "*debunk* scientists' hegemony" (1992, 365, emphasis added). They simply talk at cross-purposes: the first group talks like sociologists, whereas the second group talks like engineers.

So choose your discipline and your social group: I choose sociology, not to "debunk," but to understand the complex and changing relationships between science, technology, and society *from a different point of view* from the one taken by engineers. What sociologists have to recognize to do their job is simply that the various distinctions they make in order to understand the dynamic they investigate are *analytical* constructions and not natural kinds and that only the empirical analysis will show the role played by each of them in the different case studies. And these concepts must be clearly set out in relation to (and distinguished from) the categories used by the actors themselves.

6. Structural Constraints and the Dynamics of Scientific Change

A striking consequence of the wave of microanalysis of scientific practice has been the tendency to describe actors as if they were absolutely free to move in any direction, to elaborate any argument, or to reject any kind of objection. Callon and Latour, for example, frequently suggest that scientists and engineers move continuously from the lab to the industry or to the minister's office. Their notion of "translation" is supposed to help the analyst follow the creation of associations, but it does not explain why

8. For a typical example of such an easy strategy, see Cambrosio et al. 1991.

associations fail or succeed and why some scientists (or engineers) do go into the minister's office while others do not. In more concrete terms: why do engineers graduating from a French *grande école* or from a *faculté* not seem to have the same professional trajectories or the same access to ministries?[9]

Arguments that scientists *could* have done otherwise than they did are often adduced to suggest this freedom of movement. Most of the time, however, we are never told if the possible counterargument has effectively been voiced by the actor or if the argument is an after-the-fact rationalization offered by the analyst. Constantly repeating that "things could have been otherwise" does not explain why in fact things were as they were and not otherwise. This abstract logicism is in good part due to the fact that constructivists direct their arguments more at philosophers than at sociologists in order to counter a kind of positivist determinism by insisting on the contingency of action. Up to a point, this was a good strategy to destabilize old philosophical models of science, but from a sociological point of view it is not very illuminating. If we agree to play the game of sociology seriously, we will have to develop a *model* of scientific practice that can explain the fact that in given circumstances, scientists and engineers probably could not have acted *much* differently than they did. The constant use of counterfactuals does not show that *they* could have done otherwise but that *we* can now imagine other solutions than the one achieved by the actors in given historical circumstances.

In addition to being an effect of debating with philosophers, the insistence on the contingency of action is also in direct relation to the scale of observation. At the level of microanalysis, attention to explanation has often been neglected in the face of the complexity of the interactions: observing and describing many games of chess or Go, one cannot escape the feeling that each game is different and can only be understood in the interactions between the players and their relatively arbitrary decisions. At the microlevel it is inevitable that the open-endedness of these games strikes the analyst as fundamental. In the end, however, accepting the total contingency of action is an extreme form of phenomenalism which leaves no place for any explanation of the dynamics of knowledge production; the analyst has then no other choice but to follow each case empirically to observe the result, like so many different and unique games of chess or Go. However, as these games impose minimal constraints on the movements of the pieces and yet still leave an infinite number of different possible outcomes, so I think that all the empirical studies stemming from the constructivist and ethnographic approach only make sense within a model of

9. For those interested in this question, see Bourdieu 1989.

scientific activity that explicitly recognizes structural constraints and is reflexively applicable to sociology of science itself. But far from leading to a kind of self-destructive solipsism,[10] our conception of reflexivity embodies a practical ethic of discussion and exchange as a *social* condition for the growth of knowledge.

In order to understand why and in which circumstances scientists or engineers can move from the laboratory to the minister's office, one must start from the observation that scientists are subjected to a disciplinary training that gives them a set of tools that define an intellectual horizon. Social actors socialized to live in a particular field can rarely transfer easily their skills and knowledge directly to another field (Bourdieu 1991).[11] To use Pierre Bourdieu's concept, their habitus is the product of a trajectory in a particular field and is best "adjusted" to function inside it. Each of these fields and the relations between them are the product of a past history of social relations and are, in this sense, a social construction. Thus, far from being homogeneous, the social space must be seen as composed of many relatively autonomous fields having their own logic: the plurality of fields is a plurality of worlds. It is this heterogeneous social structure that forces actors who want to circulate outside their native field to adapt, and thus transform, their discourse and practices to the implicit rules of the new field in which they want to circulate. In the case of science, this means that scientists who must appeal to the political field to get money to realize their projects must argue in a way that refers to the specific stakes of this field by talking, for example, of the "national interest" or "economic impact" of their projects in order to convince politicians and administrators. This suggests a strong sense in which to talk of "translation of interests," but this sense completely reverses the use of this term by Callon and Latour, who tend to smooth the passage from one field to another as if there were no "discontinuities" between fields (Latour

10. The best example of an approach leading to such solipsism is provided by Woolgar (1988). It is no surprise that within this kind of narcissistic reflexivity the author concludes his book by suggesting: "Self [should become] a strategic target for social science" (108). Having (textually!) "deconstructed" everything, Woolgar is finally left contemplating his own image in a mirror that is itself the projection of his own Self.

11. For those who prefer to talk about "forms of life" instead of "fields," let us note that Barnes's definition of "forms of life" is equivalent to Bourdieu's conception of practice defined as the relation between a habitus and a field. According to Barnes, "To participate competently and successfully in a form of life requires a profound and comprehensive training at the level of practice, and where the form of life is highly standardized and uniform the associated training must be highly ordered, intense and systematic . . . to provide shared perceptions and shared experiences" (1988, 79). The advantage of the notion of "field" over that of "form of life" is that the former is much more explicitly structured, the latter being, in the hands of too many writers, hardly more than a vague motto. For a comparison between Bourdieu and Wittgenstein, see Taylor 1990.

1987b, 132–44). In other words, they use the notion of translation to insist on the continuity of action without ever taking into account the heterogeneous fields that make those very "translations" possible and necessary—thereby explaining them. The existence of distinct subcultures corresponding to different fields and habitus generating particular kinds of cultural and social capital suggests that there is a barrier (and a cost) to entry in any field (Bourdieu 1979).

The heterogeneity of skills needed to circulate in more than one field also helps explain, for example, the fact that in the era of "big science" the personality of the "manager" and man of public relations took precedence over the myth of the shy and socially misfit scientist. Whereas the latter could easily survive and hide himself in the field of science, only the former could introduce himself into the field of politics (Gingras and Trépanier 1993). The transformation of the structure of the field was thus accompanied by a transformation of the habitus required to play in the field.[12]

In addition to the limits imposed on action and strategies by the habitus and the amount of social and intellectual capital possessed by the actors, the dynamics of knowledge production *in the scientific field* is also guided by criteria of communicative action. These minimal conditions can be established by starting reflexively from our own practice of sociology of science and are in fact grounded in the principle of nonperformative contradiction as enunciated by Karl-Otto Apel (1980, 1990).

Sociologists (constructivists included) who agree to play in the field of sociology by writing papers, presenting communications, and submitting arguments in favor of (or against) a given theoretical position try to be as convincing as they can. They do their best to avoid contradictions or non sequiturs in their arguments (though they do not always succeed) and track these flaws in the presentations of their opponents. In so doing they *already* accept as a practical a priori two tenets: (1) the principle of noncontradiction and (2) a rule of inference like "*p* implies *q*," which makes it possible to move from statements to statements and link some of them together. These minimal conditions are *sociologically* necessary for having meaningful communications between human actors, but we do not have to invoke their "universality" (as some philosophers do by using them as Kantian "regulative ideals"), for they in fact do not characterize all fields of activity to the same extent (for a logical analysis of these conditions, see Cherniak 1986). Though "minimal," these conditions are not trivial, for there exist different fields in which they do not operate in the same way—if at all. Though this is not the place to go into any detail,

12. For examples of historical studies of the constitution of fields, see Bourdieu 1971, 1987; Viala 1985; and Gingras 1991.

the field of politics offers a good contrasting example of a field in which the constraints of performative noncontradiction and self-consistency are very weak. Whatever the "rules" specific to a field, one thing is clear: "The only absolute freedom the game leaves is freedom to withdraw from the game, by a heroic renunciation which—unless one manages to set up another game—secures tranquility only at the cost of social death, from the point of view of the game and the *illusio*" (Bourdieu 1981, 316).

It should be clear here that talking of noncontradiction is not invoking some nonsociological criterion. As the Polish logician Jan Lukasiewicz already observed in 1910, "the principle of contradiction has no *logical* value . . . but it possesses a *pratico-ethical* value which is even more significant. *The principle of contradiction is the only weapon against error and lie*" (1991, 30, emphasis in the original).[13] Its *explicit* formulation is thus necessary as part of the sociological understanding of the scientific field. Moreover, it does not forbid arguments between actors about a *given* contradiction or deduction, for *those very debates already presuppose the acceptance of the principle of noncontradiction*. It is this distinction that is often forgotten in suggestions that logic is not a constraint in scientific debates because an actor can decide not to see a given argument as contradicting his own position. This is what Trevor Pinch, for example, suggests when he writes that the existence of a contradiction was not a constraint in the debate on neutrinos, because one actor (Bahcall) maintained his confidence in his theoretical model of neutrino production for more than a year despite experimental results that were seen by others as conflicting (1986a, 207–11). However, Pinch confuses the logical sense of the term "contradiction" with the larger sense of "disagreement" between theory and data. In fact, far from proving Pinch's argument about the flexibility of logic, the central point about Bahcall's story is that he had arguments for resisting, but after a while, confronted with new argu-

13. It is worth noting here that this concern with the practico-ethical aspects of life could be brought to bear on the discussions concerning the distinctions between human and nonhuman actors and the attribution of agency. An excellent reason why humans in general attribute special agency to humans as opposed to nonhumans—thus committing the sin of being "asymmetric"—is that they can then recognize the social responsibility of those who by their very actions save nonhuman actors like trees, whales, and so on. Everything suggests that it is not the scallops or the whales that "enroll" the humans but rather humans who for idiosyncratic reasons choose to dedicate part of their lives to becoming spokespersons for trees, whales, or birds, which, in all probability, will never know they may owe their lives to them. Though Latour believes that "the belief in causes and effect is always, in some sense, the admiration for a chain of command or the hatred of a mob looking for someone to stone" (1988b, 162), I think on the contrary that the search for causes is intimately related to an ethic of social responsibility, equity, and justice.

ments, instruments, and experiments, he finally admitted the problems faced by his model. The notion of interpretive flexibility introduced by Pinch is very useful and perfectly consistent with our criterion of communicative action. It is by constantly putting forward new arguments (theoretical or experimental) that actors try to diminish the interpretive flexibility of data and theory and thereby limit the possibility of alternative interpretations.

Even to reject new data one has to have arguments. What if Bahcall had continued to hold his position? Given the dynamics of the scientific field, one can easily predict that the rest of the community would have reacted by marginalizing him, with talk about his "silliness," his being "older," and the like. Such a social marginalization is well described by Rudwick (1985) in the case of the Devonian controversy. It is significant that in his criticism of that book Trevor Pinch has argued that "if it is the case that a scientist can still argue for a significantly different position from that embodied in the consensus, then it can be said that the empirical evidence does not *unproblematically* lead in one direction" (1986b, 711–12, emphasis added). The significance of this statement hinges of course on the word "unproblematically." It is clear from the analysis provided by Rudwick that the evidence in favor of the Devonian was far from being "unproblematic," for it took years to create a consensus. In order to insist on the contingent aspect of the consensus that emerged on the nature of the Devonian system, Pinch adds that "as far as Williams and Weaver [the two marginalized geologists in the debate over the nature of the Devonian] were concerned their own arguments were perfectly respectable," and he uses this argument to support the conclusion that since their arguments "were available when the Devonian interpretation was reached, any view in which the 'pull of the evidence' is seen as being decisive is unwarranted." What is striking about Pinch's analysis is that it is implicitly based on a subject-centered epistemology that does not take into account the sociological context. The sociological question here is: *for whom* were the arguments respectable? There can be little doubt that *their own* arguments were perfectly respectable *to themselves*—this is a tautology. The problem is that they could not convince *other geologists.* And this conclusion is not based on any general philosophical preconception about the a priori role of evidence in scientific research but on an analysis of the dynamics of the scientific field of the time. Curiously, a "philosopher" like Gaston Bachelard was already more sociological than the "sociologist" Trevor Pinch in his analysis of science when he wrote: "We propose to ground objectivity in the behavior of others [the peers]. . . . any doctrine of objectivity always submits knowledge of the object *to the control of others*" (1972, 241, emphasis added). As Mary Tiles aptly summarizes

Bachelard's view for the English-speaking reader, objectivity has an "essentially social dimension; objective knowledge is not the unique experience of an individual but that which can be agreed upon by all similarly placed rational subjects" (1984, 53). From this truly sociological point of view, there is no such thing as a private science. Ceasing to exchange arguments or to produce new data about experimental or theoretical results or procedures in the scientific field is ceasing to do science. A scientist can remain convinced for the rest of his life that he is right but his views have no social existence in the scientific field if they are not debated and accepted or rejected.

7. The Temporality of Scientific Debates

A brief analysis of a historical case study will help illustrate the principles discussed above. The tragic history of the French geologist Jacques Deprat dramatically illustrates the role of argumentation and the effects of its *time-situated* character in the acceptance or rejection of knowledge claims by a given scientific community.

In June 1919, a jury of French geologists, meeting in the geological laboratory of the Collège de France, declared one of their most esteemed and brilliant colleagues, Jacques Deprat, guilty of forgery. Expert in the geology of the south of China and the north of Vietnam, he had collected, between 1909 and 1916, a large number of fossils. However, in 1917, his close collaborator, the autodidact geologist Henri Mansuy, announced that Deprat had added fossils of European origin to his Asiatic collection. The crux of the argument put forward by the committee was that some of the trilobites were typically of European origin and current theory made it impossible that they be found in Asia. Though there was much personal acrimony between the protagonists of this affair and private interests were at stake, the crucial point is that current knowledge made it difficult for Deprat to explain the presence of these trilobites, and he could only repeat that he had never added these specimens to his collection and that they were really found during his fieldwork. With no convincing argument other than his own integrity, Deprat lost his job, was banned from the geological society, and wrote novels for the rest of his life, some of them even winning prizes (Durand-Delga 1990, 1991).

The interesting point in this affair is that Deprat's "honor" was recently restored by a historian who argued that over the last ten years geologists have reported the discovery in Asia of many trilobites of the same species that were in Deprat's collection (Durand-Delga 1991, 1346). What was thought impossible in 1917 now made sense in view of the theory of plate tectonics, according to which 400–500 million years ago Southeast

Asia and meridional Europe were closer together than they are today, thus explaining the similarity of the fossils. In view of this historical research, the French Geological Society posthumously reintegrated Deprat as a member on 10 June 1991.

The question is: was a mistake made in 1917 when Deprat was condemned? I think that the sociological answer must be no, for no convincing explanation could be offered at the time of the controversy to make sense of the presence of the contested fossils. Only now, over 70 years later, can one suggest that Deprat was not after all guilty. As this example suggests, debates, discussions, and decisions *are always located in time and thus bounded by a horizon of what is thinkable.* Only later in time can something become "obvious" and old "mistakes" be corrected. And even then, nothing prevents these "corrections" from being in turn corrected and leading, for example, to another condemnation of Deprat. The case of Deprat is not unique. The death and resurrection of Michael Polanyi's potential theory of adsorption also illustrate the crucial role of temporality in science (Polanyi 1969, 87–96).

This model of scientific change based on a dynamic of communicative action inside a structured field has the advantage of explicitly formulating principles that act as a priori constraints on practices—and are thus most of the time implicit—instead of accepting them *implicitly* by having them play important roles in discussions (among the scientists observed as well as in the texts produced by the sociologists) while at the same time trying to prove that they are not at play. In my view, argumentation is the motor of scientific advance. Alone on his island, Robinson Crusoe would not develop science and would be limited to personal opinions and practices that would rapidly stagnate. As Bachelard put it in his poetic manner: "Solitary science is qualitative. Socialized science is quantitative" (1972, 242). As G. E. R. Lloyd has admirably shown, the emergence of Greek science is closely tied with the importance of oral debates in Greek society, and experiments were used, at the beginning, more as a rhetorical device against competing theories than as an effective practice (Lloyd 1979). And a reading of *Polarity and Analogy* (Lloyd 1966) suggests that the very constitution and codification of logical rules are themselves the product of these debates.

Of course, the passage from experiment as rhetoric to full-fledged experimentalism in seventeenth-century Europe was made possible historically only by major and unforeseeable demographic, economic, technical, and institutional innovations like the growth of cities, the printing press, and the formation of scientific societies. We still lack a complete study of the formation and transformation of the scientific field over the last four centuries (but see Ben-David 1971), but even with its modern institutional and technical trappings, the scientific field can still be considered to have something in common with its Greek origins: the dynamic role of public debates. As Lloyd suggests in his analysis of the relationships between science and society in ancient Greece, "this very paradigm of the competitive debate may have provided the essential framework for the growth of natural sciences" (1979, 267).[14]

The dynamics of knowledge production is a historical product not defined in epistemological terms (as the recurrent debates between realism and relativism so often suggest) but in essentially sociological terms: to be right *at a given time* is to have arguments which, given the structure of the field and the current context of experimental and theoretical knowledge, cannot be convincingly contested and replaced by others which would win the assent of the majority of the scientists active in the field.[15] In experimental sciences, the arguments are most of the time *about* experimental data, procedures, or instruments, and these arguments are made within a certain structure of accepted knowledge and procedures (black boxes) that result from previous debates. One could even argue that the development of an effective experimental practice—as opposed to the rhetorical invocation of experiment analyzed by Lloyd (1979)—was a good strategy to oppose conflicting views: new experiments *do* modify the existing consensus and help change accepted theories or *force* opponents to produce new experiments. As Barry Barnes wrote, "natural knowledge is always learned in conjunction with the operations of manipulating and controlling material objects and physical processes and its terms are used to refer to such objects and processes in specific situations" (1988, 55–56).

The argumentative nature of the dynamics of scientific change gives time a central role, for *it always takes time* to experiment, argue, and counterargue. All this is a practical achievement that generates new experiments (acting on the world) and new theory (talking about the world). This "process of discursive rectification" gives rise to a "discursive objectivity" which for Bachelard grounds objectivity in the social control of the members of the "*cité savante*" (1972, 241–42). It should come as no surprise that in the neutrino story Bahcall could maintain for more than a year his point of view in the face of counterarguments. Interpretive flexi-

14. By labeling "persuasive argument" "a statics of logic," Pickering completely misses the point that the dynamic element comes from *debates with others*. This model is thus a far cry from his "dialectic of resistance and accommodation," which is still based on the subject-centered Cartesian epistemology in which a lonely scientist confronts nature with his instruments, leaving in the dark the collective aspect of knowledge growth (Pickering 1990, 720).

15. The study by Kim (1991) of the reception of Johannsen's genetic theory exemplifies clearly the role of argumentation in the rejection of Pearson's biometry.

bility is an aspect of this time-ordered dynamic of scientific exchange. Arguments trying to show that logic or experimental data do not constrain belief can seem convincing only by freezing time.

This importance given to time and arguments leads one to a historicist conception of knowledge which has no place for any kind of absolute criteria of truth. There is no need to reject any explicit reference to an external reality that constrains (or resists, if Pickering so prefers) knowledge claims about it in order to accept that it is the very social dynamics of a regulated exchange in the scientific field that is the condition for producing a type of knowledge that can transcend its conditions of production (Bourdieu 1991, 22–23; Bachelard 1975, 137). As Barnes put it: "we cannot know that a ball is a sphere simply by looking to other people, who will be looking to yet other people, and so endlessly. We must all first look to the ball and decide for ourselves as to its shape—and only then look to others to discern the agreed verdict, if there is one. We have practices for determining the shapes of things which we apply to the things themselves and which tell us (collectively) what shapes the things are" (1988, 179). So, without advocating any simplistic correspondence theory of truth, "there is a genuine sense in which natural knowledge can be said to have external referents," to quote Barnes again (1988, 56).

8. Conclusion

In conclusion, I would like to note that the ethic of communication built into this reflexive model of knowledge production is that in sociology of science—as in any science conceived as a practice regulated by the logic of a field—one condition for playing the game is to continue to argue and counterargue, experiment and counterexperiment (or, in history, do archival research), in order to show the shortcomings of the position of the "opponents." In other words, although, as Aristotle wrote a long time ago, "we are all in the habit of relating an inquiry not to the subject-matter, but to our opponent in argument" (quoted in Lloyd 1979, 267), we should replace straw-man rhetoric and vague reference to "sociology" or "society" as a whole by careful analysis of the actual content of the papers produced by colleagues. Only in this manner will our collective understanding of the "subject-matter" of the dynamics of science, technology, and society (to be short) make any advance. My critical analysis of some notions suggested to make sense of this dynamics only aimed at such a clarification. Faced with articulated arguments one can choose to enter the debate by addressing the specific issues raised (thus playing the game of the field), keep silent, or resort to sophistic techniques (again well analyzed by Aristotle) that avoid confronting the issues. Pace Shapin, "panache, charm and infectious wit" (1988a, 534) are characteristics that will never replace coherent argumentation and reflexive practice, for they are more attuned to the field of fashion than to that of sociology.

Acknowledgments

Since the first circulation of this paper in November 1991, I have benefited from many helpful comments, for which I want to thank Olga Amsterdamska, Donald T. Campbell, Stephen Cole, David Edgerton, Philippe Faucher, Benoît Godin, Sungook Hong, Pierre Milot, Robert Nadeau, Dominique Pestre, John Pickstone, Hans Radder, Claude Rosental, Jan Sapp, and Michel Trépanier. In order not to embarrass any one of them I insist that I am the only person responsible for the content as well as for the tone of the paper! It has also been read at the meeting of the Canadian Society for the History and Philosophy of Science in August 1992 and at the Department of History of Science, Technology, and Medicine of the University of Manchester in January 1993 and discussed in a public debate with Bruno Latour at the Cité des sciences et de l'industrie in Paris in March 1993. Thanks to Bruno Latour for his friendly discussions with me, though they have not yet convinced me to make substantial changes to the paper.

II

Stories

SEVEN

WHY HERTZ WAS RIGHT ABOUT CATHODE RAYS

Jed Z. Buchwald

1. RETROSPECTIVE DIFFICULTIES AND THE IMPORTANCE OF CONTEXT

On 29 April 1897, J. J. Thomson argued before the Royal Institution that experiments done at the Cavendish Laboratory showed cathode rays to consist of moving electric particles. Fourteen years earlier, a young and little-published German physicist named Heinrich Hertz had performed experiments which showed both that cathode rays differ in important ways from electric currents and that they do not behave like moving electric particles. Explanations for Hertz's results, which were assumed to be erroneous after 1897, spread through the literature. These events afford an excellent opportunity to examine closely how something that, at the time, seemed to its maker to constitute an unbreakably tight argument became a decade and a half later an example of a poorly performed experiment, one whose design, to quite J. J. Thomson's biographer, "was open to certain objections" (Rayleigh 1969, 78).

By the time of Hertz's premature death in 1894, the understanding of his cathode-ray experiments had already begun markedly to change, both for him and for his German contemporaries, not least because of Hertz's own laboratory production of electric waves in 1888. Electric waves in the laboratory were bound by Hertz to electromagnetic fields on paper, and the union he forged between the effect and the concept in the early 1890s affected German physics in ways that diverged from the context of his own cathode-ray experiments.[1]

I intend to use this context to probe through example some of the issues addressed by Hans Radder, Peter Galison, and Andy Pickering. I will ask first in what sense Hertz's experiment was reproduced. Did subsequent experimenters do the same (relevant) things under the same (rele-

1. A detailed analysis of the transformation will appear in volume 2 of my *The Creation of Scientific Effects*.

vant) circumstances with essentially the same kind of device that Hertz did?[2] More precisely, if we examine the conditions of apparatus and surroundings that the experimenters considered important to their success, do we find significant differences between them, and, if so, why? In Radder's terminology:

Question 1. Was there a *fixed theoretical description* across cathode-ray experiments, in the sense that the several experiments considered the same factors to be relevant? If not, what were the differences?

Second, I will ask whether, in re-creating the original context of Hertz's experiment, and in following his specific manipulations, we see him most fruitfully as *constrained* by instrumental, natural, and contextual factors (Galison), whether we do better to see these factors as *malleable resources* (Pickering), or whether there is a somewhat different way of thinking about constraints and resources, at least in cases of this kind:

Question 2. Was Hertz *constrained,* did he manipulate *resources* to achieve his ends, or should the question not be put as an opposition?

2. Purified Cathode Rays

Heinrich Hertz at 25 was frustrated. By June of 1882 he had been working as an assistant in Helmholtz's Berlin laboratory for a year and a half, but he had so far not made his mark. Having failed with a series of experiments on evaporation, he was looking for something else. Hertz decided to plunge into an area that had for nearly a decade been a major Berlin preoccupation: the beautiful, puzzling luminosity of the Geissler tube. These gas-filled glass cylinders shone with strange, patterned glows when metal termini within them were connected across powerful batteries or strong induction coils. The glows followed raylike paths that emanated from the terminus, or cathode, that was connected to the battery's negative pole. These "cathode rays" could be deflected by magnets as though they were flexible electric currents, and they had long been of considerable experimental interest to Hertz's laboratory colleague Eugen Goldstein.

In order not to violate Goldstein's property rights Hertz had to find some way of working that did not expropriate Goldstein but that nevertheless left a wide-enough opening to have an impact. He solved this social difficulty by consulting Goldstein at every step. He asked for advice and followed up leads that his cooperative colleague tendered. In this way he

2. We may ignore, in this case, Radder's *material realization* since no one followed anything quite like Hertz's specific procedures, as we shall see. We may also ignore his *replicability of the experimental result* since neither Perrin nor Thomson got Hertz's result.

eventually decided to query Goldstein's comparatively minor claim that the discharge process which produced the glow must be intermittent.

Although Goldstein did not hold strongly to a particular model, he was certain that cathode rays were some sort of ether process, and he did offer the following suggestion. The discharge in a Geissler tube, he argued, is made possible by the gas within it, but the process actually occurs in the circumambient ether. It begins with the production of a "state of labile equilibrium in the ordering of the ether's particles" (1881, 266–67) with a corresponding tension that reaches maxima and minima over certain surfaces. Discharge consists in the equilibration of that tension (due to unspecified gaseous processes), and this results in the production of "a motion" in the ether that begins at surfaces of maximal tension and flows from there to surfaces of minimal tension. These motions constitute the cathode rays proper. When the gas particles in the tube are struck by the rays, they themselves excite a different kind of ether motion, namely, the transverse waves that constitute light and that make the rays visible.

From Goldstein's perspective each of the cathode rays, or motions between extreme tension surfaces, constituted a species of open current—open because no charge is transferred between the end of one ray and the beginning of the next one. The rays are, as it were, intermittent currents that shoot out when the ether's tension equilibrates. Hertz sought first to break apart Goldstein's essential understanding of the ray as an ether process, which had become important dogma in Berlin by 1883, from this particular model. To do so he began by attacking its basis in the assumption that the discharge process must be intermittent. Since this claim was not critical to Goldstein's experimental work or to his broader understanding of cathode rays, on which his reputation was based, Hertz no doubt felt he could challenge it in an impressive display of experimental pyrotechnics without offending his coworker.

In a methodical series of carefully contrived experiments Hertz demolished intermittence. But that was not all. Having liberated cathode rays from one currentlike property (namely, of being similar to an intermittent discharge), he decided to go further and to effect a complete divorce between these rays and electric processes. He sought to establish the rays as something new, as entities sui generis that had nothing electric about them, properly speaking, at all. This was much more radical than anything Goldstein had himself ever proposed, since for him the rays were still a kind of open current.

Hertz systematically undertook two preliminary experiments—a main and a related, subsidiary one—to show that the cathode rays have little to do with the detectable current in the glowing tube (the tube current, as I will call it). The purpose of the main experiment was to trace the tube current through its magnetic effect in order to see whether it followed the path of the cathode rays. In order for the resulting map to be significant, the magnet must not be deflected by the cathode rays acting in a nonmagnetic way. The rays might, for example, actually carry the entire current but deflect the magnet in two ways, one electromagnetic and the other specific to them. The resultant map might deviate substantially from the ray paths (depending on the nature of their nonmagnetic action), in which case it would falsely indicate that the rays do not track the current. The purpose of the subsidiary experiment was to remove this possibility.

Hertz accordingly fabricated a clever device which because of symmetry ensured that neither the tube current nor the rays would have any external magnetic effect. In Hertz's apparatus (fig. 7.1) the cathode is a brass disk that just fits the diameter of the Geissler tube. A hole bored through its center carries a thermometer, through which the anode, made of nonmagnetic metal, protrudes a bit. Since the anode emerges from the center of the disk-shaped cathode, the device is completely symmetric about the disk's axis. Whatever currents exist in the tube will therefore form an ensemble whose magnetic field lines circulate about the disk's axis—and so which cannot deflect anything. Any motion of an external magnet would therefore have to be due to a nonelectromagnetic interaction between it and the cathode rays—supposing that whatever currents the rays do constitute are also closed.[3] If nothing happened, then Hertz could be confident that the rays could *at best* move magnets electromagnetically, thereby guaranteeing the capacity of his main experiment to map the lines of tube current. "The tube," Hertz wrote, "was now brought as near as possible to the magnet, first in such a position that the magnet would indicate a force tangential to the tube, then radial, and lastly, parallel to the tube. But there was never any deflection,—none amounting to even one-tenth of a scale-division in the telescope" (1883, 240).

Let us for a moment leave Hertz in his laboratory, because we have here a very clear example of why one must seek precisely to understand the original context of an experiment. Even near-contemporary readers unfamiliar with this context missed the subtlety of Hertz's work. In a review of Hertz's papers after his death, the British physicist George FitzGerald, for example, remarked:

> From experiments on kathode rays projected down a tube, and quite away from both electrodes he [Hertz] deduced that they produce no mag-

3. If (as, in fact, Goldstein believed) the cathode rays acted as open currents, then they certainly should deflect the external magnet *as* currents. Hertz's result could accordingly be taken as showing also that the rays can *at best* be closed currents.

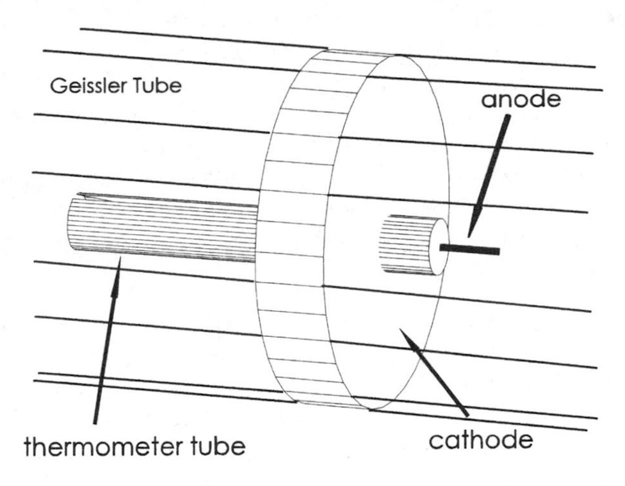

Fig. 7.1. Hertz's symmetrical device to separate rays from current

> netic action outside the tube, although they are deflected by the magnet. . . . This experiment on the magnetic action of kathode rays is quite inconclusive, and it is very remarkable that Hertz should have attributed much importance to it. Whatever current was carried down his tube by the kathode ray must have come back the tube by the surrounding gas, and these two opposite currents should have produced no magnetic force outside the tube; and this is exactly what Hertz observed. (1896, 440)

FitzGerald's remarks wonderfully miss the entire point of Hertz's trial. The Irish Maxwellian was perfectly correct in writing that Hertz got just what he should have supposing the rays to be closed currents. But he was almost certainly wrong in supposing that Hertz was unaware of the possibility, since it is entirely consistent with what Hertz was up to here, which was to show that the rays cannot affect magnets in any other fashion than by behaving *like* currents. FitzGerald had missed this critical aspect of Hertz's subsidiary experiment because he thought that all of the experiments were designed solely to demonstrate that the rays cannot be "streams of electrified particles." A later one was indeed designed (in part) to do that, but the goal of the preliminary experiments was to demonstrate that, whatever they may be, the rays nevertheless do not carry much of the current in the tube.

Confident that his laboratory work had now produced a reasonably convincing experimental argument for divorcing the rays from the tube current, Hertz began to insist on embedding the claim in appropriate language. Agreeing with Eilhard Wiedemann and Goldstein that the essence of the rays probably involves an "ether-disturbance," Hertz wished them to give up referring to the discharge as though it were the same thing as the rays. "I should," he remarked, "like to see the word 'discharge' replaced by 'cathode rays': the two things are quite distinct, although the physicists referred to do not observe the distinction." In this way Hertz staked out his claim to discovery: *he* alone had begun thoroughly to break the association between ray and current. He now began to think of ways to enforce the dissociation, perhaps even by generating the rays without using any current processes whatsoever:

> If we could prevent the production of the cathode rays, the gas would everywhere be as dark as it is in the dark intervals between the striae (although the current flows through these intervening spaces). Conversely, if we could produce the cathode rays in some other way than by the discharge, we could get luminescence of the gas without any current. For the present such a separation can only be carried out ideally. (Hertz 1883, 248)

Here was a program awaiting further realization, waiting, that is, the production of devices to generate and to manipulate rays without the troublesome presence of the inessential electric current. First, however, Hertz had to complete his argument for the entirely nonelectric character of the rays.

"If," Hertz pursued, "we admit that cathode rays are only a subsidiary phenomenon accompanying the actual current, and that they do not exert electromagnetic effects, then the next question that arises is as to their electrostatic behaviour." If we are interested in demonstrating that the rays are not currents of any kind whatsoever, then we must find a way to show this without recurring solely to the ray-magnet interaction. To clinch the issue required using an independent property of things-that-might-be-currents. A salient one was the ability of currents in metals and in electrolytes to effect concentrations of electric charge. Hertz accordingly decided to determine whether the rays had any detectable charge.

Hertz's charge-measuring device consisted of three major components: a glass tube containing the cathode and anode, a metal case that surrounded the portion of the tube that did not contain the electrodes, and a metallic mantle with a hole in it for inserting the tube (fig. 7.2). The metal case and the grounded mantle were connected to the poles of an electrometer, which therefore measured the case's potential. The device was driven by an induction coil rather than by a powerful battery, because by this time (probably late in January 1883) his self-built battery had failed.

For Hertz a charge-measuring experiment was fraught with difficulty because the cathode rays had to be carefully, thoroughly separated from the tube current that flows from the cathode to the anode. This latter current could of course produce electric accumulations, and extremely large

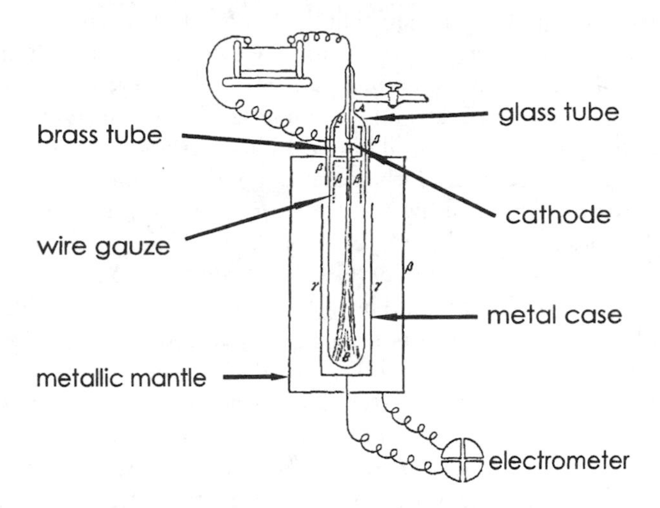

Fig. 7.2. Hertz's device for charge catching (Hertz 1883)

ones at that. Measuring *that* charge would utterly swamp whatever the rays might carry. Hertz accordingly designed an instrument to purify the cathode rays from contamination by tube current.

In Hertz's device a brass-tube anode closely surrounds the cathode. The rays flow from the cathode through a perforation in the brass, which is in metallic connection with a very important, cylindrical piece of wire gauze or mesh that hangs below it, as well as with an external mantle that surrounds the tube.[4] The mantle in turn connects to a pole of the induction coil. The space between the enclosed cathode and the anode fills with intermixed current and rays as the coil works. The rays stream perpendicularly toward the perforation; the current presumably follows a broad path to the cylindrical anode. Just past the perforation the space therefore contains primarily rays, but also some polluting tube current that has leaked through. The gauze or mesh captures this last bit of current, so that "the cathode rays are to be regarded as pure after they have passed through the opening in the metal cylinder and the wire-gauze beyond it." The space past the gauze is bathed in the current-free rays. Here, Hertz felt, and *only* here, he had produced a region within which the rays could be manipulated in their purest form. To detect—but hardly to measure—the charge of these rays was now a simple matter of surrounding the region that contains them with a metal case connected to an electrometer.

4. In Hertz's diagram (fig. 7.2) the parts marked β are in metallic connection with one another.

In two previous experiments that the young Hertz performed in Berlin he had faced the problem of having no way to calculate the magnitude of the effect that he was investigating, and this had undercut the persuasive force of his argument. Here however Hertz could determine what the charge of the purified rays would be if, in his words, "they consisted of a stream of particles charged to the potential of the cathode." Imagine with Hertz a stream of tiny metal pieces flowing out from the cathode. Such a stream, he felt, could be thought of as packed sufficiently densely to constitute in effect a single metallic object the size of the path traced by the cathode rays. This object would be in contact at one end with the cathode and would therefore have its potential. A metal rod "about the same size and position as the cathode rays" and stuck onto the cathode would be a simulacrum of the rays themselves, supposing them to consist of such a dense stream. One could then fire up the ray machine and watch the electrometric deflections produced by this thing, which is what Hertz did.

The deflection, Hertz wrote, "was too great to be measured, but could be estimated at two to three thousand scale-divisions." Here is what happened when the cathode rays replaced their metallic simulacrum:

> When the quadrants of the electrometer were connected together and the induction coil started, the needle naturally remained at rest. When the connection between the quadrants was broken, the needle, in consequence of irregularities in the discharge, began to vibrate through ten or twenty scale-divisions from its position of rest. When the induction coil was stopped, the needle remained at rest in its zero-position, and again began to vibrate as above when the current was started. As far as the accuracy of the experiment allows, we can conclude with certainty that no electrostatic effect due to the cathode rays can be perceived; and that if they consist of streams of electrified particles, the potential on their outer surface is at most one-hundredth of that of the cathode. (Hertz 1883, 250–51)

As far as Hertz was concerned, his purified cathode rays did not produce electric effects.

This was not Hertz's only, or indeed his most famous, claim concerning cathode rays. As a second test he sought to deflect the rays by passing them between electrified plates. If the rays carried charge, then they should be deflected. Hertz placed a fine wire some distance past the metal gauze. The wire cast a "sharp shadow" on the phosphorescence produced by the rays 12 cm down the tube. He then slipped the entire tube "between two strongly and oppositely electrified plates." The wire's image was not moved despite the strong electric force between the plates. There was however a possible difficulty. The extremely powerful, externally applied electrostatic force might have turned the gas in the tube into a conducting mass, in which case it would act like any conductor and completely shield the experimental space within itself from an externally applied electrostatic force. To avoid this effect Hertz placed his metal strips only 2 cm apart within the experimental space itself. In his first such experiment Hertz used 20 small Daniell's cells (so about 20 volts). "Opening and closing [the connections to the battery]," Hertz wrote, nevertheless "produced not the slightest effect upon the phosphorescent image."

Hertz did not stop here. He next investigated what would happen to the rays if they traversed a space across which a current was flowing in a direction perpendicular to their path. Using 240 Planté cells connected by a large liquid resistance produced a discharge between the plates in complete synchrony with the discharge of the tube itself. Although the "phosphorescent image of the Ruhmkorff discharge appeared somewhat distorted through deflection in the neighborhood of the negative strip [plate]," nevertheless "the part of the shadow in the middle between the two strips [plates] was not visibly displaced" (Hertz 1883, 252). *No* form of electromotive force—neither one that produces a static distribution nor one that produces a current—seemed to deflect the cathode rays.

3. Contaminated Spaces and Reproduced Experiments

In 1895 the young French physicist Jean Perrin claimed to have detected the "electrification" of cathode rays. Hertz had died shortly before, and so one cannot say with certainty what his initial response to Perrin's claims would have been. Much later, however, Perrin's experiment was seen as an improvement over Hertz's, which was said to have involved faulty apparatus:

> Another important point tested by Hertz was to try whether the cathode rays, fired into a metallic vessel (known in this connection as Faraday's cylinder or Faraday's ice pail) would carry with them an electric charge, detectable by an electrometer connected with the vessel. He failed to observe this effect, but the design of his experiment was open to certain objections which were removed in a later investigation by Perrin in 1895, directed to the same question. Perrin got definite evidence that the rays carried a negative charge. (Rayleigh 1969, 78–79)

What had Perrin done that Hertz had not, and why was this later seen as having removed objectionable aspects of Hertz's design? Was it the case, as Rayleigh implied, that Perrin had performed the *same* experiment as Hertz, only with better apparatus? Let us put Hertz's apparatus on the table next to Perrin's (fig. 7.3) to see how they fit together. These are not the same devices, even though they both use Faraday cages to detect charge. Rayleigh understood this. Whether Perrin's must necessarily be thought an improvement over Hertz's remains to be seen.

Hertz's device separated the cathode rays from the tube current by

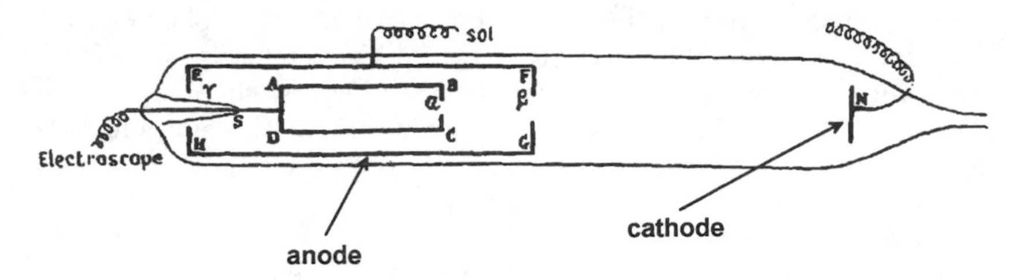

Fig. 7.3. Perrin's device for charge catching (Perrin 1895)

placing the electrodes very close together and by connecting the all-important wire gauze to the anode. Look now at Perrin's apparatus. Here the cathode rays and the current parallel one another all the way from the cathode to the anode; the hole in the anode allows both to penetrate. From Hertz's point of view Perrin's experimental space is utterly useless to probe the electric character of the cathode rays because it is irreversibly polluted with current. Worse yet, the cage lies within the tube, so that whatever current passes through the hole in the anode is likely to have a measurable effect. *Of course,* Hertz would have replied, Perrin measured charge: he measured the charge produced by the current that terminates at the anode, which has nothing to do with the cathode rays. Far from being an improvement over Hertz's design, Perrin's ruins it.

In 1897 J. J. Thomson perceived a dissimilarity between Hertz's and Perrin's experiments, writing that "the supporters of the aetherial theory do not deny that electrified particles are shot off from the cathode; they deny, however, that these charged particles have any more to do with the cathode rays than a rifle-ball has with the flash when a rifle is fired" (Thomson 1897b, 294). Thomson accordingly built a device that, like Hertz's, placed the anode close to the cathode, and he let the rays pass beyond into his experimental space. Unlike Hertz, however, Thomson placed the ray catcher out of the direct line of the rays, thereby missing whatever straight-shot particles might come from the cathode. When, and only when, a magnet deflected the rays into the catcher, he unambiguously caught charge (fig. 7.4).

Although closer in design to Hertz's experiment when Perrin's was, Thomson's nevertheless still differed from it in two major respects. Unlike Hertz, Thomson did not attach a purifying metal gauze to the anode and, like Perrin, he placed the charge catcher within the tube. From Hertz's point of view both would have been critical defects, which Hertz would certainly have noticed had he lived. They meant that the purity of Thomson's experimental space was probably ruined by residual tube current. This points to a powerful impasse between Thomson's and Hertz's ways of thinking about contamination. For Hertz the issue involved separating rays from tube current; great care had always to be taken to prevent leakage into ray space. For Thomson contamination had nothing to do with the tube current; it involved putative "electric particles" shot straight off from the cathode. He never considered, or at least admitted, the possibility of significant leakage.

Indeed, Thomson might very well have thought that Hertz had failed to measure charge with his device precisely because his purifying metal gauze pulled too much ray-producing current into the anode. In which case, whatever luminosity Hertz saw on the walls of his experimental space would have been due to residual rays that were simply too few in number to produce detectable charge. Cut the current in Hertz's fashion, Thomson perhaps thought, and there just will not be enough rays to be electrically effective even if they are charged. In this way of thinking, Thomson's contaminating electric particles had to replace Hertz's polluting current in order for charge measuring to be possible at all.

Hertz would not have had to agree with any such argument, because he was quite certain that his current-free experimental space was filled

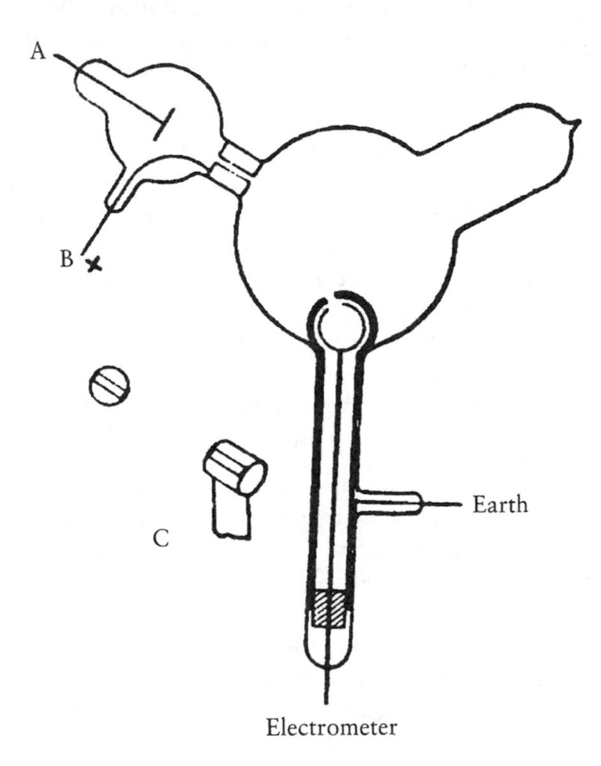

Fig. 7.4. Thomson's low-pressure charge catcher (Thomson 1897b)

with lots of rays: "[the rays past the anode trap] are none the less vivid; at low densities they cause the glass at *B* to shine with a brilliant green phosphorescence, upon which the shadow of the wire-gauze is plainly marked" (Hertz 1883, 250). This might puzzle the retrospective analyst, because the "phosphorescence" that Hertz refers to is directly proportional to the intensity of the ray beam. Hertz had sucked away most of the beam (since, we say, it must go with the current). And yet the glow on the walls is "none the less vivid." That vivid glow made Hertz certain he had an active, decontaminated space, but Thomson never mentioned Hertz's claim.

Thomson did not of course stop with charge measurement. He also tried to deflect the cathode rays electrically. He succeeded quite unambiguously in doing so with the apparatus shown in figure 7.5, but not under the same circumstances in which Hertz's rays refused to deflect. Thomson at first tried gas pressures comparable to those Hertz had used, but he also found no deflection. He, however, had available much more powerful pumps than Hertz as a result of improvements produced for the electric-light industry (Thomson 1964, 57). Even so, Thomson did not easily achieve a sufficiently low vacuum at once to produce deflection, though he did not give any sense of the difficulties he faced in his printed account for the *Philosophical Magazine*. In his autobiography Thomson recalled the problems:

> The technique of producing high vacua in those days was in an elementary stage. The necessity of getting rid of gas condensed on the walls of the discharge tube, and on the metal of the electrodes by prolonged baking, was not realised. As this gas was liberated when the discharge passed through the tube, the vacuum deteriorated rapidly during the discharge, and the pumps then available were not fast enough to keep pace with this liberation. However, after running the discharge through the tube day after day without introducing fresh gas, the gas on the walls and electrodes got driven off and it was possible to get a much better vacuum. The deflection of the cathode rays by electric forces became quite marked. (1937, 334–35)

Even with his superior pumps, Thomson had worked for "days" properly preparing his experimental space in the hope of finding something in the near absence of gas that was not there in its presence. As a means for facilitating ray deflection, Thomson's actions would (before their success) have seemed pointless to Hertz, because in his understanding gas pressure affected primarily the distance to which the rays could penetrate, and they went far enough in Hertz's experiments. Thomson thought differently because he could show that high gas pressure raised conductivity between the electrified plates, and he believed that conductivity was bad for ray deflection. Hertz had believed no such thing. On the contrary, he was apparently quite convinced that interplate conductivity

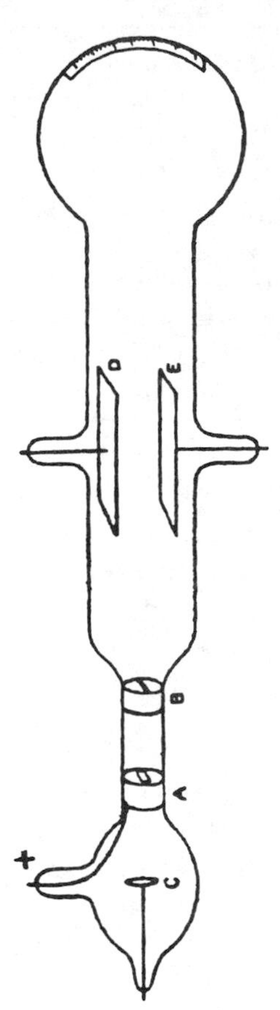

Fig. 7.5. Thomson's low-pressure ray deflector (Thomson 1897b)

(with the plates immersed in the gas) might even exacerbate ray deflection, if it occurred at all, rather than ruin it.

Thomson's successful ray deflecting, unlike his charge measuring, offered no opening to a critique based on contamination by tube current. No doubt tube current did (as Hertz would have seen it) pollute Thomson's ray-filled space. Its presence would, however, make no difference to this experiment, which concerned the visible shift in location of the rays. Novel electric-light technology, Thomson's X-ray tube work in 1896, and elaborate efforts to rid the tube of residual gas combined to elicit an effect that Hertz's high-pressure device had simply not produced 14 years previously (Falconer 1985).

Let's consider what Perrin's and Thomson's experiments have to say about Radder's notion of *experimental replication under a fixed theoretical interpretation.* Begin with two tables (one for charge catching, the

Hertz	Perrin	Thomson
Result: Rays do not carry charge	Result: Rays carry charge	Result: Rays carry charge
1. Anode close to cathode	1. Anode far from cathode	1. Anode close to cathode
2. Current trapped past anode	2. *Irrelevant*	2. *Irrelevant*
3. *Irrelevant*	3. *Irrelevant*	3. Charge catcher not on cathode-anode axis

other for charge deflecting) that briefly list several factors that were deemed necessary for success. From the table for charge-catching experiments, we see that Hertz and Thomson each deems irrelevant one of the other's requirements. Perrin deems both of their stipulations to be unimportant. Though all three consider the anode-cathode relationship to be significant, Perrin's setting differs from Hertz's and Thomson's, which do not differ from one another. Each experimenter consequently had critical requirements that the others' experiments failed to satisfy.

Most of these differences were based on the experimenters' beliefs, tacit or explicit, concerning the relationship between the cathode rays and the current in the tube. Having explicitly convinced himself through other experiments that the rays carry little of the tube current, Hertz insisted on carefully separating them. Thomson explicitly worried that the cathode would destroy the efficacy of a measuring device that faced it, but he remained tacitly unconcerned about ensuring that no tube current penetrated into the experimental space. Perrin worried about neither problem. In Radder's terms, then, there is *no* "fixed theoretical interpretation" that carries across these three experiments and that would make them comparable.

The situation with respect to the ray-deflection experiments is more complex. Hertz's experiment does not satisfy three criteria (1, 2, and 4) deemed essential by Thomson, whose first two strictures were never considered by Hertz at all. The third one was explicitly violated by Hertz because he thought that, if rays were weak currents, then interplate discharge might help, not hinder, deflection. These incompatibilities reflect very different laboratory histories and related beliefs about rays.

Thomson was convinced experimentally that a considerable amount of conduction occurs between the plates when the rays pass through at higher gas pressures. This tallied well with his belief that the rays are moving electric particles because, given that belief, it implied that the experi-

Hertz	Thomson
Result: Rays not deflected	Result: Rays deflected
1. *Irrelevant*	1. Low gas pressure
2. *Irrelevant*	2. Expunge gases from tube
3. Plates inside tube	3. Plates inside tube
4. Interplate discharge might exacerbate possible deflection	4. Avoid interplate discharge

ments had to be done at low pressures. Hertz was convinced experimentally that the rays do not carry much tube current, and this tallied well with his belief that they are sui generis. From his point of view, gaseous conduction would only test whether rays could be deflected by the kind of electromotive force that produces currents.[5] Since conduction was not something sedulously to be avoided, there was no reason to try removing more gas than was necessary to produce a satisfying glow. Using Radder's terminology, we can therefore say that Thomson did not replicate Hertz's experiment under a fixed theoretical interpretation. Yet, and despite the wrong direction of time's arrow, it perhaps makes better sense to say that *Hertz* did not reproduce *Thomson's* experiment.

4. CONSTRAINTS OR RESOURCES?

Conclusions Hertz Drew from His Cathode-Ray Experiments

1. The discharge process is not intermittent.
2. Cathode rays have at most an electromagnetic effect on magnets.
3. The ray path does not follow the path of the tube current.
4. Purified rays do not convey charge.
5. Powerfully electrified plates located outside the tube do not deflect rays.
6. Weakly electrified plates located within the tube do not deflect rays.
7. Electrified plates located within the tube and across which discharge occurs do not deflect rays.

5. Hertz certainly knew that conduction would also generate a transverse current, but such a current would not affect the rays, even if they were weak currents themselves, for precisely the same reason that currents moving in different directions in metals and in electrolytes have no effect upon one another. As far as Hertz was concerned, gaseous conduction remained just as linear a process as any other kind—there was no experimental reason to think otherwise.

Let's enter Hertz's laboratory as time-traveling physicists from the 1990s. We walk in the door and find that the room's inhabitant has left his equipment running. Snapping sparks fly about; tubes glow with curiously irregular, colorful patterns. Noxious fumes from a huge, open-plate battery mix with ozone from noisy induction machines and make the air nearly unbreathable to our environmentally sensitive noses. A primitive pump, obviously incapable of producing a good vacuum, sits in a corner. Messy notes in Hertz's hand lie near the tube and speak obscurely of things we know to be utterly irrelevant to plasma physics. Everywhere we look we see instrumental barriers, environmental impossibilities, and intellectual irrelevancies. We have entered a primitive, alien landscape, and our wonder at Hertz's failure properly to understand the nature of cathode rays fades rapidly away. We may not see just why Hertz got the results he did, but we certainly do grasp why he did *not* get the results that we want. Of the seven conclusions that Hertz drew from his experiments, the last four are understandably wrong; the first three are at best trivialities.

Let's now enter Hertz's laboratory as time-traveling German physicists from, say, the early 1830s. We see a rather large, but identifiable, battery, as well as an odd device that produces sparks between two metal terminals that have been placed atop it. The fumes do not bother us, but one smell is unfamiliar and seems to be strongest near the curious sparking machine. Our attention rapidly turns to the glowing glass cylinder on the table. We have never seen anything at all like it, though it seems to be activated either by the big battery or by the smelly spark producer. Hertz greets us warmly and tries to explain what he is doing. Our sense of confusion grows as he talks. His purposes elude us, his manipulations are strange and complex, and the device itself—the glowing tube—is entirely outside our ken. Very little in our laboratory experience makes contact with Hertz's machines and procedures, and his talk deeply puzzles us. We do nevertheless see that he is trying to probe the nature of this strange glow, about which there are apparently many different opinions. We could ourselves not even begin to think how to do this, but we are impressed by the power and novelty of Hertz's equipment. Though puzzled, we can also see that he is able to talk long and seemingly with understanding about this equipment and about what might be going on in his tubes. Surely Hertz will be able to reach a satisfying conclusion with the powerful instrumental and intellectual means at his disposal. Indeed, as we watch him skillfully work, moving with enthusiasm and confidence from experiment to experiment, and as we listen to his explanations, our belief in him grows.

What kinds of "histories" would our two sets of time travelers write? Those from the future would undoubtedly speak of the tremendous difficulties under which Hertz labored. They would write of the inadequate pumps that precluded a good vacuum, of improperly blown tubes, of irregular power supplies, and of Hertz's insistence on drawing distinctions between cathode rays and tube current in ways that surely blocked his ability to conceive of other explanations involving ionization. To these physicists constraints bound Hertz so tightly that his failure to show the rays' electric character hardly needs any further explanation, and indeed it does not.

Our physicists from the past would tell a very different story. Unlike their future counterparts, these people do not know that cathode rays track the motions of electric particles. They have had trouble understanding Hertz's devices and his goals. His talk seemed to them to be elliptical and difficult to make cohere with their ways of thinking. They have nevertheless grasped that he was seeking to find the nature of the glows he called "cathode rays" and that, unlike some of his contemporaries, Hertz did not believe them to have anything directly to do with electricity. Our physicists from the past might return home and write a secret memoir, to be opened after, say, 1900, in which they discuss with admiration the careful (though obscure and difficult) reasoning and the intricate manipulations of bizarre devices that Hertz had performed to reach his goal—one that, they might easily feel, he had succeeded admirably in accomplishing.

The limitations that the physicists from the future see certainly do mark out important barriers that Hertz could not have transgressed. Hertz, the future physicist knows, could not possibly have deflected rays at the lowest pressures available to him, and his way of thinking about rays and currents blocked him from understanding the limitations that he labored under. We can, it seems to me, agree with our physicists-from-the-future that Hertz simply could not have deflected rays with the apparatus, skills, and techniques that were available to him. We now know how to do it; he did not.

We could write a story about Hertz and rays that gives a history of vacuum technique in Geissler-tube laboratories, of the very different abilities involved in tube blowing and fabrication, of the still different skills acquired by Hertz in assimilating paper technique under Helmholtz's direction, and indeed of the many other factors that configured the boundaries of the acceptable universe for Hertz. We could assemble all of these stories, each different from, but making contact with, the others, and from them we could tell a tale of limitation and constraint.

Our physicists from the past will tell a different story. They have no idea what boundaries cannot be transgressed, what skills must be acquired, what kinds of tubes must be made, what sorts of vacua must be produced. They do not have a clear picture of the intellectual landscape in which Hertz lived, and into which his experimental practice fit. If Hertz did not deflect rays, well then, they say, he had begun to succeed, with his strange devices and peculiar ideas, in discovering the nature of these oddly behaved things. They too might write about vacuum technique, about tube blowing and making, and about Hertz's understanding (or what they could make of it). Many elements of their story would probably match elements of the story told by our future physicists. But whereas the physicists from the future, failure ever in mind, would emphasize a constraining technique, the physicists from the past, who saw success, would write about an enabling one.

If scientists from the future talk about constraints, then scientists from the past talk about power. They enter Hertz's world filled with a sense of possibility precisely because they do not know what can or cannot be done unless the limitation comes from previous work. They follow Hertz's manipulations, developing skills, and reasoning as he tries to make things go. Here there would be no talk of limitations, practical or intellectual. There would instead be talk only of actions undertaken to make the material and intellectual objects that Hertz used cohere with one another.

What the account of the physicists from the past misses is just what the account of their counterparts from the future provides: access to an understanding of the full character of contemporary practice. The fact of the matter is that finding a path through much of past science without some sort of guideline, without an arrow that points to a course of investigation, is extraordinarily difficult. The future physicist knows that low pressures must be reached to deflect cathode rays and that ionization must be considered in understanding discharge and ray behavior at higher pressures. The historian who reads Hertz's work and imagines Hertz's laboratory inevitably thinks first of tube pressures and electric-field shielding ions. This provides a point of entry for investigation as the historian carefully and sympathetically seeks to grasp how Hertz built and understood his (retrospectively) high-pressure device. Hertz's statements and practices can be bounced against, say, Jean Perrin's or J. J. Thomson's later work in order to gain a deeper understanding of them. *Difference opens the way to comprehension,* and "constraints" provide salient ones that enter at many points in a story, providing ways to recognize character through distinction. The historian's story consequently unfolds as one in which the identification and deployment of *possibilities* are aided by the retrospective guidance provided by *constraints.*

In that sense Galison and Pickering have not, it seems to me, offered radically opposite ways to understand scientific practice, at least for situations in which constraints are not acknowledged factors at the time.[6] Each

6. By which I mean that, like low-pressure tube technology, the constraint is entirely a retrospective one. Since at the time it was not known, or even thought, that the constraint existed, it could not have played a guiding role.

has instead emphasized different but complementary elements of good narrative in the history of science for cases of this kind. With Pickering we enter Hertz's laboratory and look around for the intellectual and material tools that he used; we strive to appreciate his skilled use of them as he tries hard to make Geissler tubes give meaning to cathode rays. We may, however, find it exceedingly difficult to grasp what Hertz was up to, to understand why he did *x* rather than *y*. Since we have no intrinsic map of Hertz's terrain, we build a platform from which we can see swamps and other traps that Hertz never spied. These may help more sharply to define the ones that he did see. We sally forth from Galisonian platforms, as it were, in pursuit of Pickeringian exploration.

We have done something like this as we tracked Hertz. We did not enter his territory unguided. In retrospect we think that Hertz was strikingly limited by the gas pressures he could achieve. Because I knew this, I came to Hertz's papers with the notion that gas pressure was unlikely to have been something to which he paid much attention. This constraint directed my attention away from what Hertz could not do much about (gas pressure) and toward what he could do (ray purification). Had I been a physicist from the past, I would certainly have had difficulty even beginning to understand what Hertz was up to.

My rather elliptical answer to question 2, then, is this: the constraint that prevented Hertz from deflecting rays directed my attention to the resources that allowed him to purify them. The long-term explanation for why cathode rays were not deflected until the mid-1890s grounds itself on the constraining effect of pump technology, as well as on the skills and concepts that Hertz did not develop in using high-pressure tubes and that J. J. Thomson did develop in using low-pressure ones. But the short-term explanation for Hertz's success in divorcing ray properties from electric properties involves his technology for ray purification, and that had little to do with gas pressure.

EIGHT

IS THE IDENTIFICATION OF EXPERIMENTAL ERROR CONTEXTUALLY DEPENDENT? THE CASE OF KAUFMANN'S EXPERIMENT AND ITS VARIED RECEPTION

Giora Hon

> Kaufmann's [experimental] conclusion . . . that . . . electrons have no material mass at all . . . is certainly one of the most important results of modern physics.
>
> —*H. A. Lorentz,* The Theory of Electrons

> There was a time when dynamic rather than kinematic arguments led to the notion of electromagnetic mass, a form of energy arising specifically in the case of a charged particle coupled to its own electromagnetic field. . . . The investigations of the self-energy problem of the electron by men like Abraham, Lorentz, and Poincaré have long since ceased to be relevant. All that has remained from those early times is that we still do not understand the problem. . . . We still do not know what causes the electron to weigh.
>
> —*A. Pais,* "Subtle Is the Lord . . ."

IN 1906 THE EXPERIMENTER WALTER KAUFMANN (1871–1947) published in the *Annalen der Physik* his definitive paper on the constitution of the electron. He detailed his experiment and stated the following conclusion:

> The . . . results speak decisively against the correctness of Lorentz's theory and consequently also of that of Einstein's theory. If one were to consider

> these theories as thereby refuted, then the attempt to base the whole of physics, including electrodynamics and optics, upon the principle of relative motion would have to be regarded at present as also unsuccessful.[1]

According to current physical theories this experimental result is erroneous.

A statement may be rendered erroneous only when it is embedded within a set of statements which claims to constitute knowledge. In science an error is claimed to have occurred when a statement about the physical world does not cohere with the state of affairs which a matrix of theories and observations exhibits. Thus, a claim to knowledge may be construed as an error when it does not cohere with a set of theoretical and empirical statements. To apprehend an error one needs a vantage point from which a statement that was thought to accord with the state of affairs is in fact found not to cohere.[2] The scientist must ultimately seek this vantage point by himself or herself. He or she must decide what weight to assign to a particular set of evidence with respect to the correctness or otherwise of a certain statement. This "residue of personal judgment,"[3] which depends on one's philosophical makeup and methodological approach, determines the way experimental results are assessed.

I therefore claim that the failure of a particular experiment may be attributed to different causes, depending on one's philosophical perspective and methodological disposition. These epistemological and methodological elements constitute the context in which a failure is determined and identified. Thus, the identification of an experimental error and its characterization may depend contextually on the philosophical outlook and methodological disposition with which the experiment is approached. My intention is neither to render the concept of error relative nor to explain the occurrence of error in sociological terms; rather, I wish to argue for a close connection between, on the one hand, epistemological framework and methodological approach and, on the other hand, detection of error. The case of Kaufmann's experiment and the varied responses of

1. "Die . . . Ergebnisse sprechen entschieden gegen die Richtigkeit der Lorentzschen und somit auch der Einsteinschen Theorie; betrachtet man diese aber als widerlegt, so wäre damit auch der Versuch, die ganze Physik einschließlich der Elektrodynamik und der Optik auf das Prinzip der Relativbewegung zu gründen, einstweilen als mißglückt zu bezeichnen" (Kaufmann 1906a, 534).

2. This incoherence constitutes the crucial dilemma in discerning whether an experimental error has occurred or a phenomenon which points to a new discovery has been observed. See Hon 1989, 473; Polanyi 1964, 31.

3. "There is a residue of personal judgment required in deciding—as the scientist eventually must—what weight to attach to any particular set of evidence in regard to the validity of a particular proposition" (Polanyi 1964, 31).

Poincaré, Einstein, and Lorentz to this experiment will serve to demonstrate this claim.

I shall outline the historical background of Kaufmann's experiment and then analyze the experiment, review its results, and contrast and discuss the responses Kaufmann received from Poincaré, Einstein, and Lorentz. A central motif in the discussion will be the question: is it the case that different characteristics were attributed at the time to the vitiating element of this experiment? My answer is positive and I shall generalize this claim. I shall attempt to demonstrate that the different attributed characteristics of a failed experiment are dependent contextually on the different philosophical and methodological perspectives from which the experiment is approached.

1. The General Historical Setting

In a textbook entitled *Einstein's Theory of Relativity,* the author informs the reader that

> experiments by Kaufmann (1901) and others who have deflected cathode rays by electric and magnetic fields have shown *very accurately* that the mass of electrons grows with velocity according to Lorentz's formula [$m(v) = m_0/(1 - v^2/c^2)^{1/2}$].

Here one sees the origin of a myth. Between 1898 and 1906 Kaufmann executed several sets of experiments on the relation between the mass of a charged particle and its velocity. In his initial experiments he had indeed deflected cathode rays, but the crucial innovation came when he introduced β-rays to the experiments. However, his crowning achievement was to show, and one may follow the author and ironically add "*very accurately,*" that Lorentz's formula was wrong. The author of this myth is no other than Max Born.[4]

The question whether or not mass changes in the course of its motion was dealt with in Stokes's hydrodynamics. In 1842 Stokes analyzed the slow motion of a solid, perfectly smooth sphere through an incompressible ideal fluid and reached the conclusion that the motion of such a body can be described *as if* it had a modified mass. Clearly, a sphere of mass m_0, which requires an energy of $(1/2)m_0v^2$ to reach a velocity v in a

4. Born 1965, 278 (emphasis added). Cushing 1981 and Miller 1981 rectify the distorted historical treatment which this case has received in physics textbooks. I refer to these works in the course of drawing the historical scene of the experiments. Reference should also be made to Zahar 1978, 1989. Zahar discusses the logical and philosophical implications of Planck's response to Kaufmann's experiments.

vacuum, would require a greater amount of energy to reach the same velocity when it is immersed in a fluid; for the sphere, when it is in motion, will set the fluid around it in motion. Thus, the source of energy which drives the sphere to a velocity $\boldsymbol{v}$ will have to supply additional energy to move the fluid. Stokes argued that the true mass of the sphere, m_0, can be viewed *as if* it were changed by motion to $m_0 + m'$, where m' turns out to be a constant depending on the radius of the sphere and on the density of the fluid (Stokes 1966, 17; see also Thomson 1907, 29, 30; Lorentz 1931, 256).

If one were to suppose that the forces emanating from a moving charged corpuscle could set the ether in motion and assume further that the ether has some kind of mass, then this case would be analogous to that of the sphere moving in a fluid. Consequently, one would expect an increase in the effective mass of the charged corpuscle while it is in motion. It thus comes as no surprise that as early as 1881, on entirely theoretical grounds (which were essentially Maxwellian; see Buchwald 1985a, 269 ff.) and well before his experiments on the existence and properties of cathode-ray corpuscles, J. J. Thomson could introduce the concept of electromagnetic mass (Thomson 1881, 230, 234; see also Thomson 1907, 29–30; Buchwald 1985a, 273–74).

In response to the new researches of Crooke and Goldstein in 1880 on electric discharges, Thomson set himself the task "to take some theory of electrical action and find what, according to it, is the force existing between two moving electrified bodies, what is the magnetic force produced by such a moving body, and in what way the body is affected by a magnet" (Thomson 1881, 229; see also Buchwald 1985a, 269). On the basis of Maxwell's theory, he demonstrated that a moving charged particle is surrounded by a magnetic field which possesses energy; hence more work will have to be done to start or stop the particle than if it were uncharged.

Thus, with no novel departure—assuming that the surface charge distribution on the moving particle remains *unaltered* by the motion and that the electric field is carried forward *undistorted* with the particle moving at low velocities (Jammer 1961, 136)—Thomson calculated that the "whole kinetic energy" of the moving charged corpuscle, which he for simplicity considered a sphere, is

$$T = [(m/2) + (2\mu e^2/15a)]\boldsymbol{v}^2,$$

$$T = (1/2)(m + 4\mu e^2/15a)\boldsymbol{v}^2,$$

where μ is the coefficient of the magnetic permeability of the medium through which a sphere of radius a and charge e travels with a velocity $\boldsymbol{v}$. Hence his conclusion that "the effect of the electrification is the same *as if* the mass of the sphere were increased by $4\mu e^2/15a$" (Thomson 1881, 234, emphasis added).[5]

Using a different estimate of the field energy, Heaviside calculated in 1890 this numerical coefficient for velocities that are very small compared to that of light and found it to be $2\mu e^2/3a$. He observed that "as the speed increases, the electromagnetic field concentrates itself more and more about the equatorial plane $\theta = \pi/2$" (1889, 332), that is, in the plane perpendicular to the charge's direction of motion.[6] Heaviside then made the important point that in the limiting case $v = c$, E and H become zero everywhere except in the equatorial plane; in this limiting case the electromagnetic field becomes an electromagnetic plane (1889, 332). Thomson took cognizance of this claim, elaborated it, and stated quite specifically that, in the limit $v = c$, "the increase in mass is infinite." He thus argued that "a charged sphere moving with the velocity of light behaves *as if* its mass were infinite." The velocity therefore remains constant at this limiting case, and Thomson could conclude that "it is impossible to increase the velocity of a charged body moving through the dielectric beyond that of light" (1893, 21, emphasis added).

Since the electromagnetic mass increases with the velocity of the charged body up to the limiting case of the velocity of light, it became apparent that there might be experimental means of separating the charged body's effective mass into its "true" (mechanical) and "induced" (electromagnetic) parts.[7] It thus appears that the qualitative description of a moving charged corpuscle—the electron—including its bounded velocity, had been developed well before an appropriate experiment was carried out. The results of Kaufmann's experiment at the turn of the century on the relation between the mass of a charged particle and its velocity were

5. See also Buchwald 1985a, 273–76. Lodge referred to Thomson's paper as "epoch-making" (1907, 17).

6. On the difference between the two values obtained by Thomson and Heaviside for the self-energy of the field, see Buchwald 1985a. Buchwald argues that this is the result of the inherent ambiguities of Maxwellian theory.

7. See, e.g., Lorentz 1952, 40. See also Lorentz 1931, 256. Thomson later brought these theoretical considerations to bear on his research. Seeking to discriminate between the two views concerning the nature of cathode rays (see Hon 1987a), Thomson measured the velocity of these rays. He argued that "if we take the view that the cathode-rays are aetherial waves, we should expect them to travel with a velocity comparable with that of light; while if the rays consist of molecular streams, the velocity of these rays will be the velocity of the molecules, which we should expect to be very much smaller than that of light." His measurements yielded a velocity of 1.9×10^7 cm/s (1894, 360, 364). Having found the velocity of cathode rays to be very small in comparison to that of light, Thomson probably did not expect the mass of the claimed constituent charged corpuscles, and therefore the ratio e/m, to be affected greatly by their velocity. However, he was careful not to commit himself to the view that the ratio e/m is independent of velocity.

therefore qualitatively expected. There were, however, different expressions available for the electron's mass, based on different assumptions concerning the structure of the electron and its charge distribution. The experiment was eventually seen as discriminating between these different electron theories.[8]

Kaufmann's study can be divided chronologically into two distinct periods. From 1898 to 1903—in the wake of J. J. Thomson's 1897 experimental results concerning the corpuscular constitution of cathode rays—one finds Kaufmann deflecting negatively charged moving bodies by electric and magnetic fields. Using β-rays, or Becquerel rays as they were called at the time, Kaufmann reached some tentative conclusions in 1901 with regard to the mass of fast-moving electrons. By 1903 he claimed to have reached definite results: he increased the accuracy of the experiment and adopted a specific theory of the electron, namely, the classical theory of Max Abraham, which assumes an undeformable, rigid, spherical electron.

The second period of Kaufmann's research starts in 1905 with his invention of a rotating mercury air pump, which he used in his new set of experiments of 1905–6. These experiments were essentially similar to the previous ones, but in the face of mounting criticism, they were designed with greatly increased accuracy to discriminate between several theories of the electron: those of Abraham, Bucherer, Lorentz, and Einstein (whose theory agrees formally with Lorentz's). Kaufmann concluded that his experiments excluded the theory of Lorentz and Einstein. He thereafter defended this experimental result despite intensive criticism.[9]

2. The First Set of Experiments (1898 – 1903)

The origin of Kaufmann's experimental method goes back to the work of H. Hertz, who, as early as 1883, tried—though unsuccessfully—to detect a deflection of cathode rays when they were made to pass through an electric field (Hon 1987a). Following up this idea of subjecting the rays to electric and magnetic fields, in 1897 J. J. Thomson evacuated a cathode-ray tube to a higher degree of exhaustion than that used by Hertz and succeeded in obtaining and indeed measuring such deflections. According to Thomson, this experimental result confirmed the view that a stream of identical corpuscles would behave exactly like cathode rays if the corpuscles were charged with electricity and projected from the cathode by

8. For an overview of the historical development of the theory of the electron see Pais 1972.

9. After 1908 Kaufmann ceased to publish on β-ray deflection experiments (Miller 1981, 377 n. 10).

an electric field. He went far beyond the idea of regarding the rays as particles to conclude that the experiment would "give a value of *m/e* [*sic*] which is independent of the nature of the gas and its pressure, for the carriers are the same whatever the gas may be" (1897b, 311).

Kaufmann, however, was puzzled; conducting similar cathode-ray experiments, he assumed that the moving particles were ions and expected that the ratio *e/m* should have a different value for each substance. The deflection in electric and magnetic fields should therefore depend on the nature of the electrodes or on the nature of the gas in the cathode tube. He found that neither was the case (Kaufmann 1897, 544–45; see also Pais 1983, 156). Notwithstanding, he concluded that "the hypothesis accepting that cathode rays are emitted particles is not sufficient for a satisfactory clarification of the relations observed by me" (1897, 552; quoted by Miller 1981, 107). It was thus left to J. J. Thomson to inform the Royal Institution that his experimental results "favour the hypothesis that the carriers of the charges are smaller than the atoms of hydrogen" (1897a, 109). He thereby ushered in the era of the physics of new particles.

Responding to these new discoveries, in 1898 Kaufmann focused his attention upon the ratio *e/m*. He decided to repeat with the greatest attainable accuracy those experiments which bore upon the determination of *e/m*, and he set his student, S. Simon, to work on this problem.[10] Using Kaufmann's technique of transmitting a beam of cathode rays through a magnetizing coil, Simon attempted to obtain as accurate a value as possible for the ratio *e/m*. He realized that knowledge of the variation of the magnetic field along the path of the cathode-ray beam is central to this measurement; he therefore designed a special magnetometer and was particularly careful to obtain as uniform a field as possible. Simon eventually obtained the value 1.865×10^7 cgs units for the ratio *e/m* for cathode rays.[11] Kaufmann referred approvingly to this determination and indeed relied upon it in his own later researches.[12]

However, at that time the attention of the scientific community was directed somewhere else: to the discoveries of new forms of radiation. Experimenting with these, apparently spontaneous, forms of radiation, Becquerel, Giesel, Meyer, and Schweidler demonstrated that Becquerel rays—a form of radiation that Rutherford had called β-rays (Rutherford 1899)—were deflected by a magnetic field. Subsequent studies by the Curies showed that these rays carry negative electric charge; and when in

10. See Kaufmann 1898a; see also, *Sci. Abs., Phys. Elec. Eng.*, 1 (1898): 549, art. 1049. Kaufmann, 1898b; see also *Sci. Abs., Phys. Elec. Eng.*, 2 (1899): 114, art. 282. See also Miller 1981, 107–8.

11. See Simon 1899; see also, *Sci. Abs., Phys. Elec. Eng.*, 3 (1900): 42, art. 118.

12. E.g., in his definitive paper Kaufmann 1906a, 533.

1900 Becquerel succeeded in deviating the rays by an electric field, he was giving substance to the suggestion that β-rays are of the same nature as cathode rays, with the important difference that the former are much swifter than the latter (Whittaker 1953, 3).

Kaufmann was quick to capitalize on this important and significant discovery. Looking for variations in the ratio e/m, he studied, among other things, the relation between mass and velocity. He was interested in measuring the ratio e/m for charged corpuscles with different velocities: slow as well as swift. The discovery of Becquerel rays gave him the tool he was seeking: swift particles which approach the velocity of light. He therefore abandoned his experiments on the relatively slow cathode rays and concentrated on new investigations: the measurements of the velocity of Becquerel rays and the ratio of the electric charge of the carriers to their mass. In his preliminary investigations in 1901 he outlined a photographic method by which the deflections due to magnetic and electric fields could be measured. Kaufmann was at pains to stress that with his design he had attempted to overcome two major experimental difficulties: the first difficulty was due to the fact that the velocity spectrum of the β-rays was inhomogeneous, the constituent particles having different velocities; the second difficulty arose from the possibility of attaining too poor a vacuum, in which case the rays might make the surrounding gas a conductor (Kaufmann 1901c, 1901b, 145 n. 1).[13] These explicit cautious considerations may explain why Kaufmann was regarded as a trustworthy experimenter.

To overcome the problem posed by the velocity distribution of the β-rays, Kaufmann let a collimated beam of β-rays pass normally through *parallel* electric and magnetic fields. This arrangement spreads the inhomogeneous beam into a curve on the photographic plate. Whereas with J. J. Thomson's technique one obtains a *spot* on the fluorescent screen by letting the beam of radiation pass normally across electric and magnetic fields which are mutually *perpendicular*, with Kaufmann's technique one obtains a *curve*, providing thereby a visible testimony to the inhomogeneity of the rays.[14]

13. See also *Sci. Abs., Phys. Elec. Eng.*, 4 (1901): 934, art. 2111. See also Miller 1981, 48.

14. Kaufmann borrowed here a technique from optics: Kundt's method of crossed spectra. Seeking to demonstrate the phenomenon of anomalous dispersion, Kundt—in a series of experiments performed in 1871–72—let a collimated beam of sunlight pass through two prisms so oriented that the planes of refraction were mutually perpendicular; hence the name crossed spectra (Kaufmann 1901b, 144). Miller points out that by opting for an analogue of Kundt's method over Thomson's crossed-field arrangement, Kaufmann had made "a tactical error because analysis of the resulting complicated curve was fraught with too many possibilities for error" (1981, 51, see also 49–50).

In Kaufmann's 1901 published account of these investigations, he defines the problem as follows: "to determine the speed as well as the ratio *e/m* as accurately as possible for Becquerel rays and . . . from the degree of dependence of *e/m* on *v* to determine the relation between 'actual' [*wirklich*] and 'apparent' [*scheinbar*] mass" (1901b, 144).[15] Having argued that Becquerel rays are in fact swift cathode rays (Kaufmann 1901b, 143–44; Boorse and Motz 1966, 506),[16] Kaufmann delineated the method he had employed. The idea was to use a diaphragm to produce a pencil of Becquerel rays and to record the pencil's undeformed image on a photographic plate placed perpendicularly to the beam. A simultaneous application of an electric field oriented normally to the direction of the rays and, *parallel* to this field, a magnetic field would change the point image into a curve since the deflection due to the latter field is in a direction at right angles to that due to the former field. Each point on the resulting curve corresponds to definite values of *v* and *e/m*. A whole series of observations is thus obtained on a single photographic plate from which, as Kaufmann remarked, "the dependence of *e/m* on *v* can be read off directly" (1901b, 144; see Boorse and Motz 1966, 507).

To obtain these observations Kaufmann placed a speck of radium bromide at C in a glass vessel from which the air was extracted (fig. 8.1*a*). He secured a fine pencil of β-rays by letting the particles pass through a small aperture, D, about half a millimeter in diameter, in a lead diaphragm. He collected the rays on a photographic plate, E, placed about 2 cm from the diaphragm. Kaufmann subjected the particles in their passage from C to D (a length of about 2 cm) to an electric field maintained at a potential difference of several thousand volts across the parallel plates P_1P_2 (fig. 8.1*b*), which were about 1.5 mm apart. He introduced the evacuated glass vessel between the poles of a stack of permanent magnets NS, so that during the whole of their flight from C to E the particles were subjected to a magnetic field of the order of hundreds of Gauss which was oriented, as required, *parallel* to the electric field. The undeflected γ-rays marked on the photographic plate the direct "line of fire"; its point image, B, acted as the geometrical origin for the resulting curve (fig. 8.4) (Kaufmann 1901b, 145–50; Boorse and Motz 1966; see also Thomson 1906, 653–55; Lodge 1907, 138–45; Miller 1981, 48–54; Cushing 1981, 1138 ff.).

15. See also *Sci. Abs., Phys. Elec. Eng.*, 5 (1902): 296, art. 685. For an English translation see Boorse and Motz 1966, 507.

16. Since he assumed that Becquerel rays are swift cathode rays, Kaufmann began his paper with some arguments in support of this view. He observed that "experiments on cathode rays have shown that with increased speed the deflectability decreases and the penetrability increases" (1901b, 143–44), which is entirely in keeping with the phenomena that Becquerel rays exhibit: the magnetic deflection of Becquerel rays is much smaller and their ability to penetrate solids much greater than that of cathode rays.

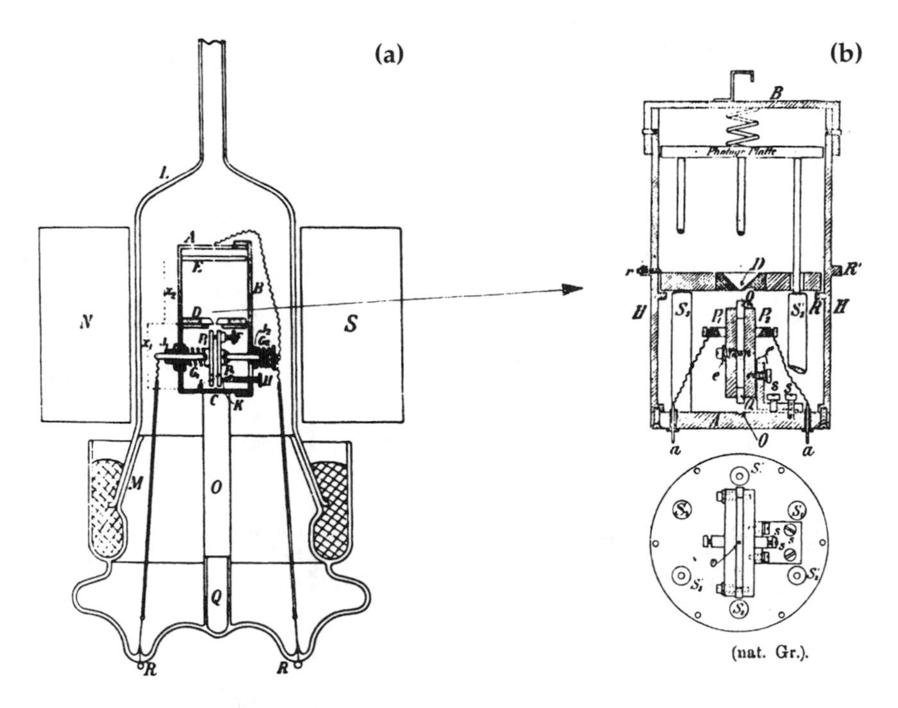

Fig. 8.1. An outline of the apparatus Kaufmann used (Thomson 1906; Kaufmann 1906a)

This arrangement of the fields and the diaphragm acted as a velocity filter. Since the particles were electrically deflected in their passage from C to D, only those with high enough velocities for passing through the aperture in the diaphragm could reach the photographic plate; all the others struck the bounding surfaces of either the condenser or the diaphragm and did not reach the plate. Kaufmann could thereby control the minimum spread that the rays must have to reach the plate.

The experiment requires analyzing the electric and the magnetic deflections of moving charged particles within the confinement of the geometrical arrangement of the apparatus. In this case the trajectory of a charged particle subject only to an electric field E is parabolic between C and D and rectilinear in the rest of its flight in the x-y plane (i.e., in the plane formed by a line normal to the plates and the ray source C; fig. 8.2). When the particle is subject only to a magnetic field H, its trajectory describes a circle CDF in the x-z plane perpendicular to the direction of the field (i.e., in the plane parallel to the condenser plates and containing the ray source C; fig. 8.3).

Clearly, if all the particles had the same velocity, they would have produced on the photographic plate a single dot; but because they have dif-

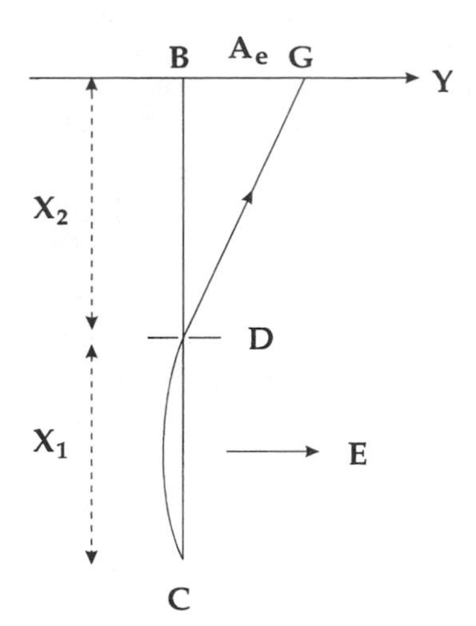

FIG. 8.2. The trajectory in the Kaufmann setup of a charged particle subject only to an electric field *E*

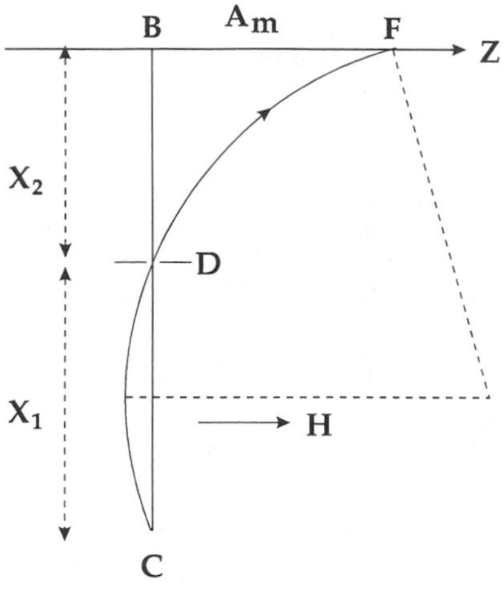

FIG. 8.3. The trajectory in the Kaufmann setup of a charged particle subject only to a magnetic field *H*

ferent velocities, one obtains, under the combined fields, a series of dots which jointly form a curved line in the *y-z* plane (i.e., in the place of the photographic plate). One can then measure with a micrometer the coordinates (y_0, z_0) of a point P on the curve and determine thereby the electric deflection, A_e, and the magnetic deflection, A_m, respectively (fig. 8.4).

In the analysis one assumes an electron theory: a particular theory for the charge distribution in a particle of a given shape. This theory yields an expression Φ for the ratio m/m_0 of the mass m at speed $\boldsymbol{v}$ to the mass m_0 for a particle at rest.

$$\Phi(\boldsymbol{v}/c) = m/m_0,$$

where c is the speed of light. This function, which depends on $\boldsymbol{v}$, can be used to construct a constant quantity, k_2, for a particular experiment in the following way. First of all, one shows that the velocity $\boldsymbol{v}$ is equal to $k_1\boldsymbol{A}_m/\boldsymbol{A}_e$, where $\boldsymbol{A}_m$ and $\boldsymbol{A}_e$ are, respectively, the magnetic and electric deflection, and k_1 depends only on the relative effective field strengths:

$$\boldsymbol{v} = k_1\boldsymbol{A}_m/\boldsymbol{A}_e.$$

Furthermore, the mass m is also equal to the product $e\boldsymbol{A}_e/k\boldsymbol{A}_m^2$, where e is the particle's charge and k depends only on the apparatus:

$$m = eA_e/kA_m^2.$$

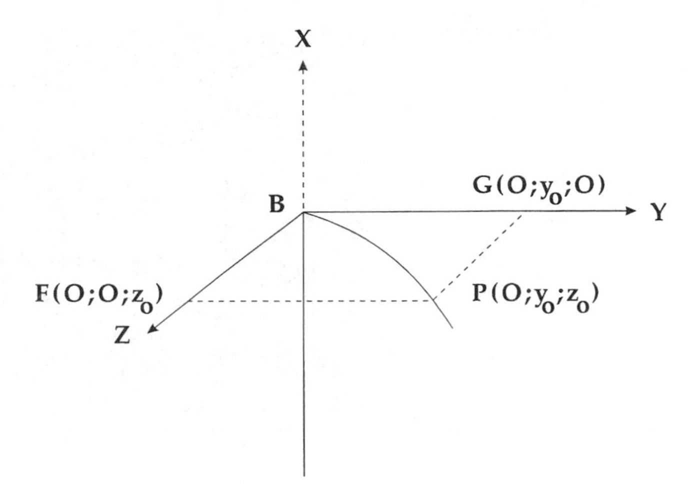

FIG. 8.4. The resulting curve on the photographic plate in the *y-z* plane under the combined electric and magnetic fields

Since by definition m is also equal to $m_0\Phi$, one can combine these expressions to produce the constant $k_2 = e/m_0k$ for the particular setup, because k appears in it. This expression for k_2 is, however, equal to $\Phi A_m^2/A_e$. Since one can compute Φ from v (i.e., from k_1A_m/A_e), one can draw up a table of values for the constant k_2 for a given specification of the fields **E** and **H**. The analysis consists in determining whether the calculated value of k_2,

$$(A_m^2/A_e)\Phi(k_1A_m/cA_e) = k_2,$$

falls within the limits of experimental error. If for a given curve this value of k_2 were to remain constant within the limits of experimental error, then the form of the hypothetical function $\Phi(v/c)$ would be to that extent verified. As a consequence, the underlying assumed structure of the particle and its charge distribution should be to that extent verified as well (Kaufmann 1901b).[17]

Kaufmann's setup remained essentially the same throughout his researches. He did, however, change the method of reducing the data.[18] To increase the accuracy, he made some of the dimensions slightly smaller and used a more active radium bromide supplied by the Curies. He could

17. This is a general outline of the experimental setup and its theoretical underpinning; for a detailed and in-depth analysis, see Lodge 1907, 140–42; Lorentz 1931, 272–74; Miller 1981, 48–54, 61–67; Cushing 1981.

18. Miller 1981, 51–54 (1901 data), 61–67 (1902–3 data), 228–32 (1905–6 data); Cushing 1981, 1138–41 (1901–3 data), 1142 (1905–6 data).

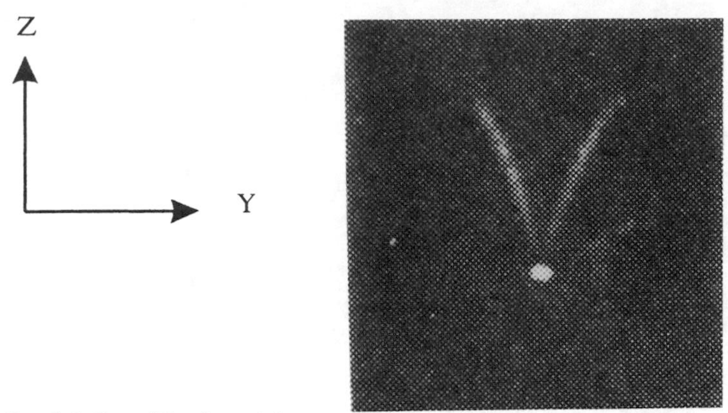

Fig. 8.5. One of Kaufmann's best runs of the 1902–3 series of experiments (Kaufmann 1903)

thus apply fields of lesser strength, which facilitated the maintenance of the required homogeneity.

Figure 8.1*b* is a diagram of the core of Kaufmann's apparatus as it appears in his definitive paper of 1906 (Kaufmann 1906a, 496). Some 48 hours of exposure were needed to produce a perceptible curve suitable for analysis. Figure 8.5 is one of Kaufmann's best run of the 1902–3 series of experiments (Kaufmann 1903, 401; see Miller 1981, 65). By reversing the electric field Kaufmann obtained two curves whose symmetry gave an indication of the uniformity of the field. Such a photographic plate made G. N. Lewis caution physicists in 1908 not to forget that Kaufmann's measurements—though taken with extreme care and delicacy—"consisted in the determination of the minute displacement of a somewhat hazy spot on a photographic plate" (Lewis 1908, 713). One should not, however, belittle these minute displacements; such is the empirical evidence of modern physics.

Kaufmann was the first experimenter to establish qualitatively that, independent of a specific form for Φ, the experimental calculations show that as $k_1 A_m / A_e$, or v, increases, the ratio e/m decreases "very markedly." Kaufmann argued that from this experimental result "one may infer the presence of a not inconsiderable fraction of 'apparent mass' which increases with speed in such a way as to become infinite at the speed of light" (1901b, 153; Boorse and Motz 1966, 509).

To arrive at a quantitative result, Kaufmann had to employ a formula for the field energy of a rapidly moving electron. He used the formula developed by the British physicist Searle in 1897 (Kaufmann 1901b; Searle 1897; see also Jammer 1961, 140; Cushing 1981, 1140–41). This formula presupposes an infinitely thin spherical shell over which the charge is evenly distributed. Kaufmann then calculated that "the ratio of apparent to true mass for speeds that are small with respect to the speed of light is . . . about 1/3" (1901b, 155; Boorse and Motz 1966, 511). Thus, the so-called true mass had to be considered, at least for this range of velocities, three times greater than the "apparent" mass.

Appropriately, the paper ends with a cautious remark: having employed Searle's formula, Kaufmann found himself obliged to point out that his conclusion concerning the ratio of apparent to true mass

> depends on the assumption that the charge of the electron is distributed over an infinitely thin spherical shell. Since we know nothing about the constitution of the electron and we are not justified *a priori* in applying to the electron the laws of electrostatics which we seek to derive from the properties of the electron itself, it is quite possible that the energy relationships of the electron can be derived from *other charge distributions,* and that there may be *distributions* which, when applied to the above analysis, give a zero true mass. (1901b, 155; Boorse and Motz 1966, 512; emphases added)

Notice that Kaufmann used the plural form. It is worth stressing that here Kaufmann made explicit the possible limitation of the experiment's argument. The claim is that if one were to endow the electron a priori with some definite structure and charge distribution and proceed to analyze on this basis the physical properties which the electron exhibits, one might not then be entitled to consider one's results conclusive: the experiment being only a comparative best fit. Indeed, this experimental method can only decide between competing theories. For each theory the experimental data are evaluated, and the "correct" theory is the one for which the values derived from the experimental data are in the best agreement with those expected theoretically. In other words, this experiment cannot refute or confirm a specific electron theory; it can only show for which theory the calculated values of m/m_0 plotted against v/c best fall on the theoretical curve. Kaufmann therefore found himself free to envisage a different structure and charge distribution which would render the true mass zero.

Kaufmann estimated that the accuracy of his 1901 results was about 5%. He obtained a value of 1.95×10^7 cgs units for the ratio e/m_0, a value which agrees, in Kaufmann's words, "quite well with that found for cathode rays [1.865×10^7]" (1901b, 152, 154; Boorse and Motz 1966, 509, 510). Although he appeared to be satisfied with his experimental method, he certainly was not happy with its results. He was a strong proponent of the electromagnetic conception of nature and he presumably considered the experimental result (which presupposes Searle's formula) that the ratio m_0/m is about 1/3 a stumbling block.

At the 1901 meeting of the German Natural Scientists and Physicians, Kaufmann outlined his expectations for an electromagnetic synthesis of the physical sciences. He concluded that before such a view could be endorsed, one would have to show that the electron mass is *entirely* electromagnetic, reduce mechanics to electromagnetism, prove that matter is composed *solely* of electrons, relate chemical periodicities to the stable dynamic arrangements of assemblies of electrons, and experimentally confirm Wien's electron theory of gravitation (Kaufmann 1901a; quoted by McCormmach 1970b, 481; emphases added). Kaufmann's 1901 result that the "apparent" mass of swift cathode rays is three times smaller than their "true," mechanical mass obstructed the very first step required by his plan for attaining a comprehensive electromagnetic view of nature. Little wonder that he sought in 1901 a structure and a charge distribution other than that suggested by Searle, which would show the "true," mechanical mass of the electron to be zero.

It did not take much time for Kaufmann's expectation to materialize. Max Abraham was intrigued by Kaufmann's experimental results. Influenced by Wien, who (after withdrawing in 1900 from Hertz's program of a unifying mechanical physics) had converted to the idea of an electromagnetic foundation for mechanics (Wien 1901), Abraham sought an electron theory that would provide the basis for Wien's idea and account for Kaufmann's results. In 1902 Abraham published his first paper on this subject. In this paper, entitled "Dynamics of Electrons" (1902a),[19] he attempted to deduce a system of electrodynamics from the differential equations of the electromagnetic field. As pointed out by Kaufmann, the central question was whether the mass of an electron could be explained *entirely* in terms of the dynamical actions of its electromagnetic field without resorting to a mass that exists independently of the electric charge. "Only when this question is affirmatively answered," Abraham remarked, "can the possibility of a purely electromagnetic foundation for mechanics be recognized" (1902a, 21; quoted by Goldberg, 1970–71a, 12).

Abraham first examined critically Searle's theory, upon which Kaufmann had relied. Abraham distinguished between two types of mass: the resistance to acceleration in the line of motion is the "longitudinal mass," whereas the resistance to acceleration perpendicular to the line of motion is the "transverse mass." Since these two masses are different functions of the electromagnetic momentum which the field energy determines, different accelerations would result from the same force depending on whether it is exerted in a direction parallel or perpendicular to the line of motion. For slowly moving electrons, when the momentum is proportional to the velocity, the two masses are equal. This is the case in Searle's theory; Searle missed here the general result that the two masses are in fact different. It was Abraham who demonstrated that mass is not a scalar, as classical

19. See also *Sci. Abs., Phys. Elec. Eng.*, A6 (1903): 238, art. 644.

mechanics has it, but rather a vector quantity of the nature of a tensor with the symmetry of an ellipsoid of revolution (Abraham 1902a, 28; see also Goldberg 1970–71a, 13–14; Jammer 1961, 151). A sacrosanct Newtonian principle—that mass was invariant with velocity—was contradicted.

On the basis of this analysis it became clear to Abraham that Kaufmann had not applied the correct formula. Since the forces in Kaufmann's experiment were exerted perpendicularly to the electrons' direction of motion, the transverse, and not the longitudinal, mass had to be considered. Drawing upon Searle's theoretical investigation, Kaufmann had actually calculated the wrong mass.

In order to arrive at an exact expression for the electromagnetic transverse mass, Abraham had to put forward assumptions concerning the structure of the electron and its charge distribution. He assumed that the electron is a rigid sphere whose charge is distributed uniformly either on its surface or throughout its volume. Calculating the electromagnetic transverse mass on this basis, he obtained

$$\begin{aligned} m_t(\text{Abraham}) &= \frac{3}{4}\, m_0\, \frac{1}{\beta^2}\left[\frac{1+\beta^2}{2\beta}\ln\left(\frac{1+\beta}{1-\beta}\right)-1\right] \\ &= m_0\left(1+\frac{2}{5}\beta^2+\frac{9}{35}\beta^4+\ \ldots\right), \end{aligned}$$

where m_0 is the mass of slowly moving electrons (such as cathode rays); β is v/c and the power series represents an expansion for $\beta \ll 1$.[20] The velocity function is, therefore,

$$\text{Abraham: } \Phi(\beta) = \frac{3}{4}\frac{1}{\beta^2}\left[\frac{1+\beta^2}{2\beta}\ln\left(\frac{1+\beta}{1-\beta}\right)-1\right].$$

On the basis of this formula it is possible to calculate the ratio s between the electromagnetic transverse masses of two different electrons with the known velocities v_1 and v_2; namely,

$$s = \frac{m_t(v_1)}{m_t(v_2)}.$$

The ratio r of the effective (total) transverse masses for such electrons can be found from Kaufmann's experimental results: this ratio is

$$r = \frac{m + m_t(v_1)}{m + m_t(v_2)},$$

20. For a detailed analysis of Abraham's theory of electrons, see Miller 1981, 55–61, 109–12; Cushing 1981, 1135–37.

where m is the mechanical mass. By eliminating $m_t(v_2)$, one obtains

$$\frac{m}{m_t(v_1)} = \frac{s - r}{s(r - 1)}.$$

If the experimental ratio r were to coincide, within the limits of experimental error, with the theoretical ratio s, then the mechanical mass m could be considered zero (Jammer 1961, 149–50; Lorentz 1952, 41). Thus, if Kaufmann's experimental data were to reveal for the effective transverse mass, $m + m_t(v)$, the same velocity dependence as for the electromagnetic transverse mass, $m_t(v)$, then the mechanical mass m would necessarily be zero (Jammer 1961, 149).

In the light of this theoretical analysis, Abraham examined afresh Kaufmann's data and stated that there was an agreement between his predictions and the experimental results. He confidently concluded that "the inertia of the electron is caused exclusively by its electromagnetic field" (Abraham 1902a, 40; quoted by Goldberg, 1970–71a, 14).

Kaufmann immediately took up this new theoretical perspective. Working closely with Abraham at Göttingen, he redid the experiment in an attempt to secure more accurate results; he presumably hoped that the new results would verify Abraham's theory and thereby render Searle's formula inadequate. In his 1902 experimental paper "Electromagnetic Mass of Electrons" (Kaufmann 1902b), he abandoned Searle's formula and adopted Abraham's theory. On the basis of this new analysis, the experiment resulted, perhaps not surprisingly, in the conclusion that the mass of the electron is purely an electromagnetic phenomenon.

The collaboration of Abraham, the theoretician, and Kaufmann, the experimenter, came to the fore in September 1902 when they read sequential papers at the 74th meeting of the Natural Scientists in Karlsbad. Kaufmann disclosed in his paper that the previous experiments, particularly those of 1901, were not in very good agreement with Abraham's theory (Kaufmann 1902a, 54). He, however, reported that his recent experiment, accurate to about 1%–1.4%, was in very good agreement with the theory. Although, as he himself pointed out, a 2% error in the measurement of the velocity of the electron would amount to 19% error in the determination of its mass,[21] he confidently concluded that the dependence of the mass on its velocity is "exactly represented by the Abraham formula" (Kaufmann 1902a, 56; quoted by Goldberg 1970–71a, 15). He then reiterated his view that "the mass of the electron is of a purely electromagnetic nature" (1902a, 56; Cushing 1981, 1141). In the ensuing discussion Kaufmann referred theoretical questions to Abraham. Subsequent to that

21. For β close to unity, $\Phi(\beta)$ varies very rapidly.

discussion Abraham presented the paper "Principles of the Dynamics of Electrons" (1902b),[22] in which he maintained with Kaufmann that the data strongly suggested that the mass of the electron was entirely electromagnetic. Naturally, in his turn, Abraham referred questions on design and experimental technique to Kaufmann.[23]

The conspicuous atmosphere of success that Abraham and Kaufmann were creating was quite understandable. At that time, in 1902, Abraham and Kaufmann could *quantitatively* predict and account for the dependence of mass on velocity in such a way that the first step toward an electromagnetic synthesis of nature could be accomplished. Indeed, so confident was he that in 1903 Abraham revealed that his initial results had not agreed too well with Kaufmann's 1902 experimental findings. He reported that Kaufmann had since then found a mistake in his calculation and improved the method of measuring the electric field strength (Goldberg 1970–71a, 17; Miller 1981, 61–66; Cushing 1981, 1141). According to Abraham the new results confirmed the view that "the mass of the electron is purely of electromagnetic nature" (quoted by Goldberg, 1970–71a, 17). For his part, Kaufmann extended the experimental conclusion—obtained only for Becquerel rays—to include cathode rays as well (1903, 103),[24] a result which H. Starke claimed later in that year to have confirmed experimentally (Starke 1903).[25]

The success of Kaufmann and Abraham came to a halt in 1904 when rival theories started to emerge. Exploring the possibility of a deformable electron which would conform to the Lorentz-FitzGerald contraction hypothesis, Lorentz arrived at a different, and mathematically simpler, formula from that of Abraham for the dependence of the transverse mass of the electron on its velocity:

$$m_t(\text{Lorentz}) = m_0(1 - \beta^2)^{-1/2} = m_0\left(1 + \frac{1}{2}\beta^2 + \frac{3}{8}\beta^4 + \ldots\right),$$

where β is v/c and the power series represents an expansion for $\beta \ll 1$ (Lorentz et al. 1952, 24; see also Miller 1981, 73–74; Cushing 1981, 1138). The velocity function is, therefore,

$$\text{Lorentz: } \Phi(\beta) = (1 - \beta^2)^{-1/2}.$$

Abraham was furious. In his view such a theory would undermine the possibility of obtaining an electromagnetic foundation for mechanics. He

22. See also *Sci. Abs., Phys. Elec. Eng.,* A6 (1903): 239, art. 645.
23. Goldberg referred to this collaboration as a symbiosis (1970–71a, 7).
24. See also *Sci. Abs., Phys. Elec. Eng.,* A6 (1903): 521, art. 1411.
25. See also ibid., A7 (1904): 95, art. 344; Miller 1981, 64–67.

argued that, owing to the deformation of the electron in Lorentz's theory, some mechanical work would have to be performed to preserve the electron's stability; that in turn would require a nonelectromagnetic inner force, quite apart from the internal electric force of the rigid electron theory. Consequently, a purely electrodynamic interpretation of Becquerel rays would be rendered untenable, and with it the electromagnetic view of nature would have to be discarded. Abraham conceded that Lorentz's formulae for the electron mass were mathematically simpler than his own; however, he stressed that the requirement of Lorentz's theory for an additional, nonelectromagnetic force would make that theory, from the physical point of view, far more complicated than his own theory of rigid electrons (Abraham 1904, 576–79 esp. 578; see also Goldberg 1970–71a, 20–21; McCormmach 1970b, 480; Jammer 1961, 147; Miller 1981, 75–79, 117–18). It is an indication of the spirit of the time that Abraham considered this argument—namely, that Lorentz had failed to consolidate the electromagnetic viewpoint—a valid form of criticism.[26]

In order to obviate the need for an additional force and yet to maintain a theory of deformable electrons, A. H. Bucherer assumed in 1904—as P. Langevin independently did a year later (Bucherer 1904, 57–58; 1905; Langevin 1905)—that if the electron were to contract in the course of its motion in accordance with the Lorentz transformation, its volume would remain invariant under these deformations and thus no mechanical work would have to be performed (McCormmach 1970b, 480). On the basis of this view that the electron is deformable but incompressible, Bucherer obtained another set of formulae, entirely different from those of Abraham and Lorentz, for the dependence of mass on velocity:

$$m_t(\text{Bucherer}) = m_0(1 - \beta^2)^{-1/3}$$
$$= m_0\left(1 + \frac{1}{3}\beta^2 + \frac{2}{9}\beta^4 + \ldots\right),$$

where β is v/c and the power series represents an expansion for $\beta \ll 1$ (Cushing 1981, 1138). The velocity function is, therefore,

$$\text{Bucherer: } \Phi(\beta) = (1 - \beta^2)^{-1/3}.$$

To these three different theories yet another one was added in 1905, which agrees formally with that of Lorentz. Einstein's 1905 paper "On the

26. Lorentz responded by noting Abraham's objection and conceding that it was certainly true that he, that is, Lorentz, had not shown that the electron, when deformed to an ellipsoid by its translation, would be in a stable equilibrium. However, in Lorentz's view the hypothesis of the deformable electron "need not be discarded for *this* reason" (Lorentz 1952, 214, emphasis in the original).

Electrodynamics of Moving Bodies"[27] laid down first principles which obviated the need to assume some electron structure and charge distribution. The two principles (the principle of relativity and the constancy of the velocity of light in empty space independent of the state of motion of the emitting source) eventually transformed electron theory—the arguments being kinematic rather than dynamic.

Einstein's theory is a comprehensive theory whose generality by far exceeds that of the theories of Abraham and Bucherer. But as late as 1910 many physicists did not perceive the general character of Einstein's theory and did not take up the fundamental issues it had raised. At that time the topical question was not whether the two principles of Einstein's theory and their implications were valid or not, but rather could one construe mass as an entirely electromagnetic phenomenon and, moreover, attain a coherent electromagnetic conception of nature? From that perspective, Einstein's theory was considered conservative: a mere generalization of Lorentz's theory (Hirosige 1968, 41–42; see also McCormmach 1970a, 60–61).[28]

3. The Second Set of Experiments (1905 – 1906)

The new theoretical developments stimulated Kaufmann to conduct a new set of experiments. While Abraham was engaged in defending the results on the theoretical front (Abraham 1904; see also Pyenson 1979, 70–74), Kaufmann further improved upon the experimental technique. He investigated new methods of high-vacuum technique and eventually succeeded in designing the first rotary high-vacuum pump.

In 1905 Kaufmann published a description of the new pump. Essentially, it consisted of a glass cylinder around which two spiral tubes, connected to the cylinder, coiled. Rotation of the cylinder round its axis caused the mercury in the two spirals to rise and fall alternately and thus to act as in the usual fall-pump (Kaufmann 1905a).[29] Although, as Boorse and Motz pointed out, "the pump was extremely fragile, unwieldy, and

27. References are to the English translation of Einstein's paper in Lorentz et al. 1952, 37–65.

28. As Pauli observed in 1921, in his article on the theory of relativity, Einstein's theory "constituted a definite progress [in] that Lorentz's law of the variability of mass could be derived from the theory of relativity without making any specific assumptions on the electron shape or charge distribution. Also, nothing need be assumed about the nature of the mass." And Pauli concluded that "the old idea that one could distinguish between the constant 'true' mass and the 'apparent' electromagnetic mass, by means of deflection experiments on cathode rays, can . . . not be maintained" (Pauli 1967, 82–83).

29. See also *Sci. Abs., Phys. Elec. Eng.*, A8 (1905): 537; Kaye 1927, 81.

temperamental" (1966, 505), Kaufmann claimed to have used it successfully in his new set of experiments (1906a, 501).

Whereas in the first set of experiments Kaufmann had eventually tested one single theory in which he appeared to believe, namely, Abraham's theory, in the second set his task was to discriminate between the various theories that had been suggested during the years 1904–5. His experimental method remained the same in both sets of experiments.

Kaufmann reported on his new experimental work in 1905 and published the final account in 1906 (1905b, 1906a). He opened his definitive paper of 1906 with a theoretical discussion of the various models, claiming that Einstein's formulae amounted to those of Lorentz; indeed, he regarded Lorentz's and Einstein's theories as two theories which assume the same electron model: the Lorentz-Einstein theory of the electron (Kaufmann 1905b, 954; 1906a, 487–95, 530–31).[30] Concluding the theoretical introduction, Kaufmann anticipated the general result of the measurements he was about to describe; he stated that "the results of the measurements are not compatible with the Lorentz-Einstein fundamental assumption. The Abraham and the Bucherer equations depict equally well the results of the observations. For the present a decision between these two theories by a measurement of the transverse mass of β-rays appears to be impossible" (1906a, 495). He then proceeded to describe the experiment and to discuss a method of analyzing the data which in his view would force a decision between the various models.

According to Kaufmann the equations for the motion of the electron given by Lorentz and Einstein differ "very considerably" (*sehr wesentlich unterscheiden*) from that of Abraham. Kaufmann therefore expressed some surprise that, as he wrote, "an application of the equations to my earlier measurements by Herr Lorentz led to the . . . result that my observations could be represented by him with the same accuracy as by the Abraham equations for the rigid electron" (1906a, 493). Given, however, sufficient experimental accuracy, the various theories could be distinguished. Indeed, Kaufmann pointed out that in the new experiments there was a 5%–7% difference between the velocities yielded by the theories of Abraham and Lorentz for each measured curve-point. Hence, "a way was provided for differentiating between the two theories." But from the outset he acknowledged that there had been the possibility of neither theory providing sufficiently accurate agreement (1906a, 493).

Kaufmann determined several constants of the curve that had been recorded on the photographic plate; he compared these constants with the corresponding constants obtained from the various theories. Clearly, the

30. Kaufmann considered Einstein's theory to be based on just one principle, namely, the principle of relative motion. See Miller 1981, 226 ff., 329.

z'	$y'_{\text{beob.}}$	p	$y'_{\text{ber.}}$ I	II	III	δ I	II	III	β I	II	III
0,1350	0,0246	0,5	0,0251	0,0246	0,0254	− 5	0	− 8	0,974	0,924	0,971
0,1919	0,0376	1	0,0377	0,0375	0,0379	− 1	+ 1	− 3	0,922	0,875	0,919
0,2400	0,0502	1	0,0502	0,0502	0,0502	0	0	0	0,867	0,875	0,864
0,2890	0,0545	1	0,0649	0,0651	0,0647	− 4	− 6	− 2	0,807	0,823	0,805
0,3359	0,0811	1	0,0811	0,0813	0,0808	0	− 2	+ 3	0,752	0,713	0,750
0,3832	0,1001	1	0,0995	0,0997	0,0992	+ 6	+ 4	+ 9	0,697	0,661	0,695
0,4305	0,1205	1	0,1201	0,1202	0,1200	+ 4	+ 3	+ 5	0,649	0,616	0,647
0,4735	0,1405	0,25	0,1408	0,1405	0,1409	− 3	0	− 4	0,610	0,579	0,608
0,5252	0,1667	0,25	0,1682	0,1678	0,1687	−15	−11	−20	0,566	0,527	0,564

Fig. 8.6. A comparison between the observed and calculated curve in Kaufmann's definitive paper (Kaufmann 1906a): z' and $y'_{\text{beob.}}$ are the reduced coordinates z_o and y_o; p is the assigned weight, a measure of the reliability of each reading. The three columns I, II, and III, under $y'_{\text{ber.}}$ represent the calculated results according to the theories of Abraham, Lorentz, and Bucherer respectively. The quantity δ ($\times 10^{-4}$) represents the difference between $y'_{\text{beob.}}$ and $y'_{\text{ber.}}$, that is, the difference between the observed and the calculated value for y'. The quantity β is the ratio v/c associated with the coordinates (y_0, z_0) according to the respective theory.

measured constants (in Kaufmann's terminology, the *Apparatkonstanten*), unlike the calculated constants (the *Kurvenkonstanten*), were independent of any specific assumptions concerning the structure of the electron and its charge distribution. Rather, they were functions of the strengths of the electric and magnetic fields and the value of e/m_0, that is, the ratio e/m for slowly moving electrons, like those that constitute cathode rays. Such a comparison could therefore serve as a fitness test for the various theories. Experimentally as well as mathematically, it was a very delicate and complicated process. To facilitate the analysis of the resulting curve, Kaufmann introduced certain approximations which required the use of deflections substantially smaller than the dimensions of the apparatus. As the actual observational results—namely, (y_0, z_0)—did not comply with this stipulation, Kaufmann had to reduce them to "infinitely small deviations."[31]

Presenting one set of readings (in his table VIII reproduced here in fig. 8.6), Kaufmann remarked that the table indicated that "from the shape of the curves alone, without taking into consideration the absolute

31. $1/A = (e/m_0)(M/c)$; $B = (e/m_0)(E/c^2)$; $C = AB = E/Mc$; $D = (E/M^2)(m_0/e)$; $e/m_0 = (cC/D)(cC/ME)^{1/2}$; $C = \beta(y'/z')$; $y'^2 = C^2z'^2 = D^2z'^4$; where y' and z' are the reduced coordinates of y_0 and z_0, and c is the speed of light in vacuum. For the method of reduction see Kaufmann 1906a, 493, 524–30, 531, 534. See also Miller 1981, 62–65, 228–29; Cushing 1981, 1142. See also nn. 17–18 above.

values of the constants, it is not possible to come to any decision in favor of one or the other theory" (1906a, 532).

Prima facie it appeared, as Kaufmann himself put it, that "one can merely say for the present that all three theories represent equally well the relative shape of the curve" (1906a, 532–33). However, he pointed out that

> from the values for β it is gathered . . . that Lorentz's theory demands quite different velocities than the theories of Bucherer and Abraham do. For both of these the velocities are almost identical, and what is still more worthy of note: the function $\Phi(\beta)$ agrees numerically, as one can easily convince oneself, for both [Abraham's and Bucherer's theories] within the velocity interval here under consideration, with a deviation of at most 2 percent. (533)

On the basis of this method—a method that did not depend on the value of e/m_0[32]—Kaufmann calculated that Abraham's and Bucherer's theories exhibited discrepancies between calculation and observation of 3.5% and 2.8% respectively, by far smaller than the discrepancy of 10.4% which Lorentz's model yielded (533).[33]

First Comparison

Abraham	−3.5%
Bucherer	−2.8%
Lorentz	−10.4%

Apparently, Kaufmann was not satisfied with a conclusion based on just one type of comparison, for he sought a second comparison, which was dependent on the value of e/m_0. This value could be obtained, on the one hand, from the direct relation between one particular theory and the observed curve of the deflected β-rays, that is, from the "curve constants" and, on the other hand, from Simon's measurement of deflected cathode rays. Thus the calculated values of e/m_0, derived from the different theories, could be compared on independent grounds with the experimental value of e/m_0. Kaufmann, however, did not use Simon's 1899 experimental result, 1.865×10^7 emug^{-1}. Applying the three theoretical models, he extrapolated Simon's value to infinitely slow electrons and obtained a

32. By considering the product of two "apparatus constants," Kaufmann could refer to a constant which is independent of e/m_0, namely, $C = AB = E/Mc$ (see n. 31, above; Kaufmann 1906a, 531; Miller 1981, 230).

33. See also Miller 1981, 230. There is a printer's error in the presentation of the discrepancies: −2.5% should be −2.8%.

mean value of 1.878×10^7 emug^{-1} (Kaufmann 1906a, 533, 548–51).[34] Kaufmann then calculated that according to Abraham's theory the mean value of e/m_0 was 1.823×10^7, with a spread of 1.9%. On the basis of Bucherer's theory the mean value was 1.808×10^7, with a spread of 1.4%; and finally, Lorentz's theory resulted in a mean value of 1.660×10^7, with a larger spread of values around the mean: 5.5%.

Comparing these values with the extrapolated value of Simon, Kaufmann found that Abraham's and Bucherer's theories gave rise to discrepancies of 2.9% and 3.7% respectively. These results are markedly smaller than the discrepancy of 11.6% which Lorentz's theory exhibited (Kaufmann 1906a, 533).

Second Comparison (e/m_0\ x 10^{-7})

	Max. value	Min. value	Mean value	Spread	Simon's extrapolated value: 1.878
Abraham	1.858	1.788	1.823	1.9%	−2.9%
Bucherer	1.833	1.780	1.808	1.4%	−3.7%
Lorentz	1.751	1.569	1.660	5.5%	−11.6%

The possibility that Simon's extrapolated value for e/m_0 might introduce a constant error into this, second, comparison did not escape Kaufmann's expert eye. Indeed, he stressed that no complete agreement had appeared to prevail concerning the value of e/m_0 (531).[35]

Kaufmann's third and final test of the contesting theories was to calculate from the "apparatus constants"—using again the extrapolated value of Simon's result and convenient values of z'—the theoretical values of y' for each theory. A comparison of the three calculated curves with the reduced observed curve showed that Lorentz's calculated curve deviated most: the greatest difference between Abraham's calculated results and those of Lorentz being about 5% (Kaufmann 1906a, 534; see also Miller 1981, 231).

34. See also Miller 1981, 230; Cushing 1981, 1142 (for $v = 0$, $e/m_0 = 1.878 \times 10^7$ emug^{-1} and not 1.885×10^7 as it appears in Cushing's paper).

35. It is worth mentioning that the accepted value of e/m_0 is $(1.75890 \pm 0.00002) \times 10^7$ emug^{-1}, that is, 6.3% lower than Kaufmann's extrapolated value of Simon. (See, e.g., Cohen, Crowe, and DuMond 1957, 267.) If one were to compare these values with the accepted value for e/m_0, one would naturally arrive at different discrepancies: Abraham's and Bucherer's theories would then show discrepancies of 3.6% and 2.8% respectively, whereas Lorentz's theory would reduce its discrepancy by more than half to 5.6%.

Third Comparison

z'	y'		
	Abraham	Bucherer	Lorentz
0.1	0.0191	0.0190	0.0196
0.2	0.0413	0.0407	0.0434
0.3	0.0712	0.0696	0.0745
0.4	0.1104	0.1080	0.1144
0.5	0.1595	0.1568	0.1642

As a careful experimenter Kaufmann once again called the reader's attention to the fact that this last comparison "is based on the assumption that the Simon value of e/m_0 is correct." However, the first comparison, according to Kaufmann, remained free from such an assumption (534).

In Kaufmann's view the conclusion from these three comparisons was clear. He confidently maintained that "the above results speak decisively against the correctness of Lorentz's theory and consequently also against that of Einstein's theory." He therefore argued that "the attempt to base the whole of physics, including electrodynamics and optics, upon the principle of relative motion would have to be regarded at present as . . . unsuccessful" (1906a, 534). Kaufmann could thus uphold the assumption that "physical phenomena depend on the movement relative to a quite definite coordinate system which we call the absolute resting ether" (535). The fact that it had not been possible to demonstrate by electrodynamic or optical experiments the existence of this fixed coordinate system did not deter Kaufmann from concluding that "a decision may not be made as to the impossibility of such a proof" (535). Kaufmann did not consider his experiment a direct test of Einstein's relativity theory; rather, he intended his experiment to be a test of what he was interested in, namely, three models of the electron (Miller 1981, 226–27, 343). The purported refutation of Einstein's theory is considered, in this final conclusion of Kaufmann, a consequence of the incompatibility of Lorentz's electron model with the experimental results (Miller 1981, 226–27, 343; see also Hirosige 1968, 41–42; 1976, 75; Holton 1975, 234–35).

Kaufmann did not carry out new and more accurate experiments to test Abraham's and Bucherer's theories. In his view, the main obstacles for increasing the accuracy stemmed from the photographic plate, namely, the behavior of the emulsion under the required experimental conditions. Other problems would arise from maintaining a constant electric field in a scaled-up apparatus and accurately measuring the magnetic field. He remarked that "a further increase in the accuracy . . . would call for means which far outstrip the current possibilities of the institute" (1906a, 535).

In response to Planck's criticism of his experiments, Kaufmann was willing to entertain the possibility of small observational errors occurring in such a way as to increase the deviations between theory and experiment. But he stressed that whereas Abraham's theory deviated 3%–5% from the resulting data, that of Lorentz deviated 10%–12% (Kaufmann 1906b; see also Miller 1981, 233).

A year later, in 1907, Bestelmeyer obtained a lower experimental result for e/m_0: 7.8% less than that of Simon and 8.5% less than the extrapolated value which Kaufmann had calculated (Bestelmeyer 1907; see also Miller 1981, 335–40). This new determination of the value of e/m_0 affected the second and third comparisons in Kaufmann's analysis; it motivated Planck to resume his criticism of the experiment. He arrived at the conclusion that the electric field might have varied as a result of an ionizing effect (Planck 1907). In his reply, Kaufmann upheld his conclusion and strongly argued that an 8% variation in the electric field was not possible in his setup; at all events, any small variation would not alter the result that the data proved the Lorentz-Einstein theory incorrect (Kaufmann 1907; see also Miller 1981, 340–41).

Campbell acknowledged at the time that "the only method which is known at present for attempting to distinguish between the various formulae . . . is to compare the results predicted with the experimental results of Kaufmann." According to Campbell,

> the requisite measurements were made by Kaufmann with exquisite skill . . . but this test cannot possibly be judged decisive: it can exclude certain theories, but the limits of experimental error are necessarily so large that it must always be possible to invent new assumptions which will give results as satisfactory as the best of these pre-existing. (1907, 201, 319)

Campbell suspected that the difference between the observed and theoretical curves was greater than the limits of experimental error; that is indeed the principal difficulty of this experiment. But, Campbell argued,

> this discrepancy need not shake our belief that the mass is purely electrical. The exact form of the theoretical curve depends on certain assumptions made as to fundamental points in electro-dynamics. Different physicists have used different assumptions and the difference between their results is greater than that between the experimental and the best of the theoretical curves. All will agree that it is better to assume that the error arises from an ignorance of the right assumption than that some small fraction of the

whole mass of the electron is of a nature different from the rest. (1907, 201–2)

In his proposed revision of the fundamental laws of matter and energy, Lewis concurred in 1908 with this conclusion. He considered Kaufmann's experiments "remarkably skilful" and was of the opinion that they agreed with his own theoretical result that a distinction could be observed between absolute and relative motion (1908, 712–15). By 1910 the experiment was still considered important; it played a crucial role in Laub's paper on the experimental basis of the principle of relativity. Laub questioned the accuracy of the experiment and searched for possible errors. Following Planck, he suggested that the vacuum might not have been sufficiently high, thereby allowing an ionization process to take place. He also referred to the different values of the ratio e/m_0 but acknowledged that "it has not been ascertained why the observations of Kaufmann, which contain so much material, deviate from those of other observers." And with some trepidation he continued: "reading the available studies, the thought occurs [to me] that in this area we are at the boundaries of perception, or that relations exist here that have not yet been well enough appreciated. It would be important to repeat Kaufmann's measurements with exactly the same experimental arrangements but with improved means (better vacuum)" (1910, 443, 444, 462).[36]

Discussing in 1914 the notion of electromagnetic inertia, Richardson expressed the view that the experimental aspect of this problem had been resolved by Kaufmann. He, however, questioned whether the experiments had not agreed with the results predicted by both Abraham's and Lorentz's formulae within the limits of experimental error (Richardson 1914, 7, 239). By 1915 more experimental results from other types of experiments became available and the emerging consensus was that Kaufmann's experiments were in error and that the Abraham classical picture of the rigid electron had to be ruled out (Lorentz 1931, 274–88; Miller 1981, 345–52).

However, while analyzing in 1938 the classical experiments on the relativistic variation of electron mass, C. T. Zahn and A. H. Spees reached the conclusion that for the higher velocities "no very satisfactory experimental distinction between the two types of electron has as yet been made

36. It is a dictum of experimental science that results should be capable of duplication. In spite of the importance of Kaufmann's experiment and the persistent calls to redo this crucial experiment, no attempt (to the best of my knowledge) was done to repeat it and to check the result experimentally. However, actual duplication of an experiment using the *precise* material arrangement is rare. The common approach is to do the experiment in a different way. Indeed, in the present case other experimental techniques, notably the use of orthogonal fields, were developed and carried out.

by direct electric and magnetic deflection methods." They remarked further that "in view of the fundamental importance of such experiments it seems that much is left to be desired" (Zahn and Spees 1938, 511; see also Miller 1981, 351–52).

Some 20 years later, the situation seemed not to have changed much. In their review of the experimental evidence for the law of variation of the electron mass with velocity, P. S. Faragó and L. Janossy concluded that "it is the fine-structure splitting in the spectra of atoms of the hydrogen type which gives the only high-precision confirmation of the relativistic law of the variation of electron mass with velocity." They stressed that "this evidence . . . is a rather *indirect* one, and it does not cover a range of velocities which is wide enough." Concerning the fairly large number of *direct* experiments on the behavior of free electrons, the authors maintained that they "could hardly find such results which would prove the validity of the relativistic relation with a margin of error much less than, say, the difference between the results of the relativity theory and the theory of Abraham" (Faragó and Janossy 1957, 1436, emphasis added). This critical view of Faragó and Janossy points at the heart of the problem: to distinguish experimentally between Abraham's theory of the electron and that of Lorentz and Einstein, a margin of experimental error which is less than the difference between the respective theoretical results has to be attained. This is a considerable task since the difference is indeed very small.[37]

In conclusion, it is fitting to return to Max Born, who, it may be recalled (see n. 4, above) dressed the Kaufmann episode in a myth. Born, however, was capable of harsh, perspicacious historical observations. In 1956, he remarked that

> as a matter of fact, the velocity dependence of energy and of mass has nothing . . . to do with the structure of the body considered, but is a general relativistic effect. Before this became clear, many theoreticians wrote voluminous, not to say monstrous, papers on the electromagnetic self-energy of the rigid electron. . . . Today all these efforts appear rather wasted; quantum theory has shifted the point of view, and at present the tendency is to circumvent the problem of self-energy rather than to solve it. But one day it will return to the centre of the scene. (1956, 196)

Kaufmann's experimental findings aroused much interest and received strictures as well as support from many distinguished physicists. These varied reactions are of considerable importance since they show how differently the same experimental results can be apprehended. More-

37. For the difference between the respective theories see Abraham 1904, 578; Jammer 1961; Goldberg 1970–71a, 18.

over, these reactions throw light on the relation between what is perceived as an error and the epistemological-methodological context within which such a perception is made. The dependence of detection of error on epistemic system and methodological disposition is discernible in such reactions. It is to this issue that I now turn by juxtaposing three different responses to Kaufmann's experimental results.

4. Poincaré's Reaction

In the introduction to his book *Science and Method*, Poincaré observed that "mechanics seems to be on the point of undergoing a complete revolution. The ideas which seemed most firmly established are being shattered by daring innovators" (1914, 11). It is natural to think that by 1908, when the book was published, Einstein would have been considered one of those innovators. However, to Poincaré, as Goldberg suggests, Einstein's work appeared as a small and rather insignificant part of a much broader theory, namely, Lorentz's theory, which had been developed from 1892 under the guiding criticism of Poincaré to its completion as a theory of the electron in 1904 (Goldberg 1970–71b, 84; 1970, 97). Indeed, Einstein is not mentioned at all in this celebrated book of Poincaré.[38] Poincaré's heroes were those who sought a new theory of matter; that is, a theory which would explain the phenomenon of matter solely in terms of electrons immersed in ether: "beyond the electrons and the ether," Poincaré proclaimed, "there is nothing" (1914, 209).

Since Einstein set himself a completely different problem, he was not one of Poincaré's heroes. Einstein questioned the meaning of simultaneity and the nature of space and time; he simply was not concerned with constructing models. As Zahar remarked, Einstein had implicitly posited " a domain of events, each of which can be referred to by coordinates (t, x, y, z) in any one of infinitely many equivalent inertial frames. *Events* are therefore the constituents of the Einsteinian universe" (1973, 122–23). Hence Einstein's interest in the way things are measured; altogether a different interest from that of Poincaré, who was intent on *actually* measuring things, on finding out experimentally the electron's properties (Goldberg 1967, 944). It comes therefore as no surprise that Poincaré praised the theoretical work of Lorentz and Abraham and the experiments of Kaufmann, even when their results were not in agreement with his own

38. Shortly after they met in 1911 at the Solvay conference, Poincaré wrote of Einstein that he "is one of the most original thinkers I have ever met. . . . Since he seeks in all directions, one must . . . expect the majority of the paths on which he embarks to be blind alleys" (quoted by Miller 1981, 255). On the relationship between Poincaré and Einstein see Pais 1983, 169–72.

ideas. After all, these researches focused on the very problem Poincaré himself was interested in.

Thus, as early as 1902, Poincaré described Lorentz's theory as "the most satisfactory theory . . . ; it is unquestionably the theory that best explains the known facts, the one that throws into relief the greatest number of known relations, the one in which we find most traces of definitive construction" (1952, 175). According to Poincaré, "the only object of Lorentz was to include in a single whole all the optics and electrodynamics of moving bodies" (176).[39] This objective was much in keeping with Poincaré's maxim: "the true and only aim [of science] is unity" (177).

Referring to the theoretical work of Abraham and the experimental results of Kaufmann, Poincaré remarked in 1905 that they "have . . . shown that the mechanical mass . . . is null, and that the mass of the electrons . . . is of exclusively electrodynamic origin." This result, Poincaré continued, "forces us to change the definition of mass; we cannot any longer distinguish mechanical mass and electrodynamic mass, since then the first would vanish; there is no mass other than electrodynamic inertia" (1946, 311). It should be stressed that Poincaré did not show here any skepticism with regard to the experimental results; he considered Kaufmann's results, at least in 1905, conclusive.

Planck took a different view; his was the critical approach. He had questioned and examined extensively Kaufmann's experimental results and eventually became skeptical of their correctness. Planck criticized both the apparatus and the interpretation. Thus, according to Planck, two kinds of error were purported to occur in Kaufmann's experiment. One error pertained to the apparatus and its working: poor vacuum allowed an ionization process to take place, which in turn interfered with the electric field. The other error pertained to the theoretical conclusion: the method of reduction was inconsistent, at least in one case in which $v/c > 1$. (On different types of experimental errors and their classification, see Hon 1989.) Tackling these two sources of error with the same method of analysis is a delicate matter. Moreover, Planck used two different values of e/m_0. In his two-pronged attack, Planck adopted a new value of e/m_0 and (1906; 1907) showed that Kaufmann's data favored what he called the Lorentz-Einstein theory.[40] However, as Cushing notes, it was a "close

39. In Poincaré's view, Lorentz "did not claim to give a mechanical explanation" (1952, 176).

40. For a detailed discussion of Planck's response to Kaufmann's experimental results, see Zahar 1978; Miller 1981, 232–35, 340–41, 345; Cushing 1981, 1142–46. In his analysis of Planck's criticism of Kaufmann's results, Cushing starts his paper on a high note by claiming that "it was Max Planck who decisively reversed the interpretation of Kaufmann's data from disconfirmation of relativity to confirmation of that theory. Planck's was a classi-

call (2% vs. 5%) and scarcely overwhelming evidence in favour of relativity. The important outcome of Planck's analysis," Cushing rightly remarks, "was that Kaufmann's experiments no longer presented a stumbling block to the acceptance of relativity theory" (1981, 1146).

Poincaré, however, did not change his mind. Three years later, in 1908, when Einstein's relativity theory was beginning to be understood as a comprehensive theory of principle, Poincaré still had much to say in praise of the results of Lorentz, Abraham, and Kaufmann. Poincaré expressed the view that "Abraham's calculations make us acquainted with the law in accordance with which the *fictitious* mass varies as a function of the velocity, and Kaufmann's experiment makes us acquainted with the law of variation of the total mass. A comparison of these two laws will therefore enable us to determine the proportion of the *actual* mass to the total mass" (1914, 206, emphases in the original). The result of Kaufmann's determination of this proportion was "most surprising: *the actual mass is nil.*" And Poincaré concluded that "we have thus been led to quite unexpected conceptions. . . . What we call mass would seem to be nothing but an appearance, and all inertia to be of electromagnetic origin" (206–7, emphasis in the original).

Notwithstanding Abraham's consideration that electrons are spherical and undeformable, Poincaré asserted that "we shall have to admit that the electrons, while spherical when in repose, undergo Lorentz's contraction when they are in motion, and then take the form of flattened ellipsoids" (1914, 226). Poincaré considered the principle of relativity (as it had been expressed in Lorentz's theory) a law of nature which had been generalized from experience (1914, 226; 1952, 135–39). He therefore felt compelled to apply the contraction hypothesis of Lorentz and FitzGerald—a hypothesis that becomes meaningful under the principle of relativity[41]—to the electrons themselves. As Poincaré clearly put it, two theories presented themselves: "one in which the electrons are undeformable, which is Abraham's; the other, in which they undergo Lorentz's deformation" (1914, 228). Poincaré sided with Lorentz and, by means of what has come to be known as "Poincaré's stress," gave a theoretical underpinning to the mechanism of the Lorentz contraction and the stability

cally beautiful application of strict logic to a rather confused situation." Cushing ends, however, on a low note with the conclusion that "even after Planck's reanalysis of Kaufmann's data, the issue was still not clearly decided in favour of either the Abraham or Lorentz theory" (1981, 1143, 1146).

41. In his 1904 paper, Lorentz inferred the contraction hypothesis from considerations of a translation between two systems; in an earlier paper he had simply posited this hypothesis (Lorentz et al. 1952, 4–7, 22–23).

of the electron under such deformation.[42] But the question still persisted: which is the correct theory?

As an exponent of conventionalism who maintained that "the aim of science is not things themselves . . . but the relations between things" (Poincaré 1952, xxiv) and for whom "science is only a classification and that . . . cannot be true, but convenient" (Poincaré 1946, 352), it is difficult to understand why Poincaré set himself the very trap which a realist would gladly have laid before a conventionalist, namely, the pitfall of the *experimentum crucis.* Whereas in 1902 Poincaré had considered experiment a guide that helps us in our free choice among all possible conventions (1952, 50), in 1908 he considered its role to be that of an arbiter. "The method employed by Kaufmann," he remarked, "would . . . seem to give us the means of *deciding* experimentally between the two theories" (1914, 228, emphasis in the original). And in the concluding remarks to his account of the new mechanics, he stated that "further experiments will no doubt teach us what we *must* finally think of . . . [the new theories]. The root of the question is in Kaufmann's experiment and such as may be attempted in verification of it" (1914, 249).

Why Poincaré, a noted conventionalist, conferred so much weight on a single experiment to the point of believing that it "revolutionizes at once Mechanics, Optics, and Astronomy" (1914, 286; see also Poincaré 1946, 545) is at the center of the difficulty underlying Poincaré's reaction to Kaufmann's experiment. One way of solving this problem is to distinguish, as Goldberg does, between Poincaré the philosopher and Poincaré the physicist, and simply to stipulate an antagonism between the two. According to Goldberg, Poincaré might well have assumed a conventionalistic position while talking about the nature of physics; however, in his work in theoretical physics "he was anything but a conventionalist" (Goldberg 1967, 938). Did Poincaré ignore conventionalism when he practiced physics? Did he waver, in the case of Kaufmann's experiment, between conventionalism and realism as Goldberg suggests (Goldberg 1970–71b, 76–77)?

Poincaré, to be sure, was quite aware of such criticism; in fact, he anticipated it. "Have you not written," he invited his reader in 1905 to ask the author, that is, Poincaré himself, "that the principles [e.g., the

42. "One obtains . . . a possible explanation of the contraction of the electron by assuming that the deformable and compressible electron is subject to a sort of constant external pressure the action of which is proportional to the volume of variation" (Poincaré 1905, 1504; quoted by Pais 1983, 158). Acknowledging Poincaré's contribution, Lorentz remarked that it had made the mechanism of the contraction much clearer (Lorentz 1952, 213–14). For a detailed discussion, see Miller 1973. See also Miller 1981, 83–85.

principle of relativity and the principle of conservation of mass], though of experimental origin, are now unassailable by experiment because they have become conventions? And now you have just told us that the most recent conquest of experiment put these principles in danger" (1946, 318). Indeed, for Poincaré "principles are conventions and definitions in disguise. They are, however, deduced from experimental laws, and these laws have, so to speak, been erected into principles to which our mind attributes an absolute value" (1952, 138). In other words, the principle of relativity and the principle of conservation of mass, for example, "are results of experiments boldly generalized; but they seem to derive from their very generality a high degree of certainty. In fact, the more general they are, the more frequent are the opportunities to check them, and the verifications multiplying, taking the most varied, the most unexpected forms, end by no longer leaving place for doubt" (1946, 301). But then, new experimental results did compel one to doubt the absolute value which had been attributed to these principles. Specifically, in 1905 Poincaré himself questioned that "if there is no longer any mass, what becomes of Newton's law? . . . If the coefficient of inertia is not constant, can the attracting mass be? That is the question" (1946, 312). Furthermore, in 1908 he acknowledged that Kaufmann's experiments "*have shown Abraham's theory to be right.* Accordingly, it would seem that the Principle of Relativity has not the exact value we have been tempted to give it" (1914, 228, emphasis in the original). Poincaré's answer to this criticism is somewhat bemusing: "Well, formerly I was right and today I am not wrong. Formerly I was right," he repeated and stressed that "what is now happening is a new proof of it" (1946, 318).

A closer reading of Poincaré's writings may clarify this position and resolve the criticism it received. Poincaré's philosophy of science, as it is stated and explained in *Science and Hypothesis,* is not a rigid conventionalism. Poincaré appears to discern in science a spectrum of philosophies. At one extreme lies conventionalism, which forms the philosophical foundations of geometry, and at the other extreme lies induction, the method of the physical sciences. Poincaré located the science of mechanics in between these two extremes; this is a science in which the two methods, deduction and induction, operate in concert.

Thus, for Poincaré, "*geometrical axioms are . . . neither synthetic a priori intuitions nor experimental facts.* They are conventions. Our choice among all possible conventions is guided by experimental facts; but it remains free. . . . In other words, *the axioms of geometry . . . are only definitions in disguise*" (1952, 50, emphases in the original). Though, according to Poincaré, experiment plays a considerable role in the genesis of geometry, "it would be a mistake to conclude . . . that geometry is, even in part, an experimental science. If it were experimental, it would only be approximative and provisory. . . . Geometry would be only the study of the movements of solid bodies; but, in reality, it is not concerned with natural solids: its object is certain ideal solids, absolutely invariable, which are but a greatly simplified and very remote image of them." "Experiment," Poincaré concluded, "tells us not what is the truest, but what is the most convenient geometry" (70–71).

By contrast, at the other end of the spectrum, where according to Poincaré the physical sciences lie, inductions is the guiding method. It is here that "experiment is the source of truth" (1952, xxvi, 140). Here, as Poincaré put it, experiment "alone can teach us something new; it alone can give us certainty" (140). He qualified this strong view, however, by remarking that all that experiment affirms "is that under analogous circumstances an analogous fact will be produced" (142). And even of that claim one is never *absolutely* sure; however well founded a prediction may appear, it may prove baseless if one tries to verify it (144). Furthermore, it was clear to Poincaré that the requirement that experiments should be carried out without preconceived ideas is impossible. He observed that "every man has his own conception of the world, and this he cannot so easily lay aside" (143). An experimental law is therefore always subject to revision: "one single piece of work by a real master"—Poincaré had Pasteur in mind—"will be sufficient to sweep . . . [bad experiments] into oblivion" (142; see also 95).

"The principles of mechanics," Poincaré argued, are "presented to us under two different aspects. On the one hand, there are truths founded on experiment, and verified approximately as far as almost isolated systems are concerned; on the other hand," he continued, "there are postulates applicable to the whole of the universe and regarded as rigorously true" (1952, 135–36). Thus, crucial experiments and conventions constitute the extremes of the spectrum and in between lies the science of mechanics, a science which combines both conventionalism, a deductive method, and experimentation, an inductive method.

In the framework of this philosophical outlook, Poincaré's reaction to Kaufmann's experiment may become clearer. On the one hand, he considered it an *experimentum crucis*, an arbiter among theories, and as such he accepted, so to speak, its judgments. On the other hand, he was aware of its possible limitations, that is, its approximations and possible errors. Hence his plea that before adopting Kaufmann's result "some reflection is necessary. The question is one of such importance," he remarked, "that one would wish to see Kaufmann's experiment repeated by another experimenter" (1914, 228). However, in his view, Kaufmann was a skillful experimenter who had taken "all suitable precautions," and "one cannot well see what objection can be brought" (229).

Nevertheless, there was one measurement whose possible sources of error Poincaré considered highly significant: the measurement of the electric field. As he put it, it was "the measurement upon which everything depends" (1914, 229). In any deflection experiment in which a uniform electric field is applied, a high vacuum is required. Kaufmann's experiment was not an exception: a high vacuum had to be created between the two plates of the condenser, so that a truly uniform electric field would be obtained. "Is this certain?" asked Poincaré. "May it not be," he questioned, "that there is a sudden drop in the potential in the neighbourhood of one of the armatures, of the negative armature, for instance? There may be a difference in potential at the point of contact between the metal and the vacuum, and it may be that this difference is not the same on the positive as on the negative side." Poincaré concluded that "we must at least take into account the possibility of this occurring, however slight the probability may be" (1914, 229).

Notwithstanding, Poincaré was reluctant to suspend judgment of, let alone to dismiss, an experiment of which he was rightly suspicious but whose importance he had acknowledged. He was therefore unwilling to look persistently for possible sources of error; he did not analyze the experimental reports with a view to detecting errors that might have led Kaufmann astray. Poincaré submitted to Kaufmann's expertise on matters of experimentation and accepted the results.[43] A theoretician, for whom experiments form the building blocks of knowledge—knowledge whose cement, so to speak, is the inductive method and to whose overall architecture conventionalism attends—Poincaré relied heavily on Kaufmann's expertise and was prepared to concede, as late as 1908, that Abraham's theory had been shown to be right.

5. Einstein's Reaction

In contradistinction to Poincaré, who construed the principles of mechanics in general and the principle of relativity in particular to be essentially bold generalizations of a posteriori experience gleaned from several experiments, Einstein elevated the principles of his relativity theory to a priori postulates which stipulate the features of the laws of nature so that there are no special privileged observers. Though he spoke in his 1905 paper, "On the Electrodynamics of Moving Bodies," of "unsuccessful at-

43. Poincaré's acceptance of the experimental findings of Blondlot, the discoverer of the so-called N-rays, is another example of his submission to authority on matters of experimentation. Poincaré visited Blondlot at his laboratory in Nancy (hence, N-rays), but he saw no effects. Yet he believed that the N-ray radiation existed. Poincaré expressed great confidence in Blondlot, who was according to Poincaré a distinguished and competent physicist. Poincaré attributed his failure to observe the effects of the alleged radiation to involuntary accommodation of his eyes (Poincaré 1904).

tempts to discover any motion of the earth relatively to the 'light medium,'" (Einstein 1905, 37), Einstein did not specify them. In his opinion these experimental attempts, about which he was notoriously vague (see Holton 1975, 261–352; Zahar 1973, 231–32; Pais 1983, 172–73; Miller 1981, 143 ff.), "suggest that the phenomena of electrodynamics as well as of mechanics possess no properties corresponding to the idea of absolute rest." Hence, "they suggest," as Einstein viewed it, that "the same laws of electrodynamics and optics will be valid for all frames of reference for which the equations of mechanics hold good" (1905, 37–38). Thus far he was prepared to go in locating the sources of his theory in experiments.

Einstein's next move, about which he was quite explicit, was to raise this suggestion or, in his words, "this conjecture . . . to the status of a postulate, and also [to] introduce another postulate, . . . namely, that light is always propagated in empty space with a definite velocity *c* which is independent of the state of motion of the emitting body" (1905, 38). In Einstein's theory of relativity an empirical finding separated itself completely from its experimental origin and attained the status of a postulate which, together with another postulate, provided the axiomatic base for "a simple and consistent theory of the electrodynamics of moving bodies" (38).

By contrast, for Poincaré the principles of mechanics always remained attached to their origins in experiments. Although the principle of relativity coheres with the theoretical models of mechanics and is confirmed by experience, it is nonetheless assailed and, as Poincaré put it, "may well not have the rigorous value which has been attributed to it" (1946, 305, 507). Speculating on this issue, Poincaré raised the question as to "what would happen if one could communicate by non-luminous signals whose velocity of propagation differed from that of light? If, after having adjusted the watches by the optical procedure, we wished to verify the adjustment by the aid of these new signals, we should observe discrepancies which would render evident the common translation of the two stations." "Are such signals inconceivable," Poincaré speculated further, "if we admit with Laplace that universal gravitation is transmitted a million times more rapidly than light?" (308). Poincaré was thus prepared on both philosophical and experimental grounds for a refutation of the principle of relativity.[44]

For Einstein, however, a refutation meant more than the rejection of a single principle; it also meant the discarding of a comprehensive theory that, on the one hand, is simple and clear in terms of its logical deductions

44. In Miller's view Poincaré never elevated the principle of relativity to a convention (Miller 1981, 376 n. 7, see also 335).

and, on the other hand, embraces a wide complex of phenomena. Einstein stressed that his was not a descriptive theory of the electron or a "constructive theory" like that of Abraham or Lorentz; rather, it was a "theory of principle": a theory that stipulates procedures for attaining physical knowledge (Einstein 1981, 223).

Responding to Ehrenfest's request for clarifying the status of the relativity theory, Einstein made it amply clear that he had intended his theory to be basically a theory of principle and not, as Ehrenfest assumed, a theory of matter (Einstein 1950, 54; 1907a).[45] Einstein explained that one should not interpret the two postulates of his theory as a system, let alone as a closed system. Rather, the two postulates—the principle of relativity and the constant nature of the velocity of light in empty space independent of the state of motion of the emitting source—form a "heuristic principle which, considered by itself, contains only statements about rigid bodies, clocks and light signals" (Einstein 1907a, 206; see also Klein 1967, 515–16; Miller 1981, 235–36). In this view, anything beyond this that the theory of relativity supplies is in the connections it requires between laws that would otherwise appear to be independent of one another (Klein 1967, 515–16). Explaining how one may inquire into the motion of fast electrons by combining the known laws for slow electrons with the relativistic transformation laws, Einstein concluded that "we are by no means dealing with a 'system' here, a 'system' in which the individual laws would implicitly be contained and from which they could be obtained just by deduction." The theory can only furnish "a principle that allows one to reduce certain laws to others, analogously to the second law of thermodynamics" (Einstein 1907a, 207; see also Klein 1967, 515–16).

A refutation of the special theory of relativity by an experiment would therefore undermine, not a certain conjectured model, but a methodology which purports to yield new knowledge of the physical world. Although Einstein concluded his 1905 relativity paper by deducing from his theory the properties of the motion of the electron that could be accessible to experiments[46]—hence, in principle, refutable—he clearly intended his

45. Ehrenfest's note is on p. 204 of Einstein 1907a. Ehrenfest raised the important question of how one applies the Lorentz transformations to a rigid body.

46. Einstein obtained three relations that constitute, in his words, "a complete expression for the laws according to which . . . the electron *must* move" (1905, 64–65, emphasis added). The three relationships intended for experimental tests are as follows: (1) the ratio of the magnetic and electric deflection, A_m/A_e, as a function of v/c; (2) the potential difference traversed as a function of v; and (3) the radius of curvature of the path of the electron in the presence of a magnetic field as a function of the electron's velocity. However, Einstein did not propose a direct test of his prediction of m_t, for which the available results of Kaufmann were applicable.

theory to be situated on some kind of metalevel. From this metalevel other theories and laws of nature may be looked upon and comprehensively correlated within a broad perspective.

What does it take then to refute experimentally a theory of principle? Does Kaufmann's refutation of Einstein's relativity theory constitute such a case? Does D. C. Miller's positive result of the "ether-drift" experiment in 1925 form another refutation of this theory?[47] It seems that if an experimental refutation of the relativity theory had been forthcoming, Einstein would have required a rigorous and more profound notion of refutation than the one offered by an isolated experiment. Indeed, the question arises as to why Einstein refrained from testing his theory with Kaufmann's available data. He did not reinterpret these experimental results in the light of his new theory as Lorentz had done, but rather, he ignored them. He must have been aware of the fact that his 1905 prediction for the electron's transverse mass disagreed with Kaufmann's data.[48]

Einstein's response to Kaufmann's challenge came in 1907. In a review article, "On the Principle of Relativity and the Conclusions That Follow from It," under the heading On the Possibility of an Experimental Test of the Theory of Motion of Material Points: The Kaufmann Investigation, Einstein summarized Kaufmann's 1906 paper. In his view, "Kaufmann has determined with admirable care [*bewunderungswürdiger Sorgfalt*] the relation between A_m and A_e [the magnetic and electric deflections] for β-rays emitted by a grain of radium bromide" (1907b, 437).[49] Having explained the theoretical background and the apparatus employed by Kaufmann, Einstein juxtaposed the calculated results based on his theory and Kaufmann's experimental results in the graph (Einstein 1907b, 439) reproduced here as figure 8.7. In the graph, the points and the width of the circles are the observed relations between the electric and magnetic deflections and their respective limits of experimental error, and the crosses are the calculated values resulting from Einstein's theory.

"In view of the difficulty of the investigation," Einstein remarked,

47. For references and discussion of D. C. Miller's ether-drift experiment, see Holton 1975, 316–17, 348–50.

48. It is quite safe to assume with Miller that when Einstein wrote his relativity paper, he was aware of Kaufmann's early experiments. Miller has argued convincingly that Einstein refrained intentionally from testing his theory with Kaufmann's experimental data. He goes on to show that had Einstein used the data of those experiments, he would have obtained values which are about 13% less than those obtained empirically by Kaufmann. This is not a good agreement between theory and experiment (Miller 1981, 333–34).

49. In Einstein's nomenclature A_m and A_e are neither z_0 and y_0 nor z' and y'; rather, z' and y' reflect the relation between A_m and A_e. Thus, generally they can be regarded as the magnetic and the electric deflection respectively. For further details see Miller 1981, 344–45.

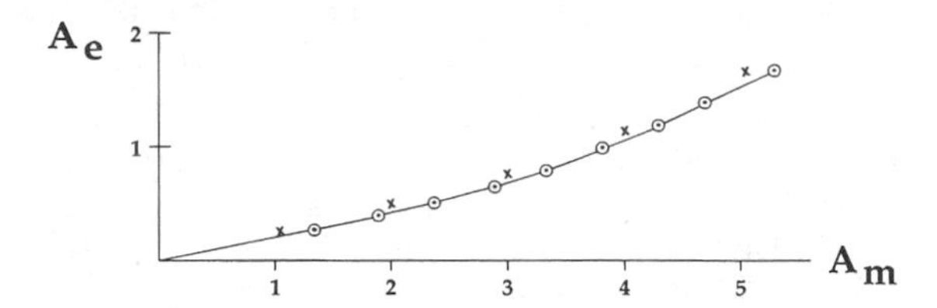

FIG. 8.7. Einstein's juxtaposition of the calculated results based on his theory (*crosses*) and Kaufmann's experimental results (*points and circles*) (Einstein 1907b)

"one would be inclined to regard the overall agreement as satisfactory" (1907b, 439).[50] However, as Einstein observed, "the existing discrepancies are systematic and considerably outside the margin of error of Kaufmann's experiment." Yet, according to Einstein, "Kaufmann's calculations are free of error [*fehlerfrei*]; it follows from the fact that Planck, using a different method of calculation, was led to results which agree entirely with those of Kaufmann" (439).[51]

Nevertheless, Einstein refused to draw a definite conclusion from these results. Rather, he maintained that "whether the systematic discrepancies are due to the existence of some so-far-not-understood source of error or whether they mean that the foundations of Relativity Theory do not correspond with the facts, can surely be decided with certainty only when a great variety of observational material is at hand" (1907b, 439). Moreover, he was not prepared to accept either Abraham's theory or that of Bucherer, although, as he himself admitted, the calculated curves that these theories had yielded fitted the observed curve considerably better than the curve obtained from relativity theory. In Einstein's opinion, "these theories should be ascribed a rather small probability because their basic assumptions about the mass of moving electrons are not made plausible by theoretical systems that encompass a wider complex of phenomena" (439; see also Pais 1983, 159; 1972, 82).

Einstein's strong view that his theory transcends the status of a theory of matter and assumes the character of a theory of principle facilitated his

50. Einstein pointed out that Kaufmann had not used the observed curve but rather a reduced curve. He made this remark in a footnote and did not elaborate on it (1907b, 438). Miller suggests that this remark could be interpreted as a caveat that was prompted by some inconsistency which Planck had found in Kaufmann's results, namely, $v/c > 1$ (Miller 1981, 233, 345; see also Planck 1906, 758; Cushing 1981, 1145; above, n. 40). Notice that Einstein did describe the results as satisfactory (*genügende*).

51. This is somewhat puzzling since Planck exposed an inconsistency in Kaufmann's method of reducing the data. See above, nn. 40, 50.

claim that Kaufmann's experiment was amiss. His response was intuitive, since he could not base his rejection of the results on either experimental or theoretical grounds. On the contrary, as we have seen, he was of the opinion that the measurements had been taken with "admirable care" and the calculations were "free of error." Einstein, it emerges, founded his case against Kaufmann's results solely on methodological grounds. What perhaps strengthened his methodological argument was the fact that according to relativity theory the calculated values for the relations between A_m and A_e lay in a consistent manner above the observed curve; a fact that could arouse his suspicion that a systematic error had vitiated Kaufmann's results. All the same, a methodological argument—that is, the available theories which could account for Kaufmann's results should nonetheless be ascribed a rather small probability since, as Einstein argued, their basic postulates had been rendered implausible by a comprehensive theory, namely, the special theory of relativity—sufficed for Einstein to substantiate his intuitive rejection of Kaufmann's results.

The consistent methodological position of Einstein motivated (one is even inclined to say, required) Einstein to suspect an error whose source he could not determine; this suspicion gave in his view sufficient ground for rejecting Kaufmann's experimental results. One sees here evidence for Einstein's growing suspicion of the epistemological priority of experiment. In the following years, Einstein expressed openly his preference for the consistency of a simple and convincing theory over the latest news from the laboratory; again and again he turned out to be right (Holton 1975, 235–36).

The contrast between Einstein's position and that of Poincaré now becomes clearer. Poincaré had questioned the correctness of a crucial measurement and proceeded to suggest a very plausible source of error; but as this correct analysis was embedded in a philosophy of science which gave heed to experimental results and dealt with their bearings upon theories of matter, it did not initiate a suspension of judgment, let alone a rejection of the experimental results at stake.

According to Einstein a physical theory can be criticized from two perspectives: the degree of its "external confirmation" and the extent of its "inner perfection," that is, the degree of its "logical simplicity" (1969, 21–23). Here we have been concerned with the former demand, which at first sight appears evident: "the theory must not contradict empirical facts" (21). However, Einstein immediately qualified this demand as he realized that its application had turned out to be quite delicate: "it is often, perhaps even always, possible to adhere to a general theoretical foundation by securing the adaptation of the theory to the facts by means of artificial additional assumptions. In any case, however, this first point of view is concerned with the confirmation of the theoretical foundation by the available empirical facts" (21–23). Since Einstein's suspicion that there was an error in Kaufmann's experiment and the consequent "blank" rejection of the experimental result were vindicated, it is surprising that Einstein did not stress or at least suggest in his review of his intellectual evolvement the possible occurrences of experimental errors.[52] It is further surprising that he preferred rather to mention the possible adaptability of a theory to empirical facts and, further, to make the obvious point that the empirical knowledge which one can have at any historical juncture is limited. Compared to the possible occurrences of experimental errors, these two qualifications are indeed very weak. Referring to his own theory of general relativity, whose great attraction is its logical consistency, Einstein remarked that "if any deduction from it should prove untenable, it must be given up. A modification of it seems impossible without destruction of the whole" (1950, 110). Flexibility and adaptability to empirical facts are not aspects of "logical simplicity," as Einstein himself indicated here and indeed in his characteristic response to Kaufmann's experimental results: he had refused to let his theory bow to these results and, to recall the words of another great scientist, would not "patch up his hypothesis."[53]

In contrast to the mellowed remarks of Einstein's "Autobiographical Notes," one can follow Dirac and imagine Einstein responding to Kaufmann's results in the following vigorous words: "'Well, I have this beautiful theory, and I'm not going to give it up, whatever the experimenters find; let us just wait and see.'" As it happened, Einstein has so far proved right; "so it seems," Dirac surmised—confirming thereby his own predilection—"that one is very well justified in attaching more importance to the beauty of a theory and not allowing oneself to be too much disturbed by experimenters, who might very well be using faulty apparatus" (Dirac 1982, 83).

6. Lorentz's Reaction

If Poincaré's and Einstein's reactions to Kaufmann's experimental results constitute the two possible extreme reactions—both suspecting an error, the former, however, favoring the acceptance of the results and the latter categorically rejecting them—then Lorentz's reaction mediates or rather

52. However, see his cautious remark in a conversation with Heisenberg (Hon 1989, 470–71).

53. Kepler: "If I had believed that we could ignore these eight minutes, I would have patched up my hypothesis accordingly." Kepler, however, unlike Einstein, did submit his theory to observations. He revolutionized astronomy by recognizing the primacy of Tycho Brahe's observations over his own theory (see Hon 1987b, 564).

vacillates between these two poles. In 1924 Lorentz remarked that "one of the lessons which the history of science teaches us is . . . that we must not too soon be satisfied with what we have achieved. The way of scientific progress is not a straight one which we can steadfastly pursue. We are continually seeking our course, now trying one path and then another, many times groping in the dark, and sometimes even retracing our steps" (1924, 608). Although Lorentz illustrated this remark with the way physicists have interpreted the phenomenon of light throughout the ages, the development of his own view with regard to the theory of the electron in general and Kaufmann's experiment in particular also provides such an illustration.

Already in his early scientific work Lorentz had sought to attain a theory of matter which would provide a consistent model as well as an elucidation of all known physical phenomena. This attempt to grasp the laws which govern matter constituted a persistent motif throughout Lorentz's scientific work. In his view it is the electromagnetic field—a field which has an independent physical reality—that provides the fundamental substance. By regarding the electromagnetic field as the state of the ether, identified when at rest with absolute space and considered an entity sui generis which cannot be affected in its structure by matter (though it can transmit action between matter), Lorentz was able to conceive a new perspective of the foundation of physics.[54]

Lorentz's scientific work in the last decade of the nineteenth century is characterized by his attempt to furnish equations for the motion of light in macroscopic dielectrics. He was much concerned with explaining several positive effects like the aberration of light, the Doppler effect, and the Fizeau experiment (Miller 1981, 18–29). But his main interest focused upon the negative result of the Michelson-Morley experiment on the motion of the earth through the ether. Lorentz was intrigued by the fact that this experiment—an experiment accurate to second-order quantities in the ratio of the earth's velocity to the velocity of light—did not demonstrate the effects of the ether, let alone its very existence as a physical entity. In order to reconcile the experimental null result with his model of the ether, Lorentz put forward in 1895 the celebrated contraction hypothesis (Lorentz et al. 1952, 3–7). However, as he himself conceded in 1904, the method of inventing special hypotheses for each new experimental result was "somewhat artificial" (Lorentz et al. 1952, 13; see also Holton 1975, 304–316). What is important to note is not so much the ad hoc character of Lorentz's early methodology (which he renounced in

54. By contrast, Hertz and Maxwell regarded the ether as just the state of dielectric. See McCormmach 1970a, 47 ff.; Hirosige 1976, 68 ff.

1904) but rather his incessant attempt to obtain a theoretical framework within which *all* experimental results would be explained. Thus, when J. J. Thomson established experimentally, just before the turn of the century, the existence of charged corpuscles, Lorentz immediately took cognizance of this experimental discovery and shifted his primary concern from the optics of macroscopic bodies to the mechanics of individual electrons. He was the founder of particle-based electrodynamics (McCormmach 1970a, 49).

The concept of the electron furnished Lorentz's theory with a physical entity whose mechanics, when immersed in the ether, could have bridged the gap between the phenomenon of light and that of matter. In Lorentz's view electrons "are extremely small particles, charged with electricity, which are present in immense numbers in all ponderable bodies, and by whose distribution and motions we endeavour to explain all electric and optical phenomena that are not confined to the free ether" (Lorentz 1952, 8). However, the notion of ether remained the fundamental assumption; "the ether not only occupies all space between molecules, atoms or electrons, but it pervades all these particles." And whereas the particles may move, "*the ether,*" Lorentz maintained, "*always remains at rest*" (11, emphasis in the original).

In Lorentz's theory, a displacement of a charged particle produces a change in the state of the ether; this perturbation propagates at the speed of light through the ether and can influence another particle at a later time. The displacements are governed, according to Lorentz, by electric and magnetic forces. He thus conceived a dynamic theory characterized by five equations: the four equations of Maxwell's electromagnetic theory and the Lorentz equation for the forces which the ether exerts on a charged particle. The last equation allows for an interaction of field and particle. As Lorentz characteristically remarked, it had been obtained by "generalizing the results of electromagnetic experiments" (1952, 14).

Indeed, Lorentz made explicit this feature of his methodology of close connection between experiment and theory. In his account of the method he had employed, Lorentz observed that the theory of the electron is "an extension to the domain of electricity of the molecular and atomistic theories that have proved of so much use in many branches of physics and chemistry" (1952, 10). The method of the molecular and atomistic theories stands in contrast to the method employed by some physicists, who, as Lorentz put it, "prefer to push their way into new and unexplored regions by following those great highways of science which we possess in the laws of thermodynamics, or who arrive at important and beautiful results, simply by describing the phenomena and their mutual relations by means of a system of suitable equations" (10). Lorentz did not deny the success of pure thermodynamics or the achievement of the equations of the electromagnetic field in their most general form. He insisted, however, that by such methods one would never have been able to attain some results to which the molecular hypotheses have given rise. The fruitfulness of these hypotheses cannot be denied, Lorentz claimed, "by those who have followed the splendid researches on the conduction of electricity through gases of J. J. Thomson and his fellow workers" (11). Lorentz was a theoretician who stood on experimental ground; he clearly saw himself as one of those who followed closely the new experimental developments at the turn of the century. As Holton observed, Lorentz was "deeply involved in constructing, step by step, a viable theory for electrodynamics, based as far as possible on existing principles and mechanisms, relying on experimental results as a guide to the detailed construction of a modification of existing theory" (1975, 306). Recalling Einstein's distinction between a constructive theory and a theory of principle, one sees clearly that Lorentz's theory is constructive.

Indeed, Lorentz's paper of 1904, "Electromagnetic Phenomena in a System Moving with Any Velocity Less than That of Light" (1904, 11–34), contains in addition to a new theory of the electron many experimental considerations. Lorentz in fact remarked that both experimental and methodological reasons had led him to reexamine the problems connected with the motion of the earth. Specifically, the two main reasons were the new experiments of Michelson, Rayleigh, Brace, and Trouton and Noble (for references, see Lorentz 1895, 3, 4; 1904, 11)—experiments in which quantities of the order v^2/c^2 could have been perceptible, though the results were all negative—and, second, the quest for fundamental assumptions which would dispense with ad hoc hypotheses (1904, 11–13).

In his 1904 theory Lorentz incorporated his dynamic equation with the four electromagnetic equations of Maxwell and assumed further a privileged frame of reference at rest in the ether that constitutes absolute space and time in the Newtonian sense. In this privileged frame all rods are of maximum length and time goes fastest with respect to rods that contract and "local times" that dilate in reference frames that move relative to this privileged frame (Whitrow 1973, 108–9; Zahar 1973, 121–22). In accordance with his methodology, Lorentz then proceeded to show how his theory could account for a large number of experimental facts. He explained the negative results of the aforementioned experiments and concluded his paper, quite appropriately, with an analysis of Kaufmann's 1902–3 results. He thus attempted to demonstrate how his theory of the electron can explain a whole range of experiments on the basis of some fundamental mechanism that lies, as he put it, "at the bottom of the phenomena" (Lorentz 1952, 8).

In this important paper of 1904, Lorentz accepted Kaufmann's results and considered them decisive; that is, he did not question their validity but rather demonstrated how his theory could account for the results: his formula for the transverse mass agreeing with the data at least as well as Abraham's formulae. According to Lorentz, Abraham's theoretical results were "confirmed in a most remarkable way by Kaufmann's measurements of the deflection of radium-rays in electric and magnetic fields. Therefore, if there is not to be a most serious objection to the theory I have now proposed, it must be possible to show that those measurements agree with my values nearly as well as with those of Abraham" (Lorentz 1904, 31). Lorentz, it appears, did not even contemplate, at least in this paper, the possibility that Kaufmann's experiment might be wrong. Moreover, he mentioned neither the small difference between the predictions of his theory and those of Abraham's theory nor the inaccuracy in Kaufmann's measurements reported by Abraham in 1903. For Lorentz it was apparently sufficient that a satisfactory agreement had been attained between the experimental results and his theory. He simply did not take the trouble to analyze the experiment itself and accepted its results, as he had accepted other experimental results, to serve as a test and a guide to his theory. Lorentz's reaction to Kaufmann's early results amounts in effect to the claim that there might have occurred an error of interpretation: Lorentz analyzed the data of Kaufmann's measurements within his own theory and showed how a satisfactory agreement with his own formulae could be obtained.[55]

As a result of Lorentz's new interpretation, Kaufmann's experiment lost its conclusiveness. Kaufmann therefore redid the experiment in an attempt to force a decision primarily between the theory of Abraham and that of Lorentz.[56] Although the new measurements did not conclusively vindicate Abraham's theory, they definitely refuted, in Kaufmann's view, Lorentz's theory.

In the spring of 1906 Lorentz delivered a series of lectures at Columbia University on the subject of the theory of the electron. The lectures were published in 1909 and have since then served, with the additional

55. Essentially, Lorentz replaced Abraham's formulae for m_t and $\Phi(\beta)$ with his own formulae. He first showed that one constant in Kaufmann's data analysis remained satisfactorily constant under this change of formulae; then, using Kaufmann's measured values z', he deduced the corresponding values of y' with which he could compare the values of Kaufmann's results. For details see Lorentz 1904, 31–34. See also Miller 1981, 74–75; Cushing 1981, 1142–43.

56. Miller reports a letter which Kaufmann wrote to Lorentz in July 1904. In this letter Kaufmann admitted that with the data at hand it was difficult to decide between the two theories; he even conceded that Lorentz's prediction agreed better. In any event, he promised, "on account of the great importance of the whole problem," to repeat the measurements with increased accuracy (Miller 1981, 75).

footnotes of the second edition of 1915, as an outstanding example of a physics which combines general principles with experimental results. From these lectures it transpires that in 1906 Lorentz admitted that, unlike his general credo, which Kaufmann's new 1905–6 results had confirmed, his own particular theory of the electron had in fact been refuted by these very results. It is worthwhile to quote Lorentz's overview at some length:

> Of late the question has been much discussed, as to whether the idea that there is no material but only electromagnetic mass, which, in the case of negative electrons, is so strongly supported by Kaufmann's results, may not be extended to positive electrons [*sic*] and to matter in general. . . . What we really want to know is, whether the mass of the positive electron can be calculated from the distribution of its charge in the same way as we can determine the mass of a negative particle. This remains, I believe, an open question, about which we shall do well to speak with some reserve.
>
> In a more general sense, I for one should be quite willing to adopt an electromagnetic theory of matter and of the forces between material particles. As regards matter, many arguments point to the conclusion that its ultimate particles always carry electric charges and that these are not merely accessory but very essential. We should introduce what seems to me an unnecessary dualism, if we considered these charges and what else there may be in the particles as wholly distinct from each other.
>
> On the other hand, I believe every physicist feels inclined to the view that all the forces exerted by one particle on another, all molecular actions and gravity itself, are transmitted in some way by the ether, so that the tension of a stretched rope and the elasticity of an iron bar must find their explanation in what goes on in the ether between the molecules. Therefore, since we can hardly admit that one and the same medium is capable of transmitting two or more actions by wholly different mechanisms, all forces may be regarded as connected more or less intimately with those which we study in electromagnetism. For the present, however, the nature of this connection is entirely unknown to us. (Lorentz 1952, 45–46)

Clearly, Lorentz believed in an electromagnetic synthesis in which ether and charged particles are the fundamental concepts; matter, in its usual sense, is just superfluous. In that respect, Kaufmann's results encouraged Lorentz to think that he was on the right track. Indeed, Lorentz cited Kaufmann's 1906 experiment with approval, arguing that "it will be best to admit Kaufmann's conclusion . . . that the negative electrons have no material mass at all." In Lorentz's view this conclusion "is certainly one of the most important results of modern physics" (43).

However, Lorentz was aware that the quantitative conclusion of

Kaufmann's experiment had undermined his own theory—the theory that could account on the basis of the contraction hypothesis for the negative results of second-order experiments. "So far as we can judge at present," Lorentz observed, "the facts are against our hypothesis. . . . Kaufmann has repeated his experiments with the utmost care and for the express purpose of testing my assumption. His new numbers agree within the limits of experimental errors with the formulae given by Abraham, but not so with . . . [my equation for the transverse mass], so that," Lorentz concluded, "they are decidedly unfavourable to the idea of a contraction, such as I attempted to work out" (1952, 212–13). It thus seemed to Lorentz that he had to relinquish the idea of a contracting electron altogether. In March 1906, Lorentz confessed in a letter to Poincaré that "unfortunately my hypothesis of the flattening of electrons is in contradiction with Kaufmann's new results, and I must abandon it. I am, therefore, at the end of my Latin. It seems to me impossible to establish a theory that demands the complete absence of an influence of translation on the phenomena of electricity and optics" (quoted by Miller 1981, 334).

Notwithstanding this conclusion, Lorentz considered it legitimate to maintain his refuted hypothesis and to pursue its implications as long as some progress in the understanding of physical phenomena could be made by employing it. He stressed that in speculating on the structure of the electron it should not be forgotten that "there may be many possibilities not dreamt of at present; it may very well be that other internal forces serve to ensure the stability of the system, and perhaps, after all, we are wholly on the wrong track when we apply to the parts of an electron our ordinary notion of force" (1952, 215). It is noteworthy that even when Lorentz tried to secure a viable position for his hypothesis, he did not resort to the possibility of rendering Kaufmann's experiment erroneous; rather, he transformed the idea of limited knowledge into an argument which was supposed to substantiate the legitimacy of developing further his refuted theory.

In the preface to the first edition of *The Theory of Electrons,* Lorentz noted that he had not treated adequately Einstein's principle of relativity. Indeed, Einstein's theory of special relativity received only a sketchy account at the end of the book. According to Lorentz, Einstein's results generally agreed with those that he had obtained; the main difference was that Einstein "simply postulates" what he, that is, Lorentz, claimed to "have deduced, with some difficulty and not altogether satisfactorily, from the fundamental equations of the electromagnetic field" (1952, 230). Lorentz remarked further that Einstein's results had led to the same contradiction with Kaufmann's experiment as his own results had done (229–30). He thus appeared to grasp quite clearly, unlike, for example, Kaufmann and

Planck, that although the two theories agree formally, they differ in their underlying principles.[57]

In the light of these views, it can be said with some force that Lorentz was not prepared on methodological grounds to doubt, let alone reject, Kaufmann's results. In 1904 he demonstrated to his own satisfaction how his theory could account for Kaufmann's 1902–3 results; however, in 1906, when Kaufmann published his definitive paper, Lorentz felt obliged to uphold the new results and, as a consequence, to relinquish his contraction hypothesis, which stood at the center of his theory. Lorentz, it appears, was caught in a dilemma: if he accepted Kaufmann's results, his general view of the physical world would be substantiated; but if he rejected the experimental results, he could perhaps save his theory of the electron but would then jeopardize his general view—the belief in the electromagnetic synthesis of nature.

Lorentz's philosophy of science, which considers experiment a guide for developing theories, constrained him, I suggest, to choose the former horn of the dilemma. He simply did not contemplate or hint at the middle way, let alone the latter horn. For Lorentz to reject an experiment, that is, to render it erroneous, would require new experimental results which contradict it conclusively with no room for doubts. Such experimental results were forthcoming in 1909 when the first edition of *The Theory of Electrons* was published, and between 1909 and 1915 it became abundantly clear that Kaufmann's experiment was indeed erroneous. The additional footnotes of the second edition of *The Theory of Electrons* bear witness to the change that took place in Lorentz's view of Kaufmann's experiment.

Whereas in 1909 Lorentz had admitted that his theory had been refuted by Kaufmann's experiment and that the contraction hypothesis had had to be relinquished, in 1915 he noted that "this can no longer be said now" (1952, 212 n. 1). He referred the reader to the experiments of Bucherer, Hupka, Schaefer and Neumann, and Guye and Lavanchy, which, in his opinion, confirmed his formula for the transverse electromagnetic mass. He therefore claimed that "in all probability, the only objection that could be raised against the hypothesis of the deformable electron and the principle of relativity has now been removed" (339). Notice that Lorentz mentioned both the hypothesis of the deformable electron and the prin-

57. Zahar observed that "before the advent of General Relativity the scientific community (e.g., Planck, Poincaré, Bucherer, Kaufmann and Ritz) spoke of the Lorentz-Einstein theory and contrasted it with the more classical theories of Abraham and Ritz: they regarded the theories of Lorentz and Einstein as observationally equivalent" (1973, 259). On Planck's misunderstanding, see Miller 1981, 254, 256.

ciple of relativity; although one can deduce within the framework of Einstein's theory the contraction phenomenon, Lorentz seems to consider here two separate theories.

It is significant that in the section where Lorentz deals with the qualitative aspect of Kaufmann's experiment, prior to the discussion of its quantitative results, he does not inform the reader that this experiment was decisively contradicted (Lorentz 1952, 42–43). Since the phenomenon that the mass of a charged particle depends upon its velocity had been confirmed by the aforementioned new experiments and, furthermore, since Kaufmann's conclusion that the mechanical mass of the electron can be regarded as zero had not been contradicted by these experiments, Lorentz presumably felt entitled to adhere to his 1909 account and did not amend this section. Thus, in the light of the new experimental results Lorentz was able to resolve his dilemma. He could now reject the quantitative results of Kaufmann's experiment, which confirmed Abraham's formula for the transverse electromagnetic mass, and accept the qualitative result, which enhanced his idea of an electromagnetic synthesis.

However, in 1915 it was not only an experiment that challenged Lorentz's theory but also a theory, namely, Einstein's theory, which was then beginning to gain wide recognition. Having rejected the experimental refutation of his theory, Lorentz now faced the simplicity and great heuristic power of Einstein's theory. He admitted that if he were to write the account of Einstein's theory again, he would have certainly given it a more prominent place (1952, 321). He further remarked that with Einstein's theory "the theory of electromagnetic phenomena in moving systems gains a simplicity that [he] . . . had not been able to attain" (321). As Lorentz viewed it, the main cause of his failure was his adherence to the idea that only the time t in the privileged rest frame could be considered as the true time; the "local time" t' being no more than an auxiliary mathematical quantity. The simplicity of Einstein's theory lies of course in its stipulation that t' carries the same physical significance as t. "If we want to describe phenomena in terms of x', y', z', t', we must work," Lorentz concluded, "with these variables exactly as we could do with x, y, z, t" (321; see also Zahar 1973, 119–20).

Since Lorentz's formulae agree formally with those of Einstein, the realization that Kaufmann's experiment was erroneous supported not only Lorentz's theory but that of Einstein as well. Lorentz was careful not to omit this important consequence, and he refers the reader, where he remarks that Einstein's results contradicted Kaufmann's experimental results, to the same footnote to which he previously alluded when he dealt with his own results (Lorentz 1952, 213, 230, 339). It is therefore no wonder that in this footnote Lorentz distinguished between the hypothesis of the deformable electron and the principle of relativity, referring no doubt to his theory and to that of Einstein respectively.

In view of the fact that in 1915 Lorentz left intact his 1909 concluding remarks, we may surmise that Lorentz did not budge from his theoretical standpoint and steadfastly held to his view of the ether as the fundamental concept. He wrote, "I think something may also be claimed in favour of the form in which I have presented the theory. I cannot but regard the ether, which can be the seat of an electromagnetic field with its energy and its vibrations, as endowed with a certain degree of substantiality, however different it may be from all ordinary matter" (1952, 230). Although Lorentz did recognize the importance of Einstein's theory, he was of the opinion that "each physicist can adopt the attitude which best accords with the way of thinking to which he is accustomed" (quoted by Hirosige 1976, 70, see also Lorentz 1931, 210). Lorentz's adherence to a classical model in which the ultimate constituents are the charged corpuscle and the ether—absolute space and absolute time—was neither wrongheaded nor idiosyncratic; on the contrary, as Zahar has convincingly argued, there was a subtle rationale behind Lorentz's obstinacy (Zahar 1973, 121–23). It is however surprising to find Zahar remarking that by 1915 Lorentz had already accepted the relativity principle (238). It seemed plausible to Zahar that "Lorentz was converted to Relativity by the realization that, while covariance was opening up new possibilities, the ether had become heuristically sterile" (1978, 80). Zahar indeed conjectured that "had Lorentz known right from the beginning that Kaufmann's experiment was not crucial, he would most probably have accepted the covariance of Maxwell's equations and joined the Relativity Programme at its inception in 1905" (80). However, in view of the fact that Lorentz carefully distinguished in 1909 as well as in 1915 between his theory and that of Einstein, and alluded in both cases to the new experimental results, it seems doubtful that Zahar's conjecture is correct. As Born pointed out (1956, 192), Lorentz never wholly accepted Einstein's theory and seemed to remain in doubt as to its physical perspective.[58]

At all events, for Lorentz the validity or otherwise of the principle of relativity was a question of particular properties of the natural forces and therefore should be decided by experiment and not by theory. "It must be possible to decide experimentally," Lorentz maintained, "whether the principle of relativity holds or does not hold" (1931, 266,

58. Elsewhere Born wrote that Lorentz "probably never became a relativist at all, and only paid lip service to Einstein at times in order to avoid argument" (1971, 198). In Pais's view, "Lorentz never fully made the transition from the old dynamics to the new kinematics" (1983, 167). See also Lorentz 1931, 208–11.

see also 255). Thus, unlike Einstein, who raised this principle to the status of a postulate, Lorentz was looking for its experimental underpinnings. It is therefore natural to find Lorentz analyzing the experimental evidence and determining, as late as 1922, the errors that vitiated Kaufmann's experiment.

In Lorentz's view the investigation of the motion of free electrons afforded the best chance for deciding whether or not the principle of relativity holds. He noted that only when the velocities of the free electrons are comparable with that of light is there an appreciable difference between the opposing theories of spherical and flattened electrons (see above, n. 37). However, Lorentz realized that in an experiment such as that of Kaufmann, "too great velocities . . . have to be avoided; . . . [since] the masses, which for β [v/c] approaching 1 tend to infinity, would then be too great, and the accelerations, and thus also the deflections produced by the applied fields, would be too small to be accurately measured" (1931, 267). Lorentz pointed out further that the glass vessel within which the electrons travel in the setup of Kaufmann's experiment "must be well evacuated not only to avoid collisions of the β-particles with the air molecules, which would give diffuse deflections, but also to prevent the formation of a current between the plates of the condenser, which would spoil the uniformity of the field" (272). Thus, as late as 1922, Lorentz thought it important to acquaint the reader with the technical difficulties of Kaufmann's experiment, an experiment whose line of investigation Lorentz considered correct and viable.

Having described the experiment, Lorentz reported that Kaufmann "believed himself to be driven to the conclusion that the theory of the spherical electron fitted better" (1931, 274). However, by 1922 Lorentz was already in a position to refute Kaufmann's claim: abundant experimental results were then available that contradicted this claim. With the backing of these new results, Lorentz searched for the errors in Kaufmann's experiment. He remarked, in the first place, that on the basis of the spherical electron theory there were great deviations between the calculated and the measured values, deviations that were, as he put it, "left behind." Moreover, "a number of sources of experimental error can be pointed out. Thus, e.g., the vacuum was not high enough. In fact, now and then a spark passed between the plates of the condenser, which shows that there was always some ionisation current left between these plates, and that therefore the homogeneity of the electric field was not above doubt. In fine, no definite verdict can be based upon Kaufmann's experiments in favour of either theory" (1931, 274). Are we being wise after the event? Lorentz maintained here that in fact Kaufmann's experiment had not even confirmed Abraham's theory; and then it was rendered erroneous anyhow because of the false assumption that the electric field was homogeneous.[59]

Lorentz did not report on the source of his evidence. The implication is that it was possible to establish the vitiating element in Kaufmann's experiment solely on the basis of a thorough examination of the experiment itself. It should be stressed that Lorentz's objections in 1922 to Kaufmann's experimental results are in fact independent of any knowledge of other experiments. The new experiments which Lorentz cited had not provided him with any critical analysis of Kaufmann's experiment. Though these experiments investigated, like Kaufmann's experiment, the motion of free electrons, they sought theoretically and experimentally different types of demonstration. These new experiments simply did not constitute in any way reruns of Kaufmann's experiment.

In view of the fact that Lorentz considered Kaufmann's experiment crucial with regard to the validity or otherwise of his own theory, it is difficult to understand why Lorentz had not subjected Kaufmann's definitive paper of 1906 to a thorough and rigorous analysis but had rather conceded that Kaufmann had shown that the contraction hypothesis was untenable.[60] I suggest that only when Lorentz was fortified with a battery of experiments that confirmed his theory and therefore contradicted Kaufmann's quantitative results, did he examine afresh the experiment at

59. Cohen, Crowe, and DuMond reported that A. E. Shaw had made an important discovery concerning the experimental technique of applying an electric field in vacuum. His discovery "revealed a source of previously unsuspected systematic error in all such work. It was shown that in vacuum there are formed, on or very close to the metal plates between which electric fields are established, polarization charges which effectively reduce the potential used to accelerate or deflect an electron beam. Furthermore, a charge seems to accumulate on a more or less permanent insulating layer which forms on metal surfaces in vacuum when these are bombarded with electrons. Such effects depend on the material of the metal plates, the residual gas pressure, the cleanliness of the vacuum pumping arrangements (freedom from organic materials), and the intensity of the electron bombardment. Residual gas such as oxygen adhering to the metal may account for the effect in part. It has been found that vaporizing a layer of gold on the interior of the entire vacuum chamber so as to cover the surfaces of all metal parts reduces these effects materially. However, the lesson from Shaw's work is one which every experimental physicist should remember: it is almost impossible to hope to define the potential of an evacuated region by means of electrically conducting metal walls, slits, or what-not with an uncertainty very much smaller than ±1 volt. Many an otherwise well-planned experiment has ended in disappointment because of ignorance of this difficulty" (Cohen, Crowe, and DuMond 1957, 139, 181).

60. For comparison, Planck suspected right from the time of Kaufmann's 1905 and 1906 publications that the experiment was erroneous. As a protagonist of the theory of relativity, Planck subjected Kaufmann's experiment to both logical and physical analyses and endeavored to establish its errors already in 1906 (see above, n. 40; Planck 1906; 1907. See also Zahar 1973, 238 n. 1, 241–42; Pais 1983, 150–51).

stake and eventually pronounce it inconclusive and indeed erroneous. Had it not been for the opposing results of many experiments, Lorentz, it seems, would not have attempted a critical reexamination of Kaufmann's experiments.

For Lorentz, the rejection of experimental results required the attainment of a vantage point from which other experimental results could be seen to replace the results in question.[61] A philosophy of science of the kind Lorentz held, which sees in experiment not only a means of testing the validity of a theory but also a guide that may indicate further developments, simply cannot sustain (unlike, for example, Einstein's philosophy of science) the "vacuum" of knowledge which a rejected experiment may leave behind if no other experimental results replace it. Furthermore, it is a contradiction between experimental results that motivates a critical reexamination of the experiments at issue.

We may thus observe a shift in Lorentz's view. Whereas in 1904 he had suggested that Kaufmann's result could be interpreted differently, implying thereby that an error of interpretation might have occurred, in 1922 he maintained that Kaufmann had made mistakes in his analysis of the data and, more importantly, that he had assumed erroneously that the electric field was homogeneous. Thus, at one time the experimental error is characterized as pertaining to problems of interpretation, and at another time as an error concerning the actual setup and its working. We may note further that the shift took place in two phases: in 1909 the experimental result was accepted, and in 1915 it was rejected with no explanation, merely on the basis of counter experimental results.

7. Conclusion

Poincaré accepted Kaufmann's experimental result though he warned against a possible error. He suggested that an error might have occurred which could pertain to the working of the apparatus. He, however, did not suspend his judgment and assented to the conclusion that classical theories of the electron had been confirmed; hence, the relativity principle could not serve as a solid foundation for physics. Einstein did not find any error; he suspected a systematic error but could not find its source. Indeed, he praised Kaufmann's expertise but all the same rejected the result outright on methodological grounds. He could not give up the relativity principle. Lorentz, on his part, vacillated until some 15 years later he finally

61. Holton also appears to hold this view. Remarking that 10 years had had to elapse before it was fully realized, through the experimental work of Guye and Lavanchy in 1916, that "Kaufmann's equipment was inadequate [*sic*]," Holton implied precisely this point (Holton 1975, 235).

rejected Kaufmann's experimental result.[62] Initially, he had thought that the interpretation was wrong; then he found the experimental result correct and relinquished his own hypothesis; and later he argued that the apparatus had not functioned as had been assumed. Moreover, in his opinion, mistakes were made in the reduction of the data.

We have seen then that the failure of Kaufmann's experiment assumed at the time different characteristics: error of interpretation, mistakes in reducing the data, and malfunctioning of the apparatus (i.e., the apparatus did not function as it had been assumed it would). Finally, it was claimed that there had been no error at all and equally maintained that a systematic error had occurred whose source was unknown. (On different types of experimental errors and their characteristics, see Hon 1989).

I have argued that these different identifications of the error in Kaufmann's experiment reflect the specific philosophical outlook and methodological disposition held by each of the protagonists. Thus, the identification of an error appears to depend on the specific context—philosophical as well as methodological—within which the experiment is examined.

A consensus eventually emerged that Kaufmann's experimental conclusion was false, and the experiment was pronounced erroneous, but then there was no agreement as to the cause of the failure. In such erroneous cases the alleged empirical findings simply fade away, and there appears to be no need for the concerned scientists to reach an agreement as to the characteristics of the error. However, it is precisely with such cases that the historian and philosopher of science may gain an insight into the practice of science.

The Kaufmann case study demonstrates that understanding failure is as knowledge dependent as understanding success. "Knowledge and error flow from the same mental sources," observed Mach, and he continued that "only success can tell the one from the other" (1976, 84). But in view of the present case, we may go further and conclude that, in their mutual relation, knowledge and error not only flow from the same sources but also determine one another.

Acknowledgments
I wish to express my debt to Jed Buchwald, Martin Carrier, Jim Cushing, Allan Franklin, Yves Gingras, Ian Hacking, Heinz Post, Saul Smilansky, and Gereon Wolters. I wish further to acknowledge the generous assistance of the Alexander von Humboldt Stiftung and the Zentrum Philosophie und Wissenschaftstheorie, Universität Konstanz.

62. It is worth noting, from a sociological point of view, that Einstein's reputation was at that time newly minted and the relativity theory was one of its central pillars. By contrast, Lorentz's reputation had already been secured, whatever the outcome of Kaufmann's experiment. He had after all founded particle-based electrodynamics.

NINE

SCIENTIFIC CONCLUSIONS AND PHILOSOPHICAL ARGUMENTS: AN INESSENTIAL TENSION

Margaret Morrison

1. INTRODUCTION

LARGELY AS A RESULT OF KUHN philosophers of science have turned increasingly to detailed case studies in order to validate their epistemological theories. Attention to foundational problems, like the nature of truth and induction, has been superseded by interest in the practice of science (past and present). Nevertheless, philosophers seem reluctant to give up their traditional epistemological role. Very often case studies are used to support philosophical conclusions arrived at either by using abstract methods like logical analysis or by adhering to traditional principles of empiricism or rationalism. But even those for whom practice is seen as a crucial *starting point* in philosophy of science, the aim is still to isolate criteria (like manipulability or causal interaction) to serve as necessary and sufficient conditions in the legitimation of belief.[1] As with any attempt to set out conditions or rules, there are usually counterexamples that undermine their descriptive accuracy and thereby call into question their normative force; and the criteria above are no exception. The result

Acknowledgments

I would like to thank Jed Buchwald and Paul Forster for discussions about this material and helpful suggestions on an earlier draft. Thanks also go to Francis Everitt, James Maffie, and David Stump, who read and commented on a version of the paper that was presented at a conference on contextualism and disunity in science held at Stanford in April 1991, as well as to Henk de Regt for his careful comments, questions, and criticisms. Support of research by the Social Sciences and Humanities Research Council of Canada and the University of Toronto Connaught Fund is gratefully acknowledged.

1. When I refer to foundationalism in this context I am also including empirical claims that are supposedly open to refutation and refinement; hence I am not equating foundationalism with absolute certainty. I use the word only to indicate philosophical views that attempt to establish epistemological priority for some group of phenomena or theoretical claims.

is a reinforcement of the idea that there must be a gap between philosophical norms and those that emerge in scientific practice.

As a way of investigating what the role of philosophy of science should be, I propose to examine the relationship between philosophical and scientific norms in a number of different cases. I begin by examining the debate between the atomists and the anti-atomists in the late-nineteenth and early-twentieth centuries and how some present-day philosophers and historians have analyzed that debate. Mach's and Duhem's reluctance to accept the kinetic theory despite its overwhelming empirical success represents a paradigm case of what I refer to as externally motivated epistemology of science—one that is static in the face of changing evidential relationships and experimental techniques.[2] For them, philosophical considerations concerning the status of explanation and the importance of observation (narrowly construed) determined what would constitute legitimate science. These factors functioned as first principles or a kind of metamethodology. Equally interesting are the recent accounts of the atomism debate by historians and philosophers; some cite what they consider to be purely philosophical reasons for the beliefs of, say, Mach, while others focus on scientific difficulties as the motivating factor in the disputes over the merits of the atomic theory. However, despite lengthy analyses no attempt is made to see how and to what extent scientific and philosophical presuppositions interact.

I will use Hertz's work on electromagnetism to illustrate the way in which scientific work can embody both philosophical and scientific norms in a way that allows for the possibility of genuine interaction. Although Hertz voiced considerable reservation about aspects of his experiments that were uncritically accepted by the British and about then current views regarding the ether, his skepticism was in principle resolvable, unlike that of Mach and Duhem. Hertz utilized specific criteria based on philosophical beliefs about the nature of good science—conditions that enable one

2. Here I want to call attention to my use of "internal" and "external." In contrast to Carnap's use of the term, I do not mean to suggest that questions external to the practice of science are somehow meaningless, nor do I want to equate my usage with the internal versus external approach to history. I use the word "external" only to characterize criteria that can supposedly function in some neutral way. They stand outside practice in the sense that they are immune to the kind of criticism that arises from within the practice. One example is Hacking's (1983) condition that we should believe in entities that can be manipulated. There is a sense in which this condition functions as an external epistemological principle because it excludes large portions of current science, yet philosophically this does not weaken its normative force. In his 1989 paper Hacking (1989a) appeals to manipulability as a key feature in his argument against realism about extragalactic phenomena like quasars and gravitational lenses. A similar example is van Fraassen's emphasis on observability as a condition for realism.

to judge the merit of what he called "scientific representations." However, the way these rules apply is not fixed and can vary from case to case. In that sense the criteria are nonfoundational, allowing Hertz to adopt a critical stance without recourse to first principles. Finally, I discuss the impact of these accounts on contemporary debates.

My point is both philosophical and historiographical. If philosophical principles can function in a dynamic way, changing in response to practical needs, then the temptation to give explicit priority to philosophical over scientific concerns (or vice versa) or to reduce one to the other will no longer exist. I claim that scientific norms and philosophical presuppositions should be considered distinct, yet mutually informing, aspects of a holistic model of scientific practice. Separating the philosophical from the scientific need not commit one to a set of immutable first principles; philosophy can function in a methodological capacity without advocating a set of fixed methods. Moreover, there is no way to construct a set of static philosophical principles detailed enough to decide cases yet general enough to have a broad range of application. Good philosophy should be like good science, capable of accommodating changes in the domain of empirical inquiry. An interactive view of this kind retains a role for philosophy by emphasizing its reconstructive character.

2. SCIENCE AND METAPHYSICS

The controversy over atomism in the latter part of the nineteenth century highlights the ways in which scientific and philosophical constraints about what counts as an acceptable theory played a role in the debate about the relative merits of phenomenological thermodynamics and the kinetic theory. Although some argue that philosophical presuppositions (which I will discuss below) were important, Clarke (1976) and Gardner (1979) claim that *scientific* considerations, like the lack of predictive power, were responsible for the weak support for the kinetic theory and that it was Einstein's and, later, Perrin's work on Brownian motion that finally established its viability. Clarke, in an effort to explain this case using Lakatos's ideas about the progressiveness of research programs, claims that in the last decade of the nineteenth century the kinetic program had "run out of steam." Boltzmann had proposed a solution to the specific heats anomaly that violated one of the heuristic principles of the program—namely, that the collisions which give rise to molecular vibrations should be subject to the laws of mechanics. In Boltzmann's molecular model the degrees of vibratory freedom were independent of the thermal capacity of the gas.[3] In

3. The problem was that each degree of freedom (of which there might be infinitely many) required an equal portion of the total energy of the system.

addition to this problem there was no general solution for the Boltzmann transport equation; hence the calculation of the transport coefficients required arbitrary approximations. Other molecular hypotheses had been put forward to account for the specific heats ratio but all were ad hoc, designed merely to account for the observed ratio; consequently they lacked independent support and produced no novel consequences.[4] Other predictions concerning thermal and density fluctuations within a gas in thermodynamic equilibrium were not empirically confirmed. Furthermore, one of the most serious difficulties was that the statistical interpretation of the entropy theorem contradicted the phenomenological second law, which had been enormously successful in predicting novel facts.

Because of the phenomenological nature of thermodynamics there was no *systematic* way to extend its laws, yet its empirical progressiveness in the 1870s made it a reasonable option for physicists, like Planck and Ostwald, who were interested in applying it in new domains. Not only had thermodynamics provided a derivation of the gaseous dissociation laws, but these laws could not be derived from the kinetic theory without the aid of ad hoc hypotheses that could not be independently supported (see Clarke 1976, 70–72). Similarly, thermodynamics solved the problem of chemical affinity (how to predict the direction of a particular chemical reaction and determine the strengths of the affinities involved) while the molecular hypotheses provided only post hoc explanations which added nothing to the thermodynamic results (Clarke 1976, 72–74). Despite this, Clarke claims that it was equally legitimate to follow the path of the kinetic theory since its strong heuristic and its past success made it reasonable to think that its rather stagnant character in the 1890s might be temporary.[5] The early kinetic theory of the 1850s and 1860s was progressive

4. One particular hypothesis that Clarke cites is the FitzGerald-Watson hypothesis that postulated an equipartition time lag wherein the heat added to a gas would take so long to be equally distributed that the predicted ratio based on the equipartition theorem would never be observed.

5. The heuristic of the kinetic theory consisted of four methodological rules that required investigators to (1) make specific assumptions about the nature of elementary individuals and their degrees of freedom in order to guarantee the applicability of the laws of mechanics; (2) treat the chaotic aggregate motion in such a way that for every property of that motion a mean value, determined by the distribution of that property among the molecules, is assumed to exist; (3) eliminate or weaken simplifying assumptions once (1) and (2) have been introduced in order to simulate conditions obtaining in a real gas; (4) use these specific assumptions to investigate the internal properties of gases (e.g., viscosity) and derive macroscopic properties as limiting cases. These rules provided suggestions on how to fill in, elaborate, and draw consequences from theories in the kinetic program. As such, the heuristic was considered a research policy or directive with details on how one might carry out the program in an efficient manner (Clarke 1976, 45–46). For more on how a heuristic functions, see Lakatos 1970.

in that one could derive laws like Dalton's law of partial pressures and Gay-Lussac's law of equivalent volumes without the theory being specifically designed to accommodate them. It also gave a novel qualitative explanation of evaporation and equilibrium between liquid and gaseous phases. The introduction of the mean free path technique facilitated an investigation of the internal properties of gases, which in turn resulted in the Maxwell law for velocities, a law that successfully predicted other novel phenomena.[6] By 1905 the kinetic theory was beginning to regain its empirical progressiveness. Einstein's work showed that the existence and magnitude of Brownian motion could be directly predicted from the kinetic model of thermal equilibrium, and subsequent experimental confirmation by Perrin in 1910, which included a specific determination of Avogadro's number, restored faith in the theory. These results, in effect, confirmed the statistical nature of the second law as documented by the kinetic theory and established the theory as the foundation for empirical progress in the field. And, according to Clarke's Lakatosian model, it is this empirical progress which ultimately explains the success of the kinetic theory.

Although Gardner does not subscribe to the Lakatos model of research programs, he does emphasize the importance of novel prediction and argues that it was "scientific" reasons that delayed but ultimately resulted in the acceptance of the kinetic theory over thermodynamics. In addition to the theory's predictive power and its ability to specify quantities like Avogadro's number, the work on Brownian motion was crucial to the success of mechanical explanations, replacing those suggested by the energeticist program, which upheld the nonstatistical nature of the second law. Although energeticists like Ostwald had pointed to the empirical anomalies associated with the mechanical approach, they also had epistemological reasons for preferring energeticism. Chemical dynamics had rendered the atomic hypothesis unnecessary, and instead of genuine investigation into problems, Ostwald felt that the kinetic theorists tended to cover up difficulties by using "arbitrary assumptions about atomic positions, motions and vibrations" (Ostwald 1907; see also Nye 1972, 44 n. 71; quoted in Clarke 1976, 88–89). In addition to the lack of direct proof for the existence of atoms, they saw energy and the conservation law as *the* basic principles governing all phenomena. Not only did they see "energy" as free of hypothesis (thereby realizing an ultimate scientific ideal), but it is through the transmission of energy that, supposedly, we obtain all of our knowledge. Because everything we know is transmitted to us through our sense organs and because transmission of energy is re-

6. Two such phenomena were the existence of a thermal conductivity coefficient for gases and the viscosity independent of the density.

quired for our sense organs to function, all phenomena should be interpreted in terms of energy (see Ostwald 1927; Gardner 1979).

Although Ostwald shared some of Mach's epistemological concerns (especially his antimechanistic attitudes), he was eventually converted to the atomic theory by its empirical merit, especially its predictive power. He specifically cites the isolation and counting of gas ions as well as the agreement of Brownian motion with the predictions of the kinetic theory and Perrin's ensuring confirmation in 1910.[7] Whereas this was undoubtedly the turning point for some, the experimental results proved insufficient to convince epistemological conservatives like Mach and Duhem. Both Clarke and Gardner suggest that Duhem's reluctance to sanction the atomic theory stems from scientific, rather than philosophical, concerns, specifically the failure of the kinetic theory to make true predictions. Although this may have been the case prior to 1910, it does not explain Duhem's position subsequent to Perrin's work. Similarly, in the case of Mach, Clarke claims that he was critical of the heuristic and empirical degeneration of the kinetic program. In fact, Clarke points to this degeneration together with the empirical progress of thermodynamics as the catalyst in the rise of scientific positivism (1976, 45 n. 222). But again, this view fails to explain why Planck and Ostwald in particular were eventually persuaded (for essentially scientific reasons) to accept the theory while Duhem and Mach remained unconvinced. The attempt to recast this event in purely scientific terms by emphasizing predictive and heuristic power ignores the way that philosophical presuppositions functioned, especially for Mach and Duhem, in determining what could be considered legitimate science. Consequently, it not only sacrifices historical accuracy but, by ignoring the role of philosophical presuppositions, fails to explain why Duhem and Mach found the later scientific evidence uncompelling.

Brush 1976 and more recently Nyhoff 1988 have offered an alternative account. They argue that the negative reaction to the kinetic theory was based primarily on philosophical, rather than scientific, objections. Evidence for this position includes Duhem's criticism of the mechanical models associated with the kinetic theory, models that encumbered thermodynamics "like a parasitic growth . . . fastened . . . on a tree already full of life" ([1906] 1954, 95). In addition, Nyhoff (1988, 87) remarks that Mach's opposition to explanatory mechanical theories became firmly established when the kinetic theory was making good progress. He claims that Mach's criticism of atomism was philosophically based, traceable to his skepticism about metaphysics and mechanical explanations that appeal to a "reality" behind the phenomena. Hence, it was an antimeta-

7. These remarks were made by Ostwald in the fourth edition of *Grundriss der allgemeinen Chemie,* written in 1909, and are quoted in Brush 1976.

physical, antimechanical position that motivated his reaction rather than particular scientific difficulties thought to be responsible for the rise of positivism.

Although I am sympathetic to the view of the role of philosophical considerations as outlined by Brush and Nyhoff, they fail to emphasize the different *ways* in which the epistemological concerns influenced methodological presuppositions, neither of which can be separated from the scientific context in which these commitments were formed. Both the Clarke/Gardner and the Brush/Nyhoff accounts fail to describe how philosophical and scientific considerations interact. By segregating the philosophical and scientific reasons, Brush and Nyhoff ultimately lose sight of the way that these sets of concerns support each other. Yet, we must also caution against collapsing the two in the way that Clarke and Gardner suggest. Gardner argues that Brush's position presupposes an "untenable distinction between philosophical (e.g. methodological) principles and empirical considerations" (1979, 14) and claims that philosophical principles about what counts as a legitimate explanation or hypothesis are often based at least in part on empirical considerations. Although this is certainly true, it in no way validates the *reduction* of philosophical presuppositions to empirical considerations in the way the "scientific" thesis implies.

A closer look at the different attitudes toward the kinetic theory reveals the importance of distinguishing the different ways in which science and philosophy form an interactive whole. Consider, for example, the *scientific* use of atomism and the kinetic theory exemplified by Mach, Ostwald, and Planck. While it is tempting to classify into two groups those that were convinced by the new evidence and those that were not, in Planck's case it is important to point out that the transition to the kinetic theory was not simply the result of Perrin's experimental work. Before Ostwald derived the laws of constant proportions, multiple proportions, and combining weights in 1904, Ostwald viewed atomism as a convenient picture of the relations of weights and volumes; and although it was unacceptable as a literal account it did possess heuristic value (Ostwald 1900). Unlike the energeticists, Planck was a strong supporter of mechanics but he nevertheless shared their skepticism about the kinetic theory and atomism in general, not only because of the difficulties it encountered in explaining the analogy between gases and liquids but primarily because atomism contradicted his absolutist interpretation of the second law.

However, Kuhn (1978) has suggested that Planck's opposition was primarily the result of strategic considerations. Research on gas theory was not wrong in principle; rather, it was simply not a productive strategy given its difficulties and especially in light of the enormous success of thermodynamics. But, Kuhn maintains, a demonstrated need or achievement on the part of the kinetic theory would have been sufficient to change Planck's mind; and indeed this need arose in his work on blackbody radiation, in which he made use of Boltzmann's statistical methods. Consequently, Planck's use of the kinetic theory was much more than a convenient way of portraying the phenomena; the theory was a productive tool enabling him to achieve results that were unavailable using the methods of thermodynamics. Although he may have had some residual antagonism toward atomistic reasoning in the early 1900s, by 1910 he had been converted to the kinetic molecular theory with its probabilistic account of entropy. He was completely convinced of the reality of atoms and the validity of their measurements, regardless of whether they could actually be seen. So, Planck had come full circle from thinking that the kinetic theory should not be considered as even a heuristic device to a strong supporter of atoms and statistical methods.

In the early stages of his career Mach was also quite willing to *use* the kinetic theory as a *memoria technica* or formula, but his motivation and attitude were significantly different from Planck's. Planck's distrust of atomism as a research strategy was based on scientific reasons, whereas Mach's own later rejection was founded on methodological assumptions about the proper aims of science, which were in turn motivated by an epistemological account of what could be properly called a "fact." Initially, Mach saw the kinetic theory as a mathematical *model* for "facilitating the mental reproduction of facts," playing a role in physics similar to certain auxiliary concepts in mathematics ([1883] 1942, 589).[8] But even his early support of atomism was directly linked and limited to its heuristic value. As he writes in the *Compendium* in 1863: "I have allowed the atomic theory to take first place everywhere [in this work], indeed, not in the belief that it represents the last word, or no longer requires additional support, but because it brings the phenomena into a simple and visual relationship. One may accept the atomic theory, if it is permissible to express oneself in this way, as a formula which already has led to many results" (iv). A fundamental premise of his view was the claim that the results of the atomic theory would be translatable into some future the-

8. For example, we represent vibrations by the harmonic formula, falls by squares of times, etc., yet no one assumes that vibrations *themselves* have anything to do with circular functions. We have merely found that relations between certain empirical quantities are similar to relations between mathematical functions, where the latter are employed as a means of supplementing experience. Mach thought the same was true of the conception of electricity that postulates the existence of attracting and repelling fluids moving on the surface of conductors; although fitting electrical phenomena, these "mental expedients" have nothing to do with the phenomenon itself. In other words, the analogies do not imply any physical similarities. The same is true for the case of light waves, caloric, molecules, atoms, the law of refraction and energy. For more details, see Mach 1892, 199.

ory, just as formulae in polar coordinates can be expressed in parallel coordinates.

Although Mach's opposition to the kinetic theory began while it was still thought to be empirically successful, a brief look at his methodological views reveals clues to his skeptical reaction. For Mach the goal of science was to establish, in an economical way, laws that expressed functional relationships between phenomena. Once these connections were found, the theory or model from which they were derived ceased to be useful, unless of course it established further associations. "The electric fluid is a thing of thought . . . [an] element of physical science contrived for very special purposes. They are discarded . . . when the interconnection of ABC . . . has become familiar; for this last is the very gist of the affair" (Mach 1890, 396).

Moreover, atomism was seen as facilitating the reduction of all natural phenomena to mechanics, a strategy that Mach ardently opposed on the grounds that it simply made processes more familiar but not more intelligible. And this misconception was achieved at the expense of epistemological legitimacy since there was no experimental verification for the atoms themselves and no proof that all action and properties were mechanical. "The ultimate intelligibilities on which science is founded must be facts, or, if they are hypotheses, must be capable of becoming facts" (Mach [1872] 1941, 5).

Hence, it is not surprising that once results of the atomic theory could be translated into the language of thermodynamics Mach no longer saw it as a productive tool; it provided nothing more than a phenomenological approach and, furthermore, directly contradicted his own methodological principles. Unlike Planck, Mach's use of atomism would never and could never evolve into ontological commitment. Although the kinetic theory regained its priority after Perrin's work, the epistemological and methodological price for Mach was too high; and given its recent history there was no reason to assume that these new results would not be translatable into some acceptable theory in the future. Mach's limited support for the kinetic theory was, like Planck's, rooted in its scientific advantages, but philosophical differences about its ultimate interpretation and legitimacy made its use as a tool a very different activity in each case.

The nature of theorizing was also the primary factor in Duhem's rejection of atomism. This, together with his skepticism about the possibility of fully *explaining* phenomena, prevented him from making the transition to the kinetic theory, regardless of its newly established predictive power. "A physical theory is not an explanation. It is a system of mathematical propositions, which aim to represent as simply, as completely, and as exactly as possible a set of experimental laws" ([1906] 1954, 19). "To explain is to strip reality of appearances covering it like a veil, in order to see the bare reality itself" (7). And, as Duhem notes, for this kind of inquiry to be possible or even make sense we must claim to know with certainty that there is a hidden reality behind our perceptions. But since science cannot render this reality accessible to the senses, theories can only be *hypothetical* explanations.

Scientifically the views of Mach and Duhem have not prevailed, at least with respect to the kinetic theory and the existence of atoms. Yet in recent philosophical work one can find a great deal of support for *certain aspects* of the conservative approach to theorizing that they prescribed.[9] This raises a number of questions about the split between scientific and philosophical reasons for accepting experimental evidence and the extent to which these considerations influence one another, if at all. For both Duhem and Mach the philosophical problems that plagued the kinetic theory could not properly be countered simply by enumerating the reasons for its empirical success after 1910. Their objections required a different kind of evaluation, using criteria that were in some sense external to the debate about the particular merits of atomism and the kinetic theory. This contrast results in two different but related kinds of questions: straightforward scientific questions about the merit of the theory under investigation and the more philosophically oriented questions regarding the evaluation of the standards that we use to judge these theories. The first group involves considerations about whether a particular theory or piece of evidence has met the requisite conditions for approval specified by the scientific community, and the second involves an estimate of the appropriateness of the conditions used in making these judgments.

In the former case, usually a well-established set of conditions recognized by the scientific community is brought to bear (criteria such as experimental agreement, coherence with well-established theories, and successful prediction). The scientific standards that determine when a theory or piece of experimental evidence should be accepted include the credibility of the experiments themselves and the relationships between the old and new theories and the data. When results are called into question, the experiment can be repeated and refined or hypotheses can be adjusted until an acceptable solution is reached or until new sets of standards are judged to be appropriate to the case.[10] When dealing with the philosophi-

9. See, e.g., van Fraassen 1980, Cartwright 1983, and Hacking 1983, all of whom adopt different attitudes about belief in theoretical entities and theories, each of which resembles Duhemian or Machian arguments in one form or another.

10. Some might refer to this shift to new standards as theoretical, referring to a purely scientific context where philosophical considerations have little effect on decision making; hence, the kinds of evaluative methods used in the adjudication of experimental evidence are seen as determined by purely scientific concerns with no philosophical input. What I am trying to argue, however, is that the proper way to characterize the evolution of the scientific

cal considerations, the concerns center on the reliability of our standards: whether they are suitable for the task and when it is legitimate to reevaluate or replace them. One example is whether the measure of what counts as legitimate evidence for belief in a particular hypothesis is reasonable given the kind of evidence that is practically available. Is it reasonable, for instance, to demand that observation be a constraint on belief when most of the entities postulated by our theories cannot be observed in anything like the ordinary sense of "observation"? When particular standards run counter to our best available scientific evidence for a theory or when these standards render inexplicable what a majority of the scientific community agrees upon, then there is a sense in which their normative force becomes eroded.

An example of previously held standards giving way in the face of new theoretical and experimental demands is the introduction of renormalization into quantum electrodynamics. When renormalization of mass was proposed as a way around the difficulties with the Dirac electron theory in 1947, it clearly violated certain canons of acceptability in mathematical physics.[11] However, the results that precipitated that correction to the Di-

---

process is to see philosophical and scientific concerns as mutually reinforcing. Although some decisions are made for what are thought to be purely scientific reasons, it is important to note how philosophical considerations enter even such basic cases as determining when an experimental result is sufficient to confirm a hypothesis. My discussion here is somewhat vague, but below I give a more detailed account of how philosophical or evaluative standards function within scientific discourse.

11. Although quantum field theory (QFT) enjoyed much success in the 1930s, the theory suffered from a fundamental physical inconsistency. The emission of virtual photons and their reabsorption by the same electron produce a self-energy which shows up as an observable energy-level shift. The problem was that QFT predicted an infinite energy shift. Because an infinite number of electromagnetic interactions between electrons and virtual photons was possible, the probability of the interaction was infinite as well since it was determined by the sum of the contributions of all the possible events. The self-energy of the electron involves a sum over the number of ways the momentum can be carved up, as it were. This infinite self-energy appeared when the electron was both moving in an orbit and at rest. But according to special relativity, the energy of a particle at rest is related to its mass by $E = mc^2$. This suggests that the electron's mass would have to be identified with both its "bare" mass, which appears in the equations for the electron field, and its infinite "self"-mass, produced by the interaction with its virtual photon.

The observable effects calculated in QFT depend on e and m only as they are measured experimentally; and what is actually measured is not the "bare" mass of the point particle but the properties of the electron bound together with its cloud of virtual particles. Hence only the net mass and charge (the measured quantities) are expected to be finite. Very basically if you have a bound (1) and a free (2) electron, the observable mass $m_0 = m + \delta m$, where m is the bare mass and δm is the self-mass. The δm's are given by divergent integrals but can be substituted to yield $m_0^{(1)} - m_0^{(2)} = \delta m^{(1)} - \delta m^{(2)}$. If the relationship is properly calculated, we are left with a finite residue. Because the second electron is free, the difference is the contribution of the self-mass to the binding energy of the first electron. This measurable rac equation (the Lamb-Retherford experiment on the fine structure of hydrogen) had been established unquestionably, and in order to sustain the unprecedented predictive power of quantum electrodynamics, physicists sacrificed the mathematical consistency of the theory.[12] Although it was for scientific reasons, the decision to accept renormalization as a part of quantum electrodynamics also had philosophical implications insofar as it violated what had been a basic constraint on theorizing in favor of instrumental value. In this case the standards used in appraising scientific activity evolved as a result of practical concerns which effected a change in the philosophical presuppositions regarding the *kinds* of methods, theories, and experiments that were deemed acceptable given certain ends. Not only is the inability to provide a physically motivated account of mass renormalization tolerated in the light of the overwhelming success of the theory, but renormalizability has become a requirement of quantum field theories. Here the scientific and philosophical constraints interact in a way that creates a cooperative context that balances practical demands and methodological goals.

In contrast, consider the objections raised by Duhem and Mach about the kinetic theory. No amount of scientific evidence would have been sufficient to resolve their difficulties since their skepticism was motivated by concerns that were in some sense external to the practice of science. They called into question the very *nature* of science as exemplified by atomism and the kinetic theory. Both believed that the kinetic theory embodied an approach to scientific inquiry that utilized illegitimate metaphysics, and both believed that its proponents failed to realize the constraints on our ability to *know* the natural world with any degree of certainty. However, Duhem and Mach refused to consider the possibility of new forms of knowledge that seemed to be emerging at the time. Admittedly, Mach's initial scientific reasons for accepting atomism gave way to scientific reasons for rejecting it; he felt that atoms had been specially devised "for the purpose in view" ([1883] 1942, 588) and had properties that contradicted attributes observed in bodies.[13] In addition, there was the problem of re-

---

quantity can then be compared with experimental results like the Lamb shift. In order for the infinities to be eliminated, it is necessary that they arise in a limited number of ways (as corrections to mass, charge, etc.). What renormalization does is eliminate infinities by absorbing them into a redefinition of physical parameters.

12. For a discussion of the Lamb-Retherford experiment, see Morrison 1986.

13. Although this was more a methodological objection stemming from Newton's third rule, it was a problem that could be addressed from within the boundaries of science by providing a more concrete representation of atoms. In addition, atoms were also ad hoc in the sense that inelastic hard bodies were thought to produce an infinite force on impact. Obviously, this effect was not taking place; hence, Clausius and Maxwell defined atoms as perfectly elastic spheres. But, as Maxwell himself was aware, this endowed the atom with

lating the spatial concept of the atom to the spectra of chemical elements, something he had attempted while still a supporter of the heuristic value of atomism (see Hiebert 1970, 102). But this is only part of the story; without these difficulties atomism would still never attain the philosophical legitimacy required by Mach's view of science. Even when presented with spinthariscope scintillations, which some took to be direct evidence of atoms, Mach was unwilling to equate what he saw as microsensations with atoms. The latter could only be properly understood in a space of more than three dimensions, whereas sensations existed only in three dimensions.

Scientific theories, for both Mach and Duhem, involved organizing phenomena in an economical way; for Duhem, the goal was classification of laws, and for Mach, it was arrangements of facts ascertained by the senses in accordance with principles of economy of thought. This organization yielded knowledge of relations or the dependence of phenomena on one another, which was the aim of natural science (Mach [1872] 1941, 9). Although the production of new phenomena was considered important, there were certain constraints on how that knowledge was to be discovered and expressed. Consequently, thermodynamics was seen by Duhem as more scientific because its methods of observation and classification were superior to those of atomism. By contrast, the kinetic theory attempted to *explain* phenomena by postulating atoms, a program which for Mach and Duhem involved not only illegitimate methods but an unrealizable goal. For them philosophical considerations determined what science was and the way it should be practiced, and they exhibited little or no sensitivity to the changes taking place at the level of physical theory. Ironically, by having epistemological and methodological constraints that are in some sense indifferent to practice, philosophy becomes isolated in a way that threatens its importance as a crucial feature of scientific debate. As a result, philosophical disputes about the relationship between theories and evidence are sometimes seen as abstract and irrelevant to what occurs *internally* at the level of practice. This is particularly so in cases where adherence to philosophical principles inhibits the assessment of certain kinds of data because they do not conform to ontological standards presupposed by a particular philosophical tradition. This tension between the internal and external domains was nicely captured by Bridgman, who in 1927 claimed that the atom is a "construct [whose] existence [is] entirely inferential," yet we are "as convinced of its physical reality as of our hands and feet" (1927, 59). In other words, we cannot claim to have *direct*

---

the very property that required explanation in aggregate bodies, the property for whose explanation the atom had originally been postulated. See Nye 1976, 254.

knowledge that atoms exist even though we firmly believe in them. The former represents the epistemological position while the latter captures what is sometimes called the commonsense scientific attitude.

Up to this point we have seen three different characterizations of the relationship between science and philosophy. In their reading of history Clarke and Gardner neglected the importance of philosophical presupposition in the debate over the atomic theory by reducing philosophical issues to scientific ones. Brush and Nyhoff emphasized the importance of philosophical constraints, but by segregating philosophy and science, especially in the case of Mach and Duhem, we fail to understand the ways in which philosophy and science relate to each other. From their claim that their concerns are solely philosophical, one gets the impression that scientific beliefs are simply reducible to philosophical beliefs, with no clear demarcation between the two. What I have tried to highlight is the way in which scientific beliefs are influenced by philosophical presuppositions, without seeing the process simply as an eliminative reduction. Indeed, Mach's and Duhem's philosophical beliefs determined what they would consider legitimate science but they also acknowledged the kinetic theory as a productive tool, a judgment that was made for purely scientific reasons. Hence, the domains interact in ways that escape us when we focus on a single set of concerns. Properly construed, the work of Mach and Duhem illustrates an interplay between science and philosophy, yet ironically, it is a relationship that exemplifies, at one level, a detachment of philosophical constraints that renders them exempt from challenge on practical scientific grounds. Although they themselves endowed philosophical presuppositions with a form of epistemic authority that in some sense segregated science and philosophy, we must also remember that they were, after all, working scientists whose scientific beliefs were formed in the context of that practice.[14]

Undoubtedly their scientific methodology was closely tied to philosophical principles, yet it is important to recognize that the motivation behind some of the specific objections to the kinetic theory can be properly characterized only as scientific concerns. In addition to Mach's scientific objections to the kinetic theory discussed above, Duhem's preference for thermodynamics also embodied a scientific dimension. Not only does the abstract nature of a phenomenological theory like thermodynamics bring

14. Although the *relationship* was asymmetrical, this does not mean that one can simply reduce science to philosophy; rather, philosophy was understood as having implications for science but not vice versa. Consequently, to claim that the reasons for rejecting the kinetic theory was solely philosophical is technically not correct since scientific concerns did play a role. What I want to deny is the reductive claim that decisions were either purely philosophical or exclusively scientific.

economy to the theory, but the algebraic correspondence that can exist between abstract theories (e.g., thermodynamics and electrostatics) constitutes a method of discovery ([1906] 1954, 96–97). More specifically, thermodynamics furnished the relation between theoretical and practical isotherms as well as equations for osmotic pressure, whereas the isothermal equation and the law of corresponding states derived from the kinetic theory proved inapplicable in experimental contexts. According to Duhem, thermodynamics had clearly provided more fruitful results than the kinetic theory and the mechanical models associated with it. Undoubtedly he had philosophical reasons for preferring certain kinds of theories (e.g., those that do not postulate a hidden reality), but in the case of thermodynamics scientific consequences involving the resourcefulness of the theory were also crucial. Although the experimental results of 1910 provided strong support for the kinetic theory, they did not significantly undermine thermodynamics to the point where the decision to abandon it was scientifically compelling.[15] Hence, it is possible to view both Mach's and Duhem's choice as being philosophically and scientifically motivated.

Despite the scientific factors that figured in the decision, one can still criticize the intractability of the epistemological and methodological criteria invoked by Mach and Duhem. Planck, on the other hand, seemed to display little in the way of critical philosophical awareness in his evaluation of atomism and was content to embrace it so long as it provided techniques that thermodynamics could not deliver. Ostwald was clearly more circumspect, basing his decision on the rather obvious empirical superiority of the kinetic theory and forming his methodological and philosophical criteria partly in response to considerations that arose from within the practice itself.[16] The important question that arises in this context is whether and to what extent it is possible for philosophical and scientific worries to be addressed in a way that preserves a normative element, or whether a praxis-based methodology results in little more than descriptive accuracy. That is, can the apparent tension between the theory (philosophy) and practice of science be resolved in a meaningful way?

My goal is to retain a role for philosophy that highlights its importance without setting it apart in the sense of having its concerns remain static in the face of changing practice. Unlike Brush and Nyhoff, who sees Mach's and Duhem's difficulties with the kinetic theory as ultimately reducible to a set of philosophical concerns, I want to acknowledge the importance of their philosophical beliefs while emphasizing how these beliefs

15. I would like to thank Jed Buchwald for drawing my attention to this point.

16. William James considered Ostwald a fellow pragmatist, citing with approval Ostwald's remark that different views must exhibit some practical difference if they are to have any difference in meaning (James 1907).

related to the scientific aspects of their overall approach. Instead of attempting to show how scientific or philosophical beliefs are ultimately reducible to the other, I stress the interaction between the two, even when that interaction is somewhat limited in scope. Consequently, my point is a normative one that has both historiographical and philosophical implications. Not only do the "scientific" and "philosophical" versions of the history of atomism fail to provide a full understanding of how these two domains influenced each other, but the relationship between science and philosophy exemplified in the work of Mach and Duhem (as well as in some current literature) prohibits the kind of cooperation that renders philosophy and science an interactive process.

My approach however should not be equated with an endorsement of traditional scientific realism, which is often guilty of rather weak philosophical arguments (e.g., inference to the best explanation) for the acceptance of the conclusions of contemporary theories.[17] Traditional scientific realism (as well as some of its modern versions) attempts to impose philosophical and logical conditions for truth and knowledge on a practice (science) that is ill equipped to meet these standards. Both Duhem and Mach recognized that this kind of realism was the wrong model for science, yet in seeking to defeat it they extended their conservatism in a way that set their vision against a newly emerging scientific culture. What I want to urge is a compromise, the beginnings of which can be seen within the framework of some nineteenth-century work on electromagnetism, particularly that of Hertz. Here we find a form of methodological conservatism that is not only well founded on scientific grounds but is, at the same time, sensitive to the possibility of overcoming skepticism when future evidence permits. In order to understand how Hertz's philosophical position developed, some historical background to his experimental work is required.

3. From Theory to Reality

3.1 Hertz and Electromagnetic Waves

Although Maxwell's theory successfully unified electromagnetism and optics, it had little if any direct experimental support. Maxwell arrived at the velocity of propagation for electromagnetic waves in a purely theoretical manner without any experimental evidence concerning the nature or existence of these waves. In addition, his calculation required postulating a displacement current, which was also without experimental justification.

17. For excellent critiques of the practice known as inference to the best explanation, see van Fraassen 1990 and Cartwright 1983.

Support for Maxwell's theory was further weakened by the fact that similar equations could be obtained from the action-at-a-distance theories of Weber and Helmholtz. Weber's theory construed electricity as fluids of positive and negative particles that possessed mechanical inertia and interacted through a force or potential similar to gravitational attraction. Helmholtz's account treated the ether as a dielectric whose polarized parts, no matter how far apart, interacted without delay; whereas Maxwell insisted on contiguous action.[18]

In order to adequately test Maxwell's theory, Hertz needed to demonstrate the finite propagation of electromagnetic waves in air. When he initially operated the apparatus, he found that the phases of interference (the relation of phases between forces traveling through air and along the wire) were different at different distances from the primary circuit. From this he concluded that the alternation would correspond to an infinite rate of propagation in air because if the waves had the same velocity, they would interfere at all distances with the same phase. After much discouragement Hertz reexamined his results and found it necessary to assume that the velocity in air was finite but greater than along the wire, a result at odds with theory. The same outcome was confirmed in a further series of experiments, and although the difference seemed improbable, he saw no reason for doubting the validity of the experiments. At the time he speculated that the motion along the wire could have been affected by some obscure cause like the essential inertia of free electricity. His trust in the experiments was further strengthened by the fact that different kinds of wires (crooked, spiral, smooth) resulted in noticeable effects on the velocity. In addition, Hertz assumed that if the velocities of waves in air and along wires were the same, then lines of electric force would be perpendicular to the wire (1962, 10). Hence, a straight wire could not exert any inductive action upon a neighboring parallel wire. When this action was actually observed, Hertz concluded that the lines of force were parallel to the wire and thus the velocity of the waves was different from that of light. A few years later when summarizing his results in the introduction to

18. As Helmholtz remarked: "The two theories are opposed to each other in a certain sense, since according to the theory of magnetic induction originating with Poisson, which can be carried through in a fully corresponding way for the theory of dielectric polarizations of insulators, the action-at-a-distance is diminished by the polarization, while according to Maxwell's theory on the other hand the action-at-a-distance is *exactly* replaced by the polarization. . . . It follows . . . from these investigations that the remarkable analogy between the motion of electricity in a dialectric and that of the light aether does not depend on the particular form of Maxwell's hypotheses, but results also in a basically similar fashion if we maintain the older viewpoint about electrical action-at-a-distance" (1882; quoted in Woodruff 1968, 305, emphasis added).

Electric Waves, he ceased to find these factors important and instead speculated that the velocity discrepancy may have been the result of special conditions of resonance in the room that he used.

Hertz was clearly bothered by this difference between velocities in air and along wires, but because he was unsure how to interpret it, he was reluctant to see it as a decisive argument against Maxwell's theory. Hertz was careful to stress the independent nature of the experimental results despite their obvious agreement with the basic tenets of field theory:

> I have described the present set of experiments, as also the first set on the propagation of induction, without paying special regard to any particular theory; and indeed the demonstrative power of the experiments is independent of any particular theory. Nevertheless, it is clear that the experiments amount to so many reasons in favour of that theory of electromagnetic phenomena which was first developed by Maxwell from Faraday's views. It also appears to me that the hypothesis as to the nature of light which is connected with that theory now forces itself upon the mind with still stronger reason than heretofore. (1962, 136)

Having repeated the experiments under the same conditions with the same results, Hertz felt that the only way to resolve the discrepancy was to use waves that were longer than those he initially produced. Unfortunately, he had no access to rooms large enough to accommodate the experiment. Other experiments were performed using waves even shorter than Hertz's, and here the difference between the velocities tended to disappear.[19] Hertz remarked that he had "little doubt that [experiments would] decide in favour of equal velocities" (1962, 14), and in 1893, five years after his initial results, Sarasin and de la Rive carried out further experiments in the Great Hall of Rhône which showed the equality of velocities in air and along wires. Despite this result, what is particularly interesting is that the controversy over Hertz's waves has never really been solved. Because Sarasin and de la Rive used waves of a different length, it was technically a different experiment. Hertz's effect was especially troubling since it was never reproduced in a way that could facilitate further investigation; that is, it could not be manipulated or compared with longer waves produced under different conditions. One obvious problem was at the level of instrumentation; the dipoles Hertz used were uncontrollable and therefore could not be adjusted and refined in order to vary the outcome. Consequently, because the effect could not be controlled, there was

19. These were done by two Genevan physicists, Sarasin and de la Rive. Despite their result Hertz felt that a true test would require long waves.

no adequate understanding of why the phenomenon occurred, and as a canonical experiment it was not entirely convincing.[20]

In the introduction to *Electric Waves,* Hertz claimed that his experiments did prove, for the first time, the propagation in time of what was previously thought to take place through action-at-a-distance:

> This fact forms the philosophic result of the experiments; and, indeed, in a certain sense the most important result. The proof includes a recognition of the fact that the electric forces can disentangle themselves from material bodies, and can continue to subsist as conditions or changes in the state of space. The details of the experiments further prove that the particular manner in which the electric force is propagated exhibits the closest analogy with the propagation of light. (1962, 19)

One can begin to see Hertz's philosophical ideas about the relationship between theory and experiment emerge in a rather interesting way. Although Maxwell's was the only available field-theoretic approach, Hertz felt it necessary to emphasize that while his experiments confirmed that electromagnetic waves were not transmitted by distant action, they failed to explicitly confirm Maxwell's theory, violating one of its important components, namely, the equality of velocities in air and along wires. His opinions about what the experiments show is indicative of an epistemological conservatism that is critical yet, unlike Mach's and Duhem's, tractable in light of appropriate standards of evidence as defined within scientific discourse. The problematic result of the propagation experiment is troubling not only from a theoretical perspective but also because it was ideas surrounding phase relations that initially stimulated him to this work. Now the very method he saw as providing the appropriate technique for testing Maxwell's theory had resulted in an unexpected and inexplicable result. It is important to emphasize that Hertz did not see his experiments as simply establishing a "scientific" result. He refers to the field-theoretic conclusion as the "philosophical and the most important result of the experiments" (1962, 19). The outcome is philosophical in the sense that it not only determines a new way of conceptualizing electromagnetic phenomena but also functions as a precondition on what will

20. While testing his results against the assumptions of field theory, Hertz ran into some difficulty with coiled wires. When a wire was rolled into a spiral and the velocity was measured along the axis of the spiral, the wave moved much more slowly. When the velocity was measured along the wire itself, it moved more rapidly. A similar phenomenon occurred with crooked wires. Maxwell's theory was also unable to account for this; according to his theory the propagation along the axis of the spiral must take place at the velocity of light for every form of conductor.

count as an acceptable explanation and hypothesis in the future. In other words, any account of electromagnetic and optical phenomena must now subscribe to a field-theoretic format, resulting in an entirely different view of the kinds of objects that exist and the ways in which they relate to one another. In this setting philosophy functions in a role that involves the legitimation of particular kinds of theories and data, yet its connection with empirical practice fosters a kind of fluidity in the way the two levels relate to each other. As a result we have an interactive relationship between scientific discourse and philosophy, replacing the static external approach that often characterizes the kind of epistemological conservatism practiced by Mach and Duhem.[21] But the story does not end here; although Hertz's *transition* to this new way of conceptualizing the phenomena is not without its difficulties, it is in the very process of recasting his ideas about electromagnetism that some interesting philosophical considerations arise. A closer look at Hertz's methodology reveals the way in which his conservatism *links* philosophical and practical concerns rather than distancing them from one another.

One might want to argue that the emergence of the field-theoretic view is simply an instance of theoretical change that took place in light of an experimental outcome and has little if anything to do with philosophical presuppositions. Although there is a sense in which Hertz's new view is theoretical, to suggest that it has no philosophical import is to miss an important subtlety in the way philosophical considerations operate at the level of practice. Although the evidence furnished by the experiments ruled out action-at-a-distance, Hertz himself stressed its neutrality in that it failed to fully confirm a *specific* theoretical account. However, what is especially interesting from a philosophical and methodological perspective is the completely different and significantly more enthusiastic response by the British to the same data. The differences between Hertz and his British colleagues show how one can view philosophical presuppositions as having some normative role while at the same time constructing a descriptively accurate account of how and why practice evolves as it does.

21. One might object that my characterization of Duhem and Mach as having an external methodology is rather odd since both were practicing scientists themselves. I do not mean that they necessarily stood outside practice but rather that their views about the goals of scientific activity can be traced to an epistemological position that was immune from revision in response to empirical results. Even if their views were established in light of what they took to be an accurate representation of science at the time, their reluctance to accept the kinetic theory despite its overwhelming success displays a conservatism that is clearly at odds with practice. Of course, one can only speculate about whether they would have eventually been convinced—both of them died shortly after the revival of atomism.

3.2 Electromagnetic Waves: The British Response

When FitzGerald heard of Helmholtz's announcement of Hertz's success, he immediately wrote to Hertz comparing Hertz's experiments with Foucault's *experimentum crucis* and claiming that the experiments would be called "Hertz's classic experiment that decided between theories of electrodynamic action at a distance and by means of the aether."[22] Hertz's reply to FitzGerald (11 June 1888) mentioned the further production of standing waves in air and measurements of their wavelengths in addition to some cautionary remarks that FitzGerald was "supposing more than was actually the case and will be disappointed" (25). Hertz emphasized that one could not speak in terms of exact comparisons or of a measurement in the strict sense and that the experiments themselves were somewhat complicated and not as simple as one would require from an *experimentum crucis.* In contrast to FitzGerald's enthusiasm, Hertz cautioned that the experiments were "not yet a final objective, but much more the beginning and the introduction to better experiments" (25). One thing however could be concluded: that propagation from point to point takes time and is not instantaneous and at a distance. Hertz went on to remark that although he had "no doubt that the views of Faraday and Maxwell will triumph in the end," as it stood the theory needed improvement in certain areas, particularly with respect to the problem of different velocities for waves in air and along wires.

Hertz's reluctance did not dampen FitzGerald's enthusiasm, and in FitzGerald's announcement of the result at the British Association for the Advancement of Science meeting in September 1888, he claimed that the "experiments prove the aethereal theory of electromagnetism" (FitzGerald 1888). Although further experiments by Walter Thorp, carried out in FitzGerald's laboratory in 1891, confirmed Hertz's result of different velocities for waves in air and along wires, there is no evidence to suggest that FitzGerald was especially bothered by the result or that he offered a possible explanation.[23] Thorp made use of wires of different thicknesses and unlike Hertz found different velocities in different wires. Therefore, he suggested that Hertz's conclusions may have been the result of using a relatively thin wire. Thorp, however, was unable to explain this phenomenon, and his result was unhelpful to Hertz, who had observed no differ-

22. Some of the correspondence between Hertz and FitzGerald, Oliver Lodge, and Oliver Heaviside has been reprinted in O'Hara and Pricha 1987. The quotation above is from pp. 23–24. All page references in discussion of the correspondence are to this volume.

23. Prior to this, in a letter to Hertz in 1889 he expressed some uneasiness about the problem (16) of wire propagation but wanted to work out a theory of electromagnetic interactions before attempting to alter Maxwell's equations in any way.

ence between thick and thin wires. Hertz was extremely interested in these experiments, but unfortunately they were not repeated.

Although Oliver Lodge had done experimental work on waves along wires, his results could be understood without propagation and so did not really anticipate Hertz in any substantive way. In his correspondence with Lodge, Hertz again stressed that he thought of his work as a beginning and expressed his concern over the discrepancy between waves in air and along wires. It seems clear from their correspondence that FitzGerald and Lodge were far less interested in Hertz's negative result and saw it as a problem to be dealt with within the framework of Maxwell's theory rather than as a threat to one of its fundamental presuppositions. Heaviside was slightly more concerned but did not see the result as potentially harmful to the basic structure of Maxwell's account, since it might possibly be solved by adding the appropriate parameters. Hertz, whose conversion to field theory involved a very gradual and extremely troubling process, saw the difficulty as more fundamental. Consequently, he saw his experimental results as having a more circumscribed implication rather than providing substantive confirmation of Maxwell's theory. For Hertz the velocity discrepancy was a difficulty that needed to be explained or accounted for in some non–ad hoc way, whereas the British field theorists viewed it as simply an artifactual effect that did not require any substantive treatment.

Undoubtedly some of the difference in response can be traced to a difference in theoretical commitment. The British were firmly immersed in the Maxwellian program, but it is also important to stress that they advocated a rather different methodology and approach to theory construction than that favored by Hertz. Central to British dynamical theory was the principle of energy conservation, the use of Lagrangian methods, and a process of model building and the construction of mechanical images that would supplement the mathematics. Hertz's emphasis on the axiomatic approach demanded a minimum number of physical hypotheses, and any attempt to supplement the formal theory with qualitative assumptions was considered distinct from the process of theorizing. As a result the kinds of phenomena and explanations which occupied a secure place within Hertz's theoretical framework were significantly different from those allowed by the British.[24] The desire to have a physical representation of phenomena had been crucial to Maxwell's approach, and in

24. I will say more about this below when I discuss Hertz's theoretical work, but for now let me just say that one such example was energy; it was seen by the British as a foundational notion and according to Maxwell was discovered (the conservation law at least) using the "true" method of physical reasoning: deduction from phenomena. Hertz, on the other hand, saw the concept of energy as surrounded by ambiguity and therefore did not consider it fundamental.

keeping with Maxwell's rather cautious attitude toward the ontological status of the models, FitzGerald exercised restraint regarding the use of models as true representations of reality. Although FitzGerald was somewhat skeptical about *particular* ether models, he nevertheless was firmly committed to the reality of the ether itself; and it was this inability to separate field-theoretic notions from an ether-based physics that led him to conclude that Hertz's work had provided experimental *proof* of the ether (O'Hara and Pricha 1987, 43). Hence the methodological legitimacy of believing in and postulating an ether was intimately connected with the legitimacy of field theory and the possibility of providing a mechanical explanation for the transfer of energy. And since Maxwell's account was the only articulated field theory, its vindication by this new experimental evidence was not a matter of question for the British. Ironically, Maxwell himself was less enthusiastic about the possibility of establishing the existence of the ether, and although he understood energy as located in the field, this hypothesis was not equated with any particular model or physical account of the ether.

Hertz's conversion to field theory embodied particular philosophical beliefs about how natural phenomena were constituted, beliefs that further influenced his methodology and ultimately his beliefs about particular hypotheses and pieces of evidence. In order to understand this we need to briefly describe the main differences between Maxwellian field theory and the Helmholtzian view that Hertz first mastered.[25] Maxwell's theory does not postulate the direct transference of energy between material objects construed as electrodynamic entities. Instead it introduces the field as an entity which subsists in the space occupied by objects and in the space between them. When we speak of an object being charged, it is not thought of as interacting directly with the field; instead we think of this condition as a local state of the field which is dependent on the presence of matter. Hence the object's state reflects and is reflected by the structure of the field. Local changes in the state of the field affect other states in the field, giving the appearance of objects interacting with one another. Unlike Maxwell's, Helmholtz's view preserves the independence of material objects. A charged object is one that acquires this condition through its relation to other charged objects, and therefore the nature of electromagnetic interaction depends on the simultaneous states of the interacting objects. A charged body acts instantly and directly on another charged body, with every distinct interaction requiring a unique specification of the states of interacting bodies. Instead of field processes, electromagnetic processes are states of bodies, and electromagnetic interactions are unmediated re-

25. My summary is taken from Buchwald 1990.

lationships between bodies in these states. What this entails is that actions cannot be separated from their physical sources. Field theory, on the other hand, *requires* that electric and magnetic forces (fields) be independent from sources. The values of the electric and magnetic fields at a given point are sufficient to determine how an object in that space will behave.

Helmholtz's influence informed much of Hertz's thinking while Hertz was engaged in experimental activity, and as a result, his views of field theory represented a rather interesting hybrid of Helmholtzian and Maxwellian principles. Hertz viewed the field as an object and saw electromagnetic interactions as the result of a coupling between an object and the field, a view which simply transformed Helmholtz's object-object interactions into object-field interactions. As a result, Hertz's understanding of what constitutes an object and its importance for electromagnetic interactions differed from that of his British colleagues. In attempting to formulate and come to terms with this new worldview, his critical awareness was heightened by problems that impinged on that transformation—problems, like the velocity discrepancy, that were considered to be nothing more than artifactual by the British. We can see then that his conservatism is in some sense a contextualized response to what he saw as uncertainty regarding certain aspects of Maxwell's theory.

What is particularly interesting about this problem is that the indifference shown by the British to an effect that was very worrisome for Hertz cannot be simply reduced to different *interpretations* of the evidence; the British had no explanation for the discrepancy. Obviously, background philosophical beliefs about the nature of objects, beliefs that were informed by theoretical commitments, were vital in determining methodological principles. The radical reconstruction from Helmholtz's view to a new field-theoretic account involved and indeed produced a way of thinking about the phenomena that made Hertz open to future possibility yet extremely cautious about an effect that defied explanation. Immersed in their own worldview, the British reacted uncritically to an experimental result that was by no means unproblematic. But that is not to say that their contextualized response is legitimated merely because their views about field theory were well entrenched. Indeed, regardless of the different philosophical presuppositions about how electromagnetic phenomena are constituted, neither view can explain Hertz's anomalous result. So although the British reaction can be *explained* by their unqualified acceptance of Maxwellian field theory, these facts do not justify their methodological stance. The implication of this normative claim is, of course, that the British should have paid closer attention to the velocity problem, and until there was some attempt to either solve or explain it, their reaction to the supposed experimental confirmation of Maxwell's theory should have been somewhat less enthusiastic. This is especially true in light of Thorp's result. Although Hertz had no explanation for his own anomalous result, his conviction that longer waves would vindicate the theoretical prediction and his suspicions about sources of interference in the room legitimate his response to the 1893 experiments of Sarasin and de la Rive. Although Hertz's 1888–89 results have never been adequately explained, the methodological question concerns the proper epistemic attitude toward the theory-experiment relationship in the interim period. It is here that Hertz's philosophical response is superior to the British reaction.

It is this kind of uncritical assessment of evidence and willingness to accept anomalous results as artifacts that provides some of the motivation for normative philosophy; yet it seems clear from Hertz's own approach that one can exemplify the philosophical perspective without extending beyond the boundary of practice itself. His views about what exactly his results confirm exemplify a level of philosophical sophistication that can be achieved without adopting a foundational approach to the evaluation of theoretical and evidential claims. Because the domain of philosophical presuppositions includes a determination of ontological commitment taken, in part, from theoretical conclusions, we can see how both interact to contribute to the formulation of epistemological and methodological standards. Further differences between Hertz's approach and that of his British colleagues emerge even more clearly in his theoretical papers where he attempts to assimilate his experimental findings with what he takes to be acceptable aspects of Maxwell's theory.

4. A Theoretical Solution

What Hertz attempted in his theoretical papers of 1891[26] was a reformulation of Maxwell's theory using a limited number of physical conceptions and eliminating what he took to be superfluous physical and mathematical ideas.[27] He emphasized at the outset of the first paper (1962, 195–97)

26. "On the Fundamental Equations of Electromagnetics for Bodies at Rest," reprinted in Hertz 1962, 195–240; "On the Fundamental Equations of Electromagnetics for Bodies in Motion," reprinted in Hertz 1962, 241–68.

27. One such idea was the "dielectric displacement (polarization) in free aether as distinguished from the electric force which produces it, as well as the relation between the two—the specific inductive capacity of the aether" (1962, 196). These distinctions were meaningful only if it were possible to remove the ether from a space while allowing the force to remain. He also dispensed with the mathematical quantity of vector potential. The vector potential had allowed for the replacement of distance forces by magnitudes which were determined at every point in space by conditions at neighboring points. Once it was possible to identify forces as magnitudes, there seemed little point in retaining potentials. In addition, the vector potentials had been associated with displacement, which was also effectively eliminated by Hertz.

that his intention was to give the fundamental ideas and the formulae that connect them. Although explanations were added to the formulae, they were not to be considered proofs; instead, the formulae were to be regarded merely as facts derived from experience, or, in modern terminology, phenomenological laws (197).

In the second paper Hertz addressed the problem of bodies in motion, taking into account the fact that disturbances of the ether that arise with the motion of ponderable matter must exhibit some effects, even though we have no knowledge of them. Unfortunately, the failure to have such knowledge while adhering to the ether hypothesis was equivalent to saying that questions concerning motion cannot be considered without the introduction of *arbitrary assumptions* about the motion of the ether. In fact, all available knowledge seemed to indicate that the motion of the ether produced no effects (1962, 241–43). One could assume that the ether moved independently of matter (even in its interior), but to adapt the electromagnetic theory to this hypothesis, one must also suppose that the ether and matter are independent at every point in space. However, as Hertz pointed out, a consideration of the phenomena revealed that none required the hypothesis of the independent motion of ether and matter, because no indication of the magnitude of the relative displacement had been obtained. As a result these phenomena were also consistent with the hypothesis that denied relative displacement and assumed that the ether moved with matter. This latter view included the possibility of taking into account only one medium filling space. It was the view adopted by Hertz even though he admitted that "a theory built on such a foundation will not possess the advantage of giving to every question that may be raised the correct answer, or even of giving only one definite answer; but it at least gives *possible* answers to every question that may be propounded, i.e. answers which are not inconsistent with the observed phenomena nor yet with the views which we have obtained as to bodies at rest" (243, emphasis added).

In his attempt to eliminate any arbitrary assumptions from the theory, Hertz acknowledged its rather abstract and colorless appearance (1962, 28). However, he did remark that if one wanted to lend more color to it, it was possible to aid one's powers of imagination by using concrete representations of concepts such as electric current, polarization, etc. But one must remain cautious and not confuse the "simple and homely figure as it is represented to us by nature, with the gay garment which we use to clothe it. Of our own free will we can make no change whatever in the form of the one, but the cut and colour of the other we can choose as we please" (28).

He concluded the second paper by remarking that the only value he attached to his theory of electromagnetic forces in moving bodies concerned its systematic arrangement (1962, 268). Although the theory enabled one to provide a complete treatment of the electromagnetic phenomena in moving bodies, it involved certain arbitrarily imposed restrictions, restrictions that Hertz thought were "scarcely probable" in relation to the "actual facts of the case" (268).

The objection to arbitrary hypotheses can be traced to the three philosophical requirements Hertz placed on theory construction. It is clear that he wanted to distinguish between the *theory,* which for him included only the formal aspects and our understanding of what they mean, and the qualitative hypotheses like the ether drag, which is used as a *possible* explanatory framework. But in addition to distinguishing between reality and our conceptions of it, a theory must, first, be logically clear and free from contradiction; second, it must be correct in that it must agree with observed phenomena; and, finally, it must be appropriate, that is, free from ambiguity and inessential elements. It is important to point out, however, that judgments about whether or not particular assumptions were arbitrary were open to empirical investigation. Unlike the philosophical conditions imposed on theory by Duhem and Mach, Hertz was hopeful that the issue of the ether was capable of being resolved. In fact, the program outlined in *Principles of Mechanics,* which was intended to eliminate the ambiguity surrounding the notions of force and energy, was also meant partly as a clarification of the problems regarding the relationship between ether and matter.[28] Unlike Mach and Duhem, whose primary challenge could not be met in practice (without reformulating the way science gets done), Hertz recognized that the velocity discrepancy and the ether-matter connection represented significant problems, but standards of evidence determined in conjunction with scientific and philosophical considerations would ultimately settle the issue, making his challenge a practical problem. Up to that point, however, no acceptable account had presented itself.

Although it has traditionally been the goal of philosophy to seek some neutral perspective from which to judge the merit of particular theories of science or cognition, Hertz had no explicit criteria or algorithm for determining when one ought to believe a particular theory. Instead, his method exemplifies a careful, critical approach evaluating each case on its own merits. He did have what he referred to as postulates that are assigned to the scientific representations of images that we form of objects. We ascribe different properties to the images with respect to what Hertz called their appropriateness, permissibility, and correctness. Appropriateness has to do with representing the essential relations of the object with the fewest

28. Hertz believed that force, energy, and electricity were identified with such a variety of relations that inconsistencies resulted, thereby obscuring their true nature (1956, 41).

number of superfluous relations. This is done using notation, definitions, and abbreviations. With respect to appropriateness, there is no nonambiguous decision procedure; only trial and error will reveal which image is most appropriate for a particular purpose. The permissible aspect of the image is given by the nature of the human mind, that is, by the laws of thought or logic. Questions of permissibility can be determined without ambiguity and the decision is timeless. Correctness has to do with accurate representation of the relations of external things. In other words, essential relations in the image must not contradict relations in external things. Our present experience allows us to decide without ambiguity whether an image is correct, but we must always allow for new experience in the future. Although these criteria enable us to judge the merit of scientific representations, there are no necessary *rules* for their application in concrete cases. The use of rules is part of the empirical methodology of science. Herein lies the tension between what philosophy expects from an epistemological framework and what is possible given the nature of scientific activity. The imposition of static philosophical presuppositions is only valuable when the ways in which those presuppositions are translated and used are allowed to vary in different contexts. Rather than eroding the normative force of these presuppositions, some fluidity in their application retains their importance as philosophical principles that can have some input at the level of scientific practice. By recognizing that the normative element in scientific judgments can survive without the uncompromising constraints of a foundational epistemology designed to establish the priority of specific criteria, a resolution to the tension between theory and practice becomes feasible.

5. Old Problems, New Debates

These nineteenth-century debates bear a striking similarity to modern philosophical accounts of the acceptance of theoretical entities and hypotheses. Although the attempt to incorporate scientific practice and history in the formulation of criteria for belief has intensified, the project of providing normative constraints[29] designed to operate across a variety of contexts has met with some difficulties in accommodating crucial aspects of past and contemporary science. For example, van Fraassen (1980) argues that although theories are *capable* of being true or false in principle, we should *believe* only that part of the theory that makes claims about

29. I should add here that my use of the word "constraint" is intended to refer to any kind of epistemological presupposition that figures in philosophical discourse. I am not adopting the more specific use of "constraint" advocated by Galison in some of his recent work.

what is observable. All aspects that extend beyond the domain of observation are to be *accepted* on pragmatic grounds. Since we are not in a position to determine the truth of these latter claims, it is epistemologically preferable to withhold full-fledged belief. This view suggests we treat such well-entrenched entities as electrons with a skepticism that is not shared by the scientific community. This is due, in part, to the fact that observation is defined differently in van Fraassen's philosophical account than it is in science. In the latter instance acceptable methods of observation include particle detectors and electron microscopes, which are ruled out by van Fraassen since anything that is in principle unobservable by unaided human sense organs fails to qualify.[30]

Attempts to overcome this departure from practice have been made by Hacking (1983) and Cartwright (1983). Although both are sensitive to the philosophical difficulties posed by abstract theoretical hypotheses and models, they nevertheless think it possible to separate physical entities from the theories that describe them. Consequently, one can be skeptical about theories while advocating a realism about the objects postulated by the theories. According to Hacking the criterion for believing that a particular kind of entity exists is whether one can manipulate it. Hacking differentiates between experimenting on an entity and manipulating it. An example of the former is the early inquiries carried out by J. J. Thomson and R. A. Millikan to determine the existence of the electron and its properties. Manipulating an entity involves interacting with it in ways that exploit its causal powers in the production of other effects. In other words, one manipulates an entity in order to experiment on *other* more-hypothetical entities or to interfere elsewhere in nature. This kind of activity involves commitment to the existence of the entities being manipulated because they are, quite literally, tools for creating phenomena or effects in other domains.[31] One outstanding problem with using manipulation as a criterion for realism is that, like observability, which rules out belief in subatomic particles, it rules out a large body of phenomena in areas, like astrophysics, where the degree of skepticism that exists within

30. Many authors have completely misunderstood van Fraassen on this point, claiming that his conditions effectively rule out optical devices like telescopes and even eyeglasses. Instead, van Fraassen claims that anything observed with the aid of these types of instruments would be observable by humans if they were in the proper context (i.e., close enough, with proper lighting, etc.). This situation differs from the microcase, where it would be simply impossible for a human to observe an electron, regardless of the conditions. What we observe are cloudy trails or patterns on a computer screen, and based on our theoretical knowledge we *infer* that they are the results of electrons passing through the chamber or detector.

31. For criticisms of this view and a discussion of some of the problems associated with manipulation as a condition for realism, see Morrison 1990.

the scientific community is fundamentally at odds with the philosophical account. Hacking (1989a) acknowledges these difficulties but remains content to sanction an antirealist position about bodies, like gravitational lenses and quasars, that are beyond the domain of manipulation.

Cartwright, on the other hand, advocates an entity realism that is based on the causal efficacy of a particular phenomenon. Unlike Hacking, she does not require manipulation as a precondition for determining the causal role played by an entity; instead, the requirements are simply that there be no underdetermination in the causal story and that controlled experiments be performed in order to ensure as far as possible the credibility of the hypothesis. Part of the difficulty with this demand for causal explanation is that many explanations in science fail to take this form; for example, frequently one explains something by deriving the phenomenon using a differential equation together with appropriate boundary conditions. In cases of this sort, often a variety of boundary conditions can be used; hence, one cannot equate boundary conditions with causal factors. In addition to the many epistemological problems associated with the notion of causality, it is particularly important to mention that in many cases what counts as a cause will be intimately connected with high-level theoretical principles and hypotheses that describe the causal process in question. For example, in the case of a simple particle detector like the cloud chamber, we infer that it is the existence of a charged particle moving at high speeds which ionizes the gas atoms. This leaves behind a trail of ions, around which water droplets form to produce a cloudy trail. One can further determine whether the particle in the chamber is an electron by observing how it moves. Because all charged particles are deflected into curved orbits by the presence of magnetic fields, we need to rely on background theory to differentiate the negatively charged electron from other negatively charged particles, like the pion. Although this is an example of a rather unproblematic causal inference, it relies on a scaffolding of theory. Electromagnetic theory supplies the conditions under which electrons ionize, and it tells us how to use electric and magnetic fields to force electrons into the cloud chamber. It also tells us how to produce a field in the chamber and, together with mechanics, describes the shape of the electron's path and the cloud trail. Other information from heat theory and atomic theory allows us to make claims about the behavior of electrons in specific circumstances.

The problem for Cartwright's view stems from the sharp distinction she draws between entities and theories. She argues that because theories and models often provide a variety of different and sometimes mutually inconsistent ways of conceiving of the phenomena and because in many cases they do not provide us with literally true descriptions of reality (due to their level of abstraction), they cannot be considered worthy of belief.

Since entities figure importantly in the kinds of causal schemes mentioned above, they must be believed because causal interaction is a crucial part of science. However, in order to uphold a causal realism of this kind, it seems imperative that one also sanction the theories that jointly figure in the causal/theoretical story one tells.

All of these views share the primary aim of providing a condition that will legitimate belief in scientific contexts. What is particularly attractive about each of these positions is that they show, in light of science itself, why traditional arguments for realism that make use of strategies like inference to the best explanation, success of science arguments, and the like are simply unsupported. Similarly, traditional foundational problems concerning the reliability of basic knowledge claims and the problem of induction have given way to more pragmatic worries about what is reasonable to believe in particular circumstances. However, implicit in these philosophical texts is still a program for establishing epistemological standards that are in many ways immune from empirical/scientific criticism.

From a purely epistemological perspective, conditions like observability, manipulability, and causal efficacy clearly have some normative force, but they immediately encounter problems when the goal is making sense of science on its own terms. As I mentioned above, philosophical standards of evaluation like observation and manipulation are often *defined* differently in scientific contexts, making it difficult to determine their merit on empirical grounds. And, like the case of Mach and Duhem, no amount of scientific evidence short of agreement with these epistemological constraints is sufficient to produce a consensus. It is in that sense that philosophical presuppositions become *external* to practice. Although each of these conditions embodies *some* aspect of practice and therefore serves in an important descriptive capacity, they function in a way that sets them apart as static principles unresponsive to contextual concerns and practical demands, yet supposedly capable of determining belief. In addition, the emphasis on *one* criterion, like observability, as the foundation for theory acceptance will ultimately result in descriptive inaccuracies since each branch of science employs specific techniques for evaluating evidential relations; hence, conditions for acceptability should be flexible enough to accommodate this diversity.[32]

In rethinking the project of providing philosophical norms, it becomes possible to isolate an alternative to the relativism that some saw as

32. Although each of these conditions brings an important consideration to the debate as a constraint on legitimacy, the conditions need to evolve with practice in order to accommodate the changing character of science itself. If we take causes as our paradigm, we need only consider the way in which explanation using differential equations all but ruled out causation as a theoretical demand in the nineteenth century. Thus, the kinds of theories that are available will often determine what counts as an acceptable explanation.

implicit in the historical turn.[33] By retaining some features of Mach's and Duhem's critical approach together with a realization of the ephemeral aspects of scientific activity, a holistic strategy emerges that does not incorporate a reduction of philosophy to science and vice versa and yet sees each as having a role to play. In order to retain a function for philosophy it is necessary that both scientific norms and philosophical presuppositions occupy a crucial place in the debate. This was clearly the case in Hertz's work. His emphasis on clarity of concepts and on the difference between descriptive theories and qualitative explanations and his views about the relationship between theory and evidence point to philosophical constraints that inform and are informed by scientific theoretical beliefs. Neither is prior, but each inhabits a distinctive place in the overall framework. Although the work of Mach and Duhem also exemplified a role for

33. One alternative to providing normative conditions for belief that has been pursued in the philosophical literature is the naturalistic approach, which emphasizes the idea that theory acceptance is a process involving individual judgment and interaction with the environment. Most naturalistic accounts focus on "rationality" as the explanatory mechanism that provides the foundation for epistemology. However, because of the theoretical difficulties in providing a theory or even a definition of rationality, some recent attempts at a naturalistic understanding of science claim that a definition of rationality is not essential to the project. Giere (1988), for example, argues in favor of replacing such a notion with what he would call "effective goal directed action." But even this limited notion is for him ultimately grounded in evolutionary theory. Using our evolved cognitive capacities together with a feedback mechanism, we extend our knowledge of the world as well as knowledge of our own cognitive abilities. Giere sees this appeal to evolutionary theory as providing an empirical foundation for a cognitive theory of science. In that sense it establishes the *priority* of evolutionary theory as the ultimate explanatory feature in the development and understanding of science. In this particular case the problem is that there is no readily available account of the relationship between the cognitive capacities and individual decision making. Although we may learn through the kind of feedback mechanism Giere describes, no algorithm emerges from the process that will enable us to determine future cases. Nor does an empirical grounding in cognitive structures provide determinate strategies for future decision making in either science or philosophy. Although a naturalized realism can provide after-the-fact justification, this falls short of the original demand on epistemology; hence, it fails even by its own standards.

To "ground" our decisions in a theory of cognitive structure results in one of two options, both of which have similar and undesirable consequences for the naturalist. Either the cognitive approach has embedded in it a monolithic character which serves as the foundation for all rational activity (i.e., the similarity of cognitive structures implies that there is a determinate way of proceeding that is the most rational to pursue), or it is so loosely defined as to allow for a plurality of methods and a variety of conceptions of rational action. In the former case we are unable to provide an account of exactly *how* to proceed, while in the latter instance "cognitive structure" loses its explanatory force. In either case the naturalist provides little in the way of increased understanding of how philosophical norms develop in conjunction with science itself. My complaint is not that naturalism fails to provide a "theory" of scientific judgment since I am sceptical as to whether such a thing is possible. Rather, to the extent that naturalism supposedly provides an epistemology it fails in achieving its own goals.

both philosophy and science, it occurred in a context that lacked mutual support and interaction. By imposing criteria that not only are fixed but prevent us from accommodating large aspects of practice in a way that seems acceptable from a scientific point of view, we run the risk of undermining the normative function those criteria are designed to serve.[34] If standards of evidence and belief are part of an ongoing dialectical process, we engage science and philosophy in the same context, making it possible to be both sensitive to and critical of the practice in which one is immersed. Because philosophical criteria for adjudicating disputes are themselves implicated in the debate, we constantly turn to history and practice to renegotiate our standards of judgment in the same way that science renegotiates its theories in light of new evidence.

In conclusion, it is tempting to ask how this interaction between science and philosophy ought to proceed. What exactly are these scientific and philosophical norms, and how does one decide when these constraints need to be reevaluated? My answer is simple; none of these questions can be addressed in isolation from the particular context in which these issues are raised. That is, to get the appropriate answer, one needs to ask a question of the following form: what are the philosophical and scientific considerations implicit in Hertz's work that make his methodology different from that of his British counterparts? But to expect that we can develop a methodology from this or other historical case studies that can be generalized to all other cases is to simply resurrect the very problem I have tried to address. The only methodology that does emerge is the one that emphasizes the need to proceed in a localized way when evaluating the impact of philosophical and scientific norms on practical problems. My sense is that it is *only* this kind of contextual analysis that can truly uncover the philosophical presuppositions implicit in scientific practice.

34. Clearly, Mach's and Duhem's emphasis on observation and sense knowledge as crucial aspects of science would have made them incapable of embracing, in a substantial way, any of the theories that became integral parts of twentieth-century science (e.g., quantum mechanics, general relativity).

TEN

WHERE EXPERIMENTS END: TABLETOP TRIALS IN VICTORIAN ASTRONOMY

Simon Schaffer

> The work of a mere observer is the most completely horse-in-a-mill work that can be conceived. The *beau idéal* of an observer of the highest class is a compound of a watchmaker and a banker's clerk. Most of the observers that I know are far below this standard. You will see therefore that the mere observer is a person very very far below the mere chemical experimentalist.
>
> —*Airy to Harcourt, 5 September 1832*

1. WHERE EXPERIMENTS END

CAN EXPERIMENTS BE PERFORMED ON THE SUN? Many Victorian astronomers reckoned they could. The scope of "experiment" obviously needs clarification here. It has recently been common to bemoan a certain philosophical obsession with theory and a corresponding neglect of experiment. Despite obvious and noisy sectarian dissent, such reflections agree that "experiment" is a practical, material, and productive exercise. Experiment is spoken of in terms of technology, engineering, and craft skill. Andrew Pickering writes of a "symbiosis between natural phenomena and the techniques entailed in their production," in contrast to the view of experiment as the "arbiter of theory," and suggests that this symbiosis needs "more attention than it has so far received from historians and philosophers of science" (1984b, 14). Ian Hacking's "experimental argument for scientific realism" hinges on the claim that "engineering, not theorizing, is the best proof of scientific realism about entities" (1983, 274). That is, manipulation through experimental practices is a compelling ground for ontological faith. Peter Galison gives Hacking provisional endorsement. But he seeks a more flexible "openness" in our stories about experimental manipulation. He locates experimental practices in terms of their "directness" and the "stability" of their results (1987, 261, 277). All these stories deliver messages about the closure of experimental inquiries. They tell us when a trial program can be judged to have reached its end. But how general are these stories supposed to be? What of the disciplinary and historical limits of the very category "experiment"? Not when, but where, do experiments end?

Recent studies certainly raise the questions of experiment's boundaries. Pickering's "symbiosis" has limits. He usefully distinguishes between the "old physics" of the 1960s, which involved a "commonsense" attitude to laboratory phenomena, and the more theory-dominated "new physics" of theoretically privileged but rare phenomena. He suggests that "other great conceptual developments in the history of science" might also involve such "shifts in experimental practice" (1984b, 15–16). Hacking has drawn a bold line between the realm of experimental manipulation, where a scientific realism is warranted, and the realm of astrophysics, whose method allegedly remains Hellenistic and whose objects of inquiry warrant little more than a constructive empiricism: "astronomy is not a natural science at all," and so "we can get by without the full experimental method engineered into being in the seventeenth century" (1989a, 560, 577). Galison's looser demands on the practical-manipulative features of experiment relieve him of the need to restrict the realm of experiment to our own planet. But he does imply that the term "experiment" really changes its sense somewhere between "the small, tabletop experiments of earlier generations" and the industrial-scale trials of modern physics. Here "experiment" had better not be unthinkingly transferred from Maxwell's London laboratory of the 1860s to Gargamelle (1987, 29, 262). So experiments are bounded somewhere in the past (with Boyle, with Einstein, with the Manhattan Project, or with the "November Revolution" of 1974) and somewhere in space (at the edge of this sublunar sphere, at the lab's security fence, or at the edge of the lab bench).

A reasonable answer to this problem of the delimitation of experiment is to urge that the term is a typical family-resemblance one. We are highly unlikely ever to identify an essential quality all "experiments" possess (Gooding, Pinch, and Schaffer 1989, xv). Then it becomes interesting to ask why any activity gets labeled experimental. That is the burden of this paper. The case chosen is Victorian astronomy, which many of its practitioners publicly stipulated was experimental. They did so in deliberate contrast to Auguste Comte's notorious judgment that nothing experimental happened in observatories. Unlike Hacking, however, Comte and some influential predecessors, notably Laplace, argued that astronomy's natural-scientific virtue was just that laboratories and observatories were very different places. Enlightenment astronomy was construed as a perfected celestial mechanics in which experimental trials had no place. This "pattern science" was supreme because it eschewed dubious ventures beyond the solar system or into the qualities of matter. No doubt this image sustained the model of the astronomer as a nocturnal occupant of an isolated watchtower, communing directly with an unmediated heaven.

Victorian astronomers refused the Comtean gift of disciplinary mastery because the price—a severe limit on astronomy's scope—was too high. Instead, they turned the idealized watchtowers into subspecialized workshops, stocked with batteries, induction coils, magnetometers, telegraph wires, spectroscopes, reagent bottles, Bunsen burners, photographic studios, and even "artificial stars." The telescope was symbolically and practically surrounded with tabletop trials, experiments designed to calibrate and then extend the zone of astronomical operations (Schaffer 1988). Trevor Pinch uses the term "evidential context" to describe the implied scope of a trial: for example, Pinch describes challenges to the claim that the evidential context of a neutrino detector was a set of physical processes in the Sun (1985). We can say that Victorian astronomers worked to make their enterprise count as experimental so as to extend its "evidential context." Moral claims about the "unity of the universe" helped reinforce this extension. Such "unity" depended on networks connecting observatories with manipulable celestial objects. The strength of these networks hinged on the robustness of links between astronomers and other disciplinary workers: instrument makers, chemists, experimental physicists, meteorologists, geologists. The implication is that tabletop trials are linked to their evidential context through networks at once social and practical. Judgments of the scope of such trials involve judgments about the moral order of scientific disciplines.

The paper which follows moves from an examination of the structure of the Victorian astronomical community to an examination of the structure of a number of "experiments" on the Sun. In section 2 I summarize Comte's ban on speculative solar and stellar physics and indicate its wide influence in Victorian science. In section 3 I describe one strategy adopted to contest the Comtean ban: the moral reformation of the imperial observatory network centered at Greenwich. Sections 4 and 5 chart the alternative sites of astronomical production: the worlds of the grand amateur and the new public observatories at Kew and South Kensington, where experimental astronomy, in contrast to Greenwich vigilance, was promoted. Here we find the most strenuous activists on behalf of a properly experimental mapping of the Sun, and find, too, a series of techniques for making the Sun and Earth as similar as was necessary to allow solar experimentation. These techniques include spectroscopy and photography and doctrines of the plurality of worlds, of solar meteorology, and of the death of the Sun. Finally, in sections 6 and 7 I describe two key episodes in Victorian experimentation: the debates on sunspot spectra in the 1860s and on solar energy in the 1880s. Here we find rivals in dispute vigorously contesting the extension of the "experimental" label and its sense. It is concluded that the work of experiment hinges on the forging of close links between laboratories and other sites of knowledge and practice, and that these links involve the calibration of instruments in a fundamental way.

2. Auguste Comte's Embargo

Astronomers frequently told their audiences that proficient observers of the heavens saw things differently from common folk. Laplace began his *Exposition du système du monde* with this claim: "it is a very long way from the first view of the sky to the general view through which one grasps the past and future states of the system of the world." The consequences of this difference have often been very significant. Astronomy presented itself simultaneously as the most empirical of the sciences and as that which relied on the most esoteric techniques for rendering comprehensible the phenomena of the heavens. Both observation and calculation gave astronomy its status as the highest discipline. This "observation" was a gaze of a very idiosyncratic kind. Laplace summarized these techniques: "it was necessary to observe the stars for many centuries, recognize in their appearances the real motions of Earth, ascend to the laws of planetary motions, and from these laws to the principle of universal gravitation, and redescend at last from this principle to the complete explanation of all celestial phenomena even in their smallest details. This is what the human mind has accomplished in Astronomy" (1984, 13).

The ensemble of prized astronomical techniques did not merely account for its triumphant past. It also taught a portentous lesson about its future: the lesson of closure. Laplace implied, and some of his more enthusiastic interpreters insisted, that the science of astronomy was at an end. To use the phraseology of William Whewell, who shared this view, astronomy had been transformed from its progressive to its permanent phase. This transformation was especially visible in the area of what Whewell called "physical astronomy, that is the mathematical mechanics of the science" (1857, 2:221–23). An influential late-nineteenth-century astronomer commented that "there were those who began to regard it as a science which, from its very perfection, had ceased to be interesting—whose tale of discoveries was told" (Clerke 1885, 183). The subdivision was crucial, for it picked out secure from incomplete sections of the science. The celestial mechanics of this solar system was now a completed science. "Beyond the limits of our own system, all at present is obscurity," stated John Herschel on behalf of the new Astronomical Society in 1820. This mapping was important from the 1830s on, when, for example, the British Association for the Advancement of Science had to adjudicate on the division of labor in the sciences.[1] In that decade, the most celebrated, if not notorious, formulation of the thesis of astronomy's end was due to Auguste Comte.

The importance for this paper of Comte's embargo on large zones of possible astronomical work is twofold. Through his argument against the possibility of a stellar astronomy, we can understand the implications of the construal of astronomical observation as positivist vision, devoid of experimental or practical manipulation. Comte's "astronomy" corresponds closely to the watchtower version of the science, an account of astronomical practice which limits it to the calculated gaze. During the nineteenth century, however, the institutions of astronomy were workshops, highly complex sites of coordinated manipulative laboratory practices. So, second, many Victorian astronomers deliberately violated the Comtean ban, and manipulation of "tabletop" equipment was just what licensed this breach of the positivist border fence. A change in one topology, that of the astronomical workplace, sustained a change in another topology, that of the knowable universe.

Comte made celestial mechanics the center of his account of the sciences. He took Laplace's project as the model of astronomy, to which a few key improvements were to be made before it could be judged complete. He recognized the institutions of Laplacian science as the temporary ideals of scientific organization. He studied at the Ecole Polytechnique for two years before his expulsion in 1816, took courses from the master astronomer Jean-Baptiste Delambre, and each year between 1830 and 1848 gave free astronomy lectures to Paris workers in a town hall behind the Conservatoire National des Arts et Métiers. The work he embodied in these lectures appeared in a series of increasingly ambitious documents during the 1830s and 1840s. Two volumes of his *Cours de philosophie positive* appeared in 1830 and 1835, the latter of which incorporated his doctrine of the philosophy of astronomy and his model of the nebular cosmogony. This model was also prepared as an (unsuccessful) memoir for the Académie des Sciences in January 1835. Comte published the balance of his astronomy lectures together with a preliminary discourse on positivism in this *Traité philosophique d'astronomie populaire* in 1844.[2]

In Britain the principal immediate impact of this offensive was its fallacious numerical argument in defense of the nebular hypothesis of the origin of the solar system. This argument was provisionally accepted by

1. On the division of astronomical labor, see Whewell to Harcourt, 4 Nov. 1831, in Morrell and Thackray 1984, 95.

2. For Comte's work, see Carneiro and Arnaud 1970, 585–608; Merleau-Ponty 1983, 273–87; and Schweber 1990.

David Brewster and by John Stuart Mill in the late 1830s and decisively rejected by John Herschel in 1845.[3] Herschel rejected the Comte-Laplace nebular hypothesis in the name of his father, William, whose observations of nebulae had provided much apparent evidence for the existence of interstellar nebular fluid. He reckoned there was a big difference between William Herschel's model of nebular constitution and development and the local model of Laplace and Comte, which spoke of the emergence of our solar system from a spinning nebular cloud. John Herschel told his audience at the British Association that the former was secure astronomy, and the latter mere cosmogonic speculation which bred dangerous illusion (1846, xxxvi–xxxix). Comte thought just the opposite. "The interior life of our world is probably the sole one we may hope to know with certainty and thence the sole one whose knowledge is of importance to us. . . . Therefore there is in this respect, as in all other great intellectual relations, a necessary and fundamental harmony between the bearing of our real knowledge and the extent of our real needs" (quoted in Carneiro and Arnaud 1970, 606–8; and in Merleau-Ponty 1983, 170). This emphasis on the local range of legitimate astronomy amounted to an embargo on investigation of, and claims to manage, the phenomena presented by the stars.

Comte limited astronomy to a combination of pure observation and subsequent calculation. Here "observation" meant the vigilant recording of angles and times; "calculation," the exact prediction of all future states of that system which, alone, affected human conduct. So the celestial mechanics of the solar system was the most positive science. All else was dangerous illusion. Stellar astronomy was a chimera: "this blind specialty, by preventing astronomical notions from taking on a truly philosophical character, involuntarily maintains the indirect rule of habits absolutely inherent to ancient philosophies," that is, Hellenistic ones (Comte 1985, 120). Stellar distances could not be measured. The work pursued in the great German observatories by Bessel and Struve on stellar parallax, which they estimated at nearly 1/3″ for the proximate stars, was so much idleness. This was partly because of the error under which "the most perfect angular measures" languished, but it prompted Comte to state his fundamental axiom of the isolation of the solar system: "this same insufficiency of all our current means of estimating the distance of external stars yields a valuable idea for the positive philosophy which I do not fear to set up here as a truly basic dogma, since we thus confirm, in the most unchallengeable manner, the isolation of our solar system and the neces-

3. For Comtean cosmology in Britain, see Schweber 1990, 159–67; and Schaffer 1989b, 135.

sity of limiting all our rational astronomical speculations to it" (256–57). The consequences of this strict boundary went deep. Just as the recent work of the Germans on stellar parallax was a mistake, so too were the projects mounted in Britain on the mutual orbits of double stars. John Herschel stood condemned for his "servile and badly disguised imitation of studies of our own world" when he sought to apply Newtonian gravitation to stellar systems (422).[4] Part of the obstacle was clearly technical: the insurmountable errors whose noise swamped the stellar signals. Part was moral: the isolation of the solar system from the rest of the universe meant that stellar science was strictly useless and so a haven for metaphysicians. Bad astronomy licensed the wrong form of power.

3. "THE UNIVERSE IS ONE": OBSERVATORY MANAGEMENT AND ITS VISION

Comte's view was merely a more coherent and forceful account of a widespread early-nineteenth-century belief. Its appeal must not be underestimated. The doyen of observatory managers of the period, Friedrich Bessel, argued in 1832 that "what astronomy must do has always been clear—it must lay down the rules for determining the motions of the heavenly bodies as they appear to us from the Earth. Everything else that can be learned about the heavenly bodies . . . is not properly of astronomical interest" (quoted in Hufbauer 1991, 43). This came from the measurer of stellar parallax whom Comte had so roundly condemned. Even when spectroscopy and photography developed as key tools of the "new astronomy," the Comtean view remained profoundly appealing. Admiral Smyth, author of the most popular English handbook for amateur astronomers, reckoned that despite "the marvellous and extensive power of Chemistry," its techniques should not be allowed to disrupt the order of disciplines: "I really trust it will not be exerted among the Celestials to the disservice or detriment of measuring agency, and this I hope for the absolute maintenance of Geometry, Dynamics and pure Astronomy" (1864, 90).

But some Victorians decisively rejected even the most moderate form of this rule. They defined their capacity to understand and handle the phenomena of sidereal systems and to develop practices other than angle measurement, in explicit contrast to the Comtean vision. This was especially true of the ideologues of astrospectroscopy, who rarely avoided sneering at Comte when surveying the astonishing expansion of their discipline since 1850. In his autobiography, composed in 1897, William Huggins

4. Herschel's work is discussed in Williams 1984, 295–309.

reminisced about Comte's ban. He connected that failed vision with the career of the nebular hypothesis and, more significantly, with the question of experimental philosophy. "How could we extend the methods of the laboratory to bodies at distances so great that even the imagination fails to realise them?"[5] Agnes Clerke, historian of nineteenth-century astronomy, argued similarly that it was precisely the emergence of "the chemistry of the sun and stars amongst the foremost of the experimental sciences" that had shown Comte's hubris (1885, 181). "Astronomy is the oldest of the sciences; astrophysics, that is to say, spectroscopic astronomy, is the youngest. Yet they concur in their Testimony. The Universe is One"—thus the pious Greenwich astronomer Walter Maunder, for whom the enterprise of spectroscopy took its place as a coherent element in the battery of techniques deployed within the observatories (1912, 92).

Victorian piety doubtless played a crucial role in the development of techniques for escaping from the solar system. The alliance of theology, precision measurement, and spectroscopy was a powerful antipositivist force. Clerk Maxwell was a highly influential expositor of this view. In a series of lectures in the 1870s, the Cambridge professor developed a complex argument against Comte. "Not many years ago," he recalled in 1870, "if we had been asked in what regions of physical science the advance of discovery was least apparent, we should have pointed to the hopelessly distant fixed stars on the one hand, and to the inscrutable delicacy of the texture of material bodies on the other. If we are to regard Comte as in any degree representing the scientific opinion of his time, the research into what takes place beyond our own solar system seemed then to be exceedingly unpromising if not altogether illusory."[6] The key lesson of the astrospectroscopy of William Huggins and Norman Lockyer, he judged, was the identity of stellar and laboratory elements. In the 1860s, for example, he worked closely with Huggins on estimating the line-of-sight motions of stars and nebulae by means of the Doppler shifts of their spectra.[7] "We are thus assured that molecules of the same nature as those of our hydrogen exist in those distant regions, or at least did exist when the light by which we see them was emitted." The security of this identity depended on the extreme accuracy which Maxwell and his colleagues attributed to measures with spectroscopes. A caveat helped secure the claim that terrestrial and celestial molecules were truly identical, not merely similar within

5. William Huggins, "Historical Statement" (1897), in Huggins 1909, 38.

6. James Clerk Maxwell, "Address to the Mathematical and Physical Sections of the British Association" (1870), in Maxwell 1890, 2:221.

7. See James Clerk Maxwell, "On the Influence of the Motions of the Heavenly Bodies on the Index of Refraction of Light" (1867), in Huggins 1909, 201–5.

the limits of observation. "The discovery of particular lines in a celestial spectrum which do not coincide with any line in a terrestrial spectrum does not much weaken the general argument," because these lines would either be from a new element or from a dissociated one.[8]

In a celebrated paper challenging modish German models of action-at-a-distance, Maxwell argued that the physics of the electromagnetic ether showed the homogeneity of the universe and thus the connection between terrestrial and stellar trials:

> The vast interplanetary and interstellar regions will no longer be regarded as waste places in the universe which the Creator has not seen fit to fill with the symbols of the manifold order of His kingdom. We shall find them to be already full of this wonderful medium; so full, that no human power can remove it from the smallest portion of space, or produce the slightest flaw in its infinite continuity. It extends unbroken from star to star; and when a molecule of hydrogen vibrates in the dog-star, the medium receives the impulses of these vibrations; and after carrying them in its immense bosom for three years, delivers them in due course, regular order and full tale into the spectroscope of Mr Huggins at Tulse Hill.[9]

The integrity of the ether matched the integrity of electromagnetic measures. Maxwell welded this universal spectroscopic unity to a theological argument about standards. Citing John Herschel, he claimed that such omnipresent standardization was the sign of a divine creative origin for all molecules. They were "manufactured articles." Maxwell hammered home the link between the godly standards born on interstellar light and the mundane measures of the lab: "From the ineffaceable characters impressed on [molecules] we may learn that those aspirations after accuracy in measurement, truth in statement and justice in action, which we reckon among our noblest attributes as men, are ours because they are essential constituents of the image of Him who in the beginning created, not only the heaven and the earth, but the materials of which heaven and earth consist."[10] Precision measurement and celestial spectroscopy taught Victorians that the universe was one. Maxwell's was a bold bid to link the cultures of the laboratory and the observatory in common cause.

Comte was answered, therefore, with a change in the moral economy of the observatory. This change was twofold: first, the reform of observatory management, which devalued the status of the observer; second, the

8. James Clerk Maxwell, "Molecules" (1873), in Maxwell 1890, 2:375; James Clerk Maxwell, "Atom" (1875), in ibid., 478.

9. James Clerk Maxwell, "Action at a Distance" (1873), in ibid., 322.

10. Maxwell, "Atom," 483; Maxwell, "Molecules," 377.

appearance of astronomical laboratories, in explicit contrast to the traditional network of state observatories. (Schaffer 1988; Williams 1989). Technical developments such as the construction of the "personal equation," variable but measurable reaction times for different observers, demanded the regimentation of each observatory worker. Artificial stars moving at a known rate across the telescope's filar micrometer were used to calibrate each observer's performance at the eyepiece. Galvanic and mechanical stop clocks were used to effect the calibrations. Control over the act of observation was displaced from each individual observer to the collective regime of managers, clockmakers, and electricians. From the late 1840s and early 1850s, therefore, publicly funded observatory managers such as Airy (Greenwich), Quetelet (Brussels), Struve (Pulkovo), and Arago (Paris) routinely performed calibration trials on their employees. Arago wrote of his reliance on clockmakers' skill and on the division of labor of the act of observation which this prompted. Even the simplest model of the positivist gaze of the astronomer needed complex experimental manipulation to maintain it. Experimental instruments carried the burden of trust: "when one wishes in future to become independent of personal errors, it will be necessary, so to speak, to leave to the stop clock the burden of evaluating the second and fraction of a second corresponding to the passages of stars," declared Arago. An American contemporary classed an observer as "but an imperfect and variable machine." Any meridian transit "observation" now became the result of teamwork linking observatory workers and large numbers of collaborators (Arago 1859).

The public ideology of observatory management was profoundly moralistic. It distracted attention from the astronomical isolate toward observatory conduct and labor relations. Greenwich and its colonial outposts provided exemplary sites of moral order. Under Airy's immediate predecessor, Pond, the Greenwich staff were viewed as "indefatigable, hard working and above all obedient drudges" (Smith 1991, 7–10). Airy himself meditated at length on the relation between the hierarchy of the observatory and the hierarchy of the sciences. "Mere" observation was menial. So the positivist image of astronomy as pure observation was false, because its moral superiority could not lie in such a base form of life. At Oxford in 1832 he sternly told the British Association that to identify astronomy with observation was a debilitating mistake and a national sin. "I consider English observers as below foreign observers not because they are deficient in their character of mere observers (in which on the contrary they are pre-eminent) but because that character is essentially low."[11] Why low? Because "an observation is a lump of ore, requiring for its pro-

11. Airy to Harcourt, 5 Sept. 1832, in Morrell and Thackray 1984, 152.

duction, when the proper machinery is provided, nothing more than the commonest labour, and without value until it has been smelted" (1833, 184). Hence Airy's view, cited as the epigraph to this paper, that chemical experimentation was vastly superior to the common labor of the observer. The implication was that, contra Comte, astronomy's preeminence could only hinge on its regimes of experiment and manipulation.

Airy's argument was compelling. At the end of the century, it had become common for the observatory managers deliberately to tell their audiences of the startling difference between the fantasy of the positivist watchtower and the reality of the trials of the workshop. Maunder addressed the Religious Tract Society on this theme, comparing the Royal Observatory with an income tax office whose ledgers were full of "stars, planets and sunspots." Each humble observer was surrounded on all sides by lab technicians, computers, and overseers (1900, 137). In 1891 David Gill, then state astronomer at the Cape, told his audience at the Royal Institution that observation "differs very widely from the popular notion of an astronomer's occupation. It presents no dreamy contemplation, no watching for new stars, no unexpected or startling phenomena." He stressed that the popular notion was more in evidence elsewhere, and, above all, in the work of photography and spectroscopy. In a lecture chaired by Huggins, Gill's theme was apt. He showed off artificial stars, photos of nebulae and star clusters, and spectra of Sirius and of telluric iron, and thence deduced exciting lessons about nebular cosmogony and the composition of celestial matter. "It is these after all that most appeal to you, it is for these that the astronomer labours, it is the prospect of them that lightens the long watches of the night" (1970b, 1:363, 370). By displaying photographic and spectroscopic analysis as the science's ultimate goal, Gill carefully defined the map of Victorian astronomy. The major network of state-funded observatories was to be given the task of disciplining observation and permanent surveillance of the heavens. In contrast, a complementary set of astrophysical observatories sustained the new work of experimental astronomy. In neither case would a positivist model of astronomical vision be applicable. The coordination between these two networks was crucial, because it showed that observatory managers and experimental astronomers might collaboratively extend their control beyond the boundaries of celestial mechanics.

4. The Victorian Observatory as a Laboratory

The contrast between the large state observatory regimes and the enterprises of the new astronomical laboratories is crucial for our understanding of the sense of the term "experiment" in Victorian astronomy. The trials which offered astronomers the resources for an extension of control

Fig. 10.1. Spectrum analysis in action: an 1878 engraving issued by the maker, Browning, showing a spectroscope, Ruhmkorff coil, and portable metal samples. From the Whipple Museum for the History of Science.

beyond the realm of celestial mechanics included measures of magnetic strength, temperature, and optical intensity; spectral refrangibility; and "chemical rays" (as some photographic agents were known) (Huggins 1970c, 1:172). Work on sunspots promised links between magnetic and meteorological series, and in the late 1850s the work of R. C. Carrington and his amateur colleagues led to the announcement of the Sun's equatorial acceleration and sunspot structure (Carrington 1858; Hufbauer 1991, 47–49). Spectroscopic and photographic trials demanded links with chemists, photographers, and opticians. The comparison spectra, without which no set of spectral lines could be calibrated, were typically generated by spark discharges from high-power induction coils. Complex galvanic technologies and discharge apparatus were necessary for what Huggins called the "strictly fiducial" function of these comparisons (fig. 10.1). He spelled out the impact of strategies for working with these parameters on the domestic economy of his observatory:

> Then it was that an astronomical observatory began, for the first time, to take on the appearance of a laboratory. Primary batteries, giving forth noxious gases, were arranged outside one of the windows; a large induction coil stood mounted on a stand on wheels so as to follow the positions of the eye end of the telescope, together with a battery of several Leyden jars;

shelves with bunsen burners, vacuum tubes and bottles of chemicals, especially of specimens of pure metals, lined its walls.[12]

Resources for such laboratories were rapidly assembled around the midcentury, at just the period when the managerial regimes of the major observatories were also transformed. Key instruments were the camera and the spectroscope. It was claimed that with these resources astronomers could now study the phenomena of the Sun, stars, and nebulae. In the heroic myth of the "new astronomy," the announcement of dark lines in the solar spectrum by the master instrumental maker Fraunhofer in 1817 and that of the daguerreotype by Arago in 1839 were followed by the telescopic application of the daguerreotype to the Moon in 1840 and to stars in 1850. In 1854–55 John Herschel and Warren de la Rue successfully urged the initiation of a "sunspot patrol," a daily photographic record of the solar surface, while in 1859–61 the Heidelberg professors Bunsen and Kirchhoff analyzed the processes of spectral absorption and emission, interpreted Fraunhofer's lines as marks of absorption by elements above the solar photosphere, and hence proclaimed "an entirely untrodden field, stretching far beyond the limits of the Earth, or even of our solar system" (Bennett 1984, 5; see also Hufbauer 1991, 57–59). Between 1857 and 1860 the Harvard astronomer G. P. Bond inaugurated photographic astrometry by applying the new wet collodion process to the recording of precise stellar positions. Bond's enthusiastic view was that "the field for experiment is too vast to be at once occupied" and that for each stellar image "the photograph is worth three times as much as a single direct measure" (quoted in Norman 1938, 570, 573; see also Lankford 1984, 17).

To occupy Bond's "vast field for experiment," however, the tools of his trade had to be secured. Initially, these tools looked like experiments in photochemistry and spectroscopy, not trials of stars, suns, and nebulae. The evidential context was limited to the machines themselves: how was it to be extended to the heavens? A consensus had to be established that cameras and spectroscopes were untroubled instruments. Chemical and optical experiments had to be turned into black-boxed instruments in secure astronomy laboratories. This task was the precondition of the establishment of an experimental astronomy. It was a task which combined relations of trust, authority, and workplace practice. This was why the troublesome relation between established sites, such as Harvard and Greenwich, and novel institutions, such as the new astronomy labs, mattered for the character of astronomical experiment. In his astrophotography, Bond relied on "high authorities in chemistry" to judge the sensitivity

12. Huggins, "Historical Statement," 7–8.

of his plates. And he relied on photographic support staff, such as the Boston photographer J. A. Whipple, to turn trials with starlight, bromine, silver iodide, and mercury vapor into reliable representations of celestial objects:

> To be of essential service to astronomy, it is indispensable that great improvements be yet made, and these, I feel sure, will not be accomplished without a deal of experimenting. To do this properly we need for at least a year to come the services of the excellent artists who have hitherto literally given us their assistance, expensive materials and instruments. They should be liberally remunerated, and feel at liberty, when the prospect is good for a fair night, to give up their day's business and come to the work fresh and fit to spend the whole night at the telescope. As matters are at present they come to the observatory thoroughly exhausted, for it generally happens that the best nights are preceded by their busiest days. They make no charge for their time, costly chemicals and instruments, and as they are volunteers, we have no claim on them, and cannot, in conscience, require more of men utterly exhausted than they have done. But could we press this matter on, we should soon be able to say what we can and cannot accomplish in stellar photography. (Quoted in Norman 1938, 571)

The disciplinary autonomy of photographers from observatory managers was a problem for the status of astrophotography. In the third quarter of the century, photography was a massively popular, if not vulgar, craft and pastime. "In picture factories, the division of labour was so developed that production reached a thousand a day: the 'operator' never left the camera, the polisher and coater prepared the plate, the exposed plate was passed on to the mercurialiser who developed it, the gilder who coated it and the painter who tinted it" (Tagg 1988, 47–48). It was not immediately obvious, therefore, that a bulk enterprise like photographers' work was fit for the fine requirements of high-status astronomy. There were both technical and cultural issues involved in the use of the camera in the observatory. For example, collodion plates had to be kept wet during exposure, limiting exposure times to 15 minutes, and so confining use to very bright celestial objects such as the Moon. The question of artistic status famously affected debates on photography contemporarily with the accomplishments of Bond and his colleagues—thus Charles Baudelaire's celebrated statement of 1859 on *The Modern Public and Photography:* "we must see that photography is again confined to its real task which consists in being the servant of science and art, but the very humble servant like typography and stenography which have neither created nor improved literature" (quoted in Schwartz 1987, 109–10).

Collaborative work between astronomers and instrument makers was designed to stabilize the service role of photographers and spectroscopists by making robust machines which could sit alongside the telescopes. It is significant that astronomers such as Herschel and Arago played such a crucial role in new photochemical technologies, which were immediately assessed for their worth within astronomical practice. Daguerreotypes, in use from the 1840s, needed long exposure times and produced a direct, inverted, positive image, which was irreproducible and so incapable of "virtual witnessing." The process demanded heavy and complex equipment. Experimental astronomers in Britain soon abandoned this technology, but the process was clearly highly adapted to precision measurement, as Bond and Andrew Common both pointed out. From the 1850s, the wet collodion process, with faster exposures and reproducible negatives, became the technology of choice until the 1880s, when dry-plate gelatine techniques allowed exposure times almost 100 times quicker than wet collodions. At Kew Observatory, for example, solar photography using these improved methods was seen as "peculiarly suitable for ladies," being normally carried out by the daughter of the chief mechanical assistant.[13] With the adoption of the new technology, the large darkrooms of earlier astrophotographic observatories and eclipse expeditions were replaced by plates fitted at the telescope's eyepiece and tabletop developing apparatus. Common commented in 1886 on the implication of the displacement of wet- by dry-plate processing for the liberation of the astronomer from dependence on the chemist. This liberation went further: "the records that the future astronomer will use, will not be the written impressions of dead men's views but veritable images of the different objects of the heavens recorded by themselves as they existed" (1970, 1:246, 251). Part of this process was the international project for a photographic map of the heavens, in whose inauguration Common and David Gill participated. Gill's role in the stabilization of astrophotography's status was remarkable. Against severe criticism by the masters of meridional astronomy, he worked hard to make the gelatine plate an indispensable part of the equipment of the major state observatories. Indeed, when his Royal Society funding was withdrawn in 1887, he argued that a photographic star catalogue "never has been and never will be the occupation of the amateur or single-handed astronomer. . . . It can only be executed at regular Government establishments" (1970a, 323).[14] Debates like these showed the contrasting fate of experiment in observatories such as those of Common and in the work of the astronomical branches of government.

Ian Hacking rightly notes that "it is particularly true of practical in-

13. For the impact of photography on midcentury astronomy, see Rothermel 1993, esp. 138–39. For A. A. Common, see Common 1970. For Warren de la Rue, see de la Rue 1866. For solar photography at Kew, see Lankford 1984, 23–25.

14. For astrophotographic catalogues, see Clerke 1885, 429.

struments that what is read is not raw unprocessed data but a built-in deduction from a measurement" (1989b, 269). The map of Victorian observatory workers indicates that we need to understand how deductions get built into instruments and cannot assume that instruments come with ready-made messages or that all members of the community will agree on what deductions can legitimately be made. To build a stable deductive practice into instrumental work, the instrument must be made "transparent," an untroubled transmitter of nature's signals. The history of spectral experiments displays this feature rather well. Trials were calibrated in varying ways to make what might seem the "same" instrument yield different signals.[15] Crucial in this process is the distinction between calibration and experiment. This distinction shifts techniques between reliable standards and experimental trials. Such processes directly affected the security of the spectroscope. In his Munich glassworks, Fraunhofer commanded a team of 48 subordinates. His trials on the solar spectrum were a by-product of his search for calibration standards for refractive indices in his glassworks, and his instrument was not yet a recognizable spectroscope but was a modified theodolite carrying a prism table. Work on spectroscopy before 1860 was dominated by communities of physicists and instrument makers who needed good, reliable standards: thus Fraunhofer needed monochromatic light to perfect achromatic lenses; David Brewster needed a similar light source to eliminate aberration in microscopy; John Herschel needed monochromatic light for testing crystal structure. These needs show the connection betwen standardization and technology (Bennett 1984; James 1985). The heterogeneity of their suppliers was a factor which troubled spectroscopists. Huggins reminisced about the 1860s, when "a star spectroscope was an instrument unknown to the optician. . . . It is difficult for any one, who has now only to give an order for a star spectroscope, to understand in any true degree the difficulties which we met with in attempting to make such observations for the first time.[16] It was only at the start of that decade, in the collaboration between Kirchhoff, Bunsen, and the Munich maker C. A. Steinheil, especially in work on the solar spectrum, that an integrated spectroscope was produced.

Needing a constant scale against which to mark the places of spectral lines generated in the lab, the Heidelberg workers chose the Fraunhofer lines, which were judged to be unchanging. They preferred this scale to verniers or theodolite angles since they believed that the Fraunhofer lines could be reproduced in every lab. They would then be able to develop

15. For "transparency" of instruments, see Schaffer 1989a.

16. For problems of suppliers, see Bennett 1984, 3. On star spectroscopes, see Huggins, "Historical Statement," 6.

chemical analysis using spectroscopy, especially on the troublesome non-volatile alkalis. The bright D doublet was a key marker of this scale. In order to secure its stability, Kirchhoff tried shining sunlight through a sodium flame, since it was widely agreed that the D lines corresponded to the sodium spectrum. He found that as the sunlight brightened, the D lines vanished and then gradually appeared as a very clear dark doublet. This was the evidence Kirchhoff used to claim that incandescent vapors absorb light at the frequency they emit. He lacked any account of the size of the absorption or of the structure of the Sun's photosphere. In close collaboration with instrument makers in Heidelberg and Munich, and with his colleagues Bunsen and Helmholtz, whose 1854 lecture "On the Interaction of Natural Forces" discussed the link between thermodynamics and solar energy, Kirchhoff then initiated a research project on the production of the solar spectrum. Kirchhoff could eventually make the celebrated statement, published in Britain in 1861, that Steinheil's apparatus possessed "a degree of accuracy and purity which has certainly never before been reached" and his epochal announcement that "Ruhmkorff's induction apparatus produces such a rapid succession of sparks that the spectra of metals may be thus examined with as great facility as the solar spectrum." Kirchhoff's team reproduced the solar atmosphere in their lab, first with sodium, then with iron and more complex spectra. Research into the calibration technique had turned a constant scale into a variable and complex experimental site (Kirchhoff 1860; Stokes 1860; McGucken 1969, 29–34). Rather rapidly, this replication of the Sun became a textbook trial. Popular works by Lockyer (1873, 94–98) and by Maunder (1912, 27) both spell out the protocols for the generation of Fraunhofer lines with limelight and a spectroscope. These tabletop trials were just what legitimated the rival stories about the Sun's structure and composition.

Calibration mattered because spectroscopists needed an absolute scale against which to measure their absorption and emission lines (fig. 10.2). And calibration requires agreement that the test signal is sufficiently similar to the unknown phenomena instruments will be used to detect (see Collins 1985, 100–101). Kirchhoff's arbitrary numbers proved unreproducible, and as Maxwell also noted, there were major ambiguities about the appropriate reaction to spectral lines that did not match lab spectra. Venomous disputes arose about such substances as Henry Sorby's jargonium, which had a brief life during 1869; coronium, identified with Kirchhoff number 1474 in light from the solar corona in the same year; and Huggins's notorious nebulium, which he associated with the bright green "chief nebular line" in the spectra of the Orion and other major nebulae (Maunder 1912, 59–60; Meadows 1972, 60–61). Lockyer reckoned this was some dissociated form of magnesium; Huggins held it was "a

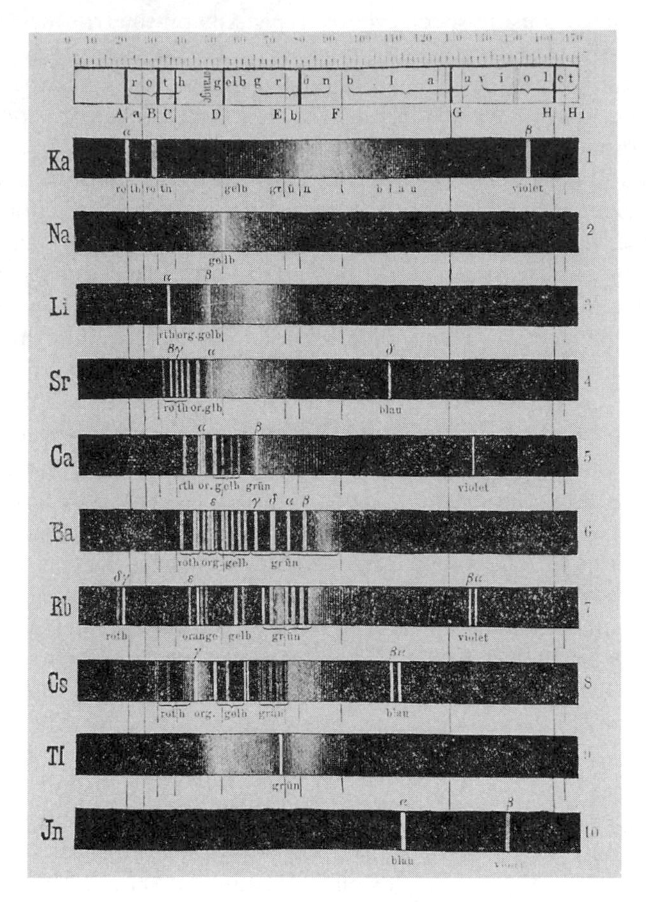

FIG. 10.2. Calibrations of line spectra by Bunsen and Kirchhoff (Stein 1886, 62)

form of matter more elementary than nitrogen which our analysis has not yet enabled us to detect."[17] Symptomatic, too, was the strong resistance to Lockyer's coinage of the term "helium" for the substance responsible for a bright yellow line seen in solar prominences during the 1868 eclipse. The lack of match between this line and any known terrestrial element prompted many chemists to doubt the basis of Lockyer's strategy for solar spectroscopy (Meadows 1972, 59–60; Maunder, 1912, 74–75). Where

17. William Huggins, "On the Spectra of Some of the Nebulae" (1864), in Huggins 1909, 116.

there was no consensus about the criterion for identification of new substances, there was dispute about competent instrument design. Huggins, Lockyer, and Secchi disagreed about the design of the collimator and the slit for producing clear line images. Differences between means of generating comparison spectra—flames, sparks, and arcs—were vital in the handling of spectrometers. Vacuum-tube and induction-coil technology was important just because these differences affected the means spectroscopists used to calibrate their instruments. Walter Maunder put the case like this, with respect to the metallic spectra detectable in the solar atmosphere and tried in laboratory replications:

> These metals thus invading the chromosphere do not muster all their lines in full on such occasions: many fail to answer the roll-call. At one time in the early history of spectroscopy, it was supposed that each element had only a single spectrum, absolutely invariable in the number of its lines, in their positions and relative brightness. Now it is known that spectra are alterable in a number of particulars according to the conditions under which they are produced; the number of lines shown may vary, their actual brightness may suffer change, and their relative brightness, their width and even to some extent their wavelength or position in the spectrum may be altered. (1912, 47)

Key features for managing these variations were the mechanisms for throwing a comparison spectrum and a photographically marked scale onto the eyepiece of the spectroscope's view-glass, and the use of a Ruhmkorff coil as the arc source. Huggins's "star-spectroscope," with which he initiated his astrospectroscopy in the 1860s, was designed at his observatory and then made and marketed by the new specialist maker John Browning. Crucial for Huggins's success was Browning's ability to make cylindrical lenses, which, placed within the telescopic focus, broadened the point image of a star anisotropically so as to visualize the spectrum at the eyepiece.[18] Browning's colleague William Ladd was equally vital for Huggins's projects, partly because he provided the astronomer with high-performance batteries with which to drive the induction coils.[19] Just as Huggins depended on his close alliance with the skilled chemist W. A. Miller of King's College London, whom he met through the Pharmaceutical Society, so Lockyer's work relied for status and skill on his collabo-

18. Huggins, "Historical Statement," 7.

19. William Huggins, "Further Observations on the Spectra of Some of the Stars and Nebulae" (1868), in Huggins 1909, 210. For Huggins's allies and his retrospection about their role, see Becker 1993, 207–22.

ration with Edward Frankland of the Royal College of Chemistry. Frankland's lab assistants were vital in Lockyer's strategies.[20]

Was work with chemists, opticians, and galvanists appropriate for an astronomical observatory? The fights that raged in the Victorian astronomical community during the later nineteenth century raised just this issue. The new moral economy of the state observatories did not help the incorporation of extradisciplinary and potentially autonomous support staffs, especially for difficult new technologies. Just as Howard Becker has urged the significance of the resources provided by support personnel in art worlds (1982, 77–92), so here we can pinpoint the conflicts and fragile alliances upon which the relation between astronomical support staffs and the "astronomers" depended. Thus, while Airy viewed overdependence on instrument makers or support staff as a source of "disgraceful difficulties," he was second to none in seeing the potential for the absorption of astrophotography, suitably managed, into the Greenwich form of life. In 1857 he praised de la Rue's wet-plate pictures of the Moon in very strong (and, in fact, ludicrously optimistic) terms: "when we obtain collodion finer in grain and still more sensitive, it will supersede hand-drawing altogether, and even now the results obtained are much more accurate than anything hitherto done by mapping or hand-drawing." As Holly Rothermel has persuasively argued, "Airy's vision of photography was inextricably bound up with his vision of observatory science."[21] His view was that such measures could be disciplined within the system, and the results of photography and spectroscopy transformed into a carefully monitored serial endeavor. Photographic star surveys and sunspot patrols would fit well into this system. The alternative pattern was the construction of alliances between experimenters, makers, and other scientists outside this system. This alternative needed its own institutions.

5. The Standards and Values of the New Astronomy

As usual, Victorian astronomers treated the work of John Herschel as a good resource in defining what such institutions should do. In 1845 he stated that "every astronomical observatory which publishes its observations becomes a nucleus for the formation around it of a school of exact practice—a standing and accessible example of the manner in which theories are brought to their extreme test." Herschel singled out the links be-

20. Huggins, "Historical Statement," 6. For Lockyer and Frankland, see Meadows 1972, 58–59.

21. For Airy on subordinate sciences, see Meadows 1975, 2, 69. For collodion, see Airy 1859, 18. For Airy's ambitions for self-registration through photography, see Rothermel 1993, 168.

Fig. 10.3. Kew Observatory, headquarters of the sunspot patrol (British Association, *Reports,* 1844, 14th report, pl. 30)

tween these nuclei and the instrumental resources of astronomy, the "continual demand for and suggestion of refinements, delicacies, and precautions in matters of observations and apparatus which react on the whole body of science" (1846, xxix). We have noted that Herschel was a promoter of spectroscopy and photography as parts of the observatories' tool kit in the mid–nineteenth century. He invented the glass-plate negative, campaigned strenuously for a sunspot patrol, and fought hard and successfully for the establishment of an experimental observatory at Kew (fig. 10.3) under the British Association's management (Cannon 1978, 228).

The memoir Herschel drafted in 1842 with colleagues such as Edward Sabine, Charles Wheatstone, and John Gassiot eloquently outlined the tasks of a "physical observatory." Based explicitly on the models established by Humboldt, Gauss, and Weber at Berlin in the 1820s and then Göttingen in the 1830s, and profiting from the collaborative enterprise of the so-called magnetic crusade they had mounted, the projected new observatory would fulfill a number of related tasks. These included verification of instrument performance, development of new astronomical and meteorological experiments and technologies, a depot for accreditation of instrument makers' work, and, above all, a site at which the new branch of experimental astronomy could be pursued (Scott 1886, 50–52). Kew Observatory, especially under the leadership of Warren de la Rue and Balfour Stewart in the 1850s and 1860s, became the key site of this experimental strategy. It was there that photographic and then spectrographic techniques were developed, including de la Rue's photoheliograph and Gassiot's improved spectroscopes.

Stewart's management was especially important, because from 1863 onward British astronomers and natural philosophers rather strenuously defended his priority over Kirchhoff in the construction of the "theory of exchanges," upon which astrospectroscopy depended. This priority fight lies outside the scope of this paper, though it is linked with contemporary discussions of Comte's ban on inferences about the chemical condition of the stars. Its relevance here is that work by William Thomson, Peter Guthrie Tait, and others on behalf of Stewart's work helped transmit spectroscopic technique. As Siegel has perceptively noted, for example, in the mid-1850s Thomson and Stokes held that the "unexpected darkness" of the solar D lines when sunlight was transmitted through sodium flames was merely a psychological effect due to optical compensation. "In the use of optical instruments, separating the phenomena under observation from instrumental and observer-related distortions is often a difficult task" (Siegel 1976, 572). However, their work for Stewart against Kirchhoff convinced Stewart's Scottish allies that the Edinburgh and Kew workers had founded the basis of astrospectroscopy and then embodied its techniques in robust experiments.

The security of programs such as those pursued at Kew hinged on their connections with existing centers of astronomical authority. Herschel, Sabine, Gassiot, and de la Rue made sure that Kew mobilized instrument makers to their cause. Thousands of meteorological, magnetic, and optical devices were tested and certified there. Large numbers of foreign observatories were supplied with standard experimental devices. Magnetism, meteorology, and solar physics formed the basis of this new discipline. Kew standards and values changed the character of experimental astronomy. These standards and values were carefully disseminated throughout the astronomical world (Scott 1886, 74–75). On the other hand, astronomical propriety had to be maintained to ensure the ethos of the major state observatories. Stewart fought fierce battles with the Kew "ancien régime," led by Sabine, over the propriety of displacing a natural-historical program of data accumulation by disciplined precision measurement. This was especially important in the Stewart and de la Rue strategy for reclaiming evidence of linkages between the Sun, the planets, and Earth-magnetic activity through their sunspot patrol.[22] Socially, ex-

22. For controversies at Kew, see de la Rue et al. (1872), 82–87; and Gooday 1989, chap. 7.

perimental astronomy was very much the province of the "grand amateurs," men such as de la Rue, Lockyer, Huggins, and Common. These workers founded their own suburban observatories, such as spectroscopic centers run by Lockyer at Hampstead and by Huggins at Tulse Hill, or the important photographic centers of Common at Ealing and of de la Rue in Cranford. De la Rue and Lockyer were also key figures in the foundation of new astrophysics labs, at Kew and the Solar Physics Observatory in South Kensington. Their practice, as historians have stressed, was devoted to hands-on technologies, either developing their own tabletop instrumentation or forging close links with the instrument makers' community.[23]

The contrast with Greenwich was that between continuous surveillance and heroic experimentation. This was an explicitly political distinction. The Devonshire Commission on the Advancement of Science was told in 1875 of the relation between expertise and these two forms of experiment: "our statesmen do not properly appreciate the value of scientific enquiry. . . . They are very much fonder of experiments made upon a large scale with no defined system, than they are of experiments which have been brought out as the result of a carefully studied previous inquiry" (Meadows 1972, 90). Controversy raged within the Royal Astronomical Society and under the aegis of the Devonshire Commission, of which Lockyer was secretary, about the formation of a solar physics lab in South Kensington (Meadows 1972, 96–100; Dreyer and Turner 1923, 174–76; Hufbauer 1991, 66–67; Smith 1991, 11–12). By 1879 Lockyer's program there was under way, but in the interim the violence of public debate about the true character of experimental astronomy prompted the resignation of Lockyer and de la Rue from the Council of the Royal Astronomical Society and involved Common in controversy on behalf of astrophysics as a legitimate object of endowment. Huggins and Lockyer disagreed on these questions; whereas Lockyer sought an independent astrophysics observatory, Huggins urged the coordination of experimentation and regimented observation under Airy's rule. Lockyer worked hard to build a counternetwork of experimental observatories: in 1871 he told an audience at Cambridge that the university should establish a solar physics observatory, and at the end of the decade he used the pages of *Nature* to support the establishment of the meteorological station atop Ben Nevis (Meadows 1972, 33, 73; Dreyer and Turner 1923, 206–11). In June 1872, the Cambridge wrangler and journalist Richard Proctor told Airy that for the Lockyer group "the 'physics of astronomy' signifies, in the main, any branch of astronomy associated with what Dr. De la Rue has done and Mr. Lockyer might do. The study of stars and nebulae, for

23. For the amateurs, see Lankford 1981; Hufbauer 1986; and Williams 1987.

instance, is only set as a subsidiary matter. Solar physics and solar photography become the leading features. . . . On the other hand, Dr. Huggins' idea is that the physics of astronomy includes in an equal degree every branch of observation by which our knowledge of the nature of the celestial bodies, individually and collectively, may be increased" (quoted in Meadows 1972, 98). As far as Proctor was concerned, the Kew lobby's interest in the "physics of astronomy," which included meteorological prediction, was little better than the astrology of Zadkiel and *Old Moore's Almanac.* It was unfit for admission to the discipline of astronomy.

The networks of great private observatories, such as those of Huggins and Common, new public institutions, such as Kew and the Solar Physics Observatory, and the imperial network of state observatories all provided different versions of what such a physics might mean. When Huggins presented his work on the solar surface to the Royal Astronomical Society in 1866, he was warned that "in the application of terrestrial physics to solar phenomena, we must bear in mind that the relations of solidity, liquidity and gaseity there must be very different to their types here" (Huggins 1866, 161). Experimental astronomers faced a tough task. In order to ply their trade, trials performed at varied observatories and labs had to be calibrated and replicated on the basis of shared standards and values. As Proctor divined, the astrophysics of Kew's sunspot patrol and the work of its spectroscopist allies were a vulnerable target. The linkage between solar and terrestrial physics was the rationale for their project. This linkage needed a set of technologies able to show plausibly that the Sun and Earth were relevantly similar. Hence the force of Proctor's reference to astrology—for the resources Kew found to back up its experiments were often drawn from themes of the plurality of worlds and the discourse of solar habitability.

Herschel, Stewart, and Lockyer were key promoters of the language linking terrestrial and solar life. Herschel backed his father's celebrated model of solar structure: sunspots, it was agreed, were openings through which the dark, solid, habitable surface of the Sun could be perceived.[24] In his influential textbooks John Herschel described spots as depressions in the gaseous cloud surrounding the Sun and then urged the connection between the "correspondence of the dynamical laws" here and elsewhere in the universe and the "provision for other races of animated beings." In response to Kirchhoff's highly controversial story of 1861 that the Sun was a hot luminous liquid above which cloudlike spots floated in the "corona," the layer where spectral reversal took place, Herschel responded

24. For sunspots and habitability, see Crowe 1986, 216–22; Schaffer 1980, 93–96; and Herschel 1851, 229.

that the photosphere was a high-density gas and the corona a layer pierced by openings (Herschel 1851, 554; 1867, 81–82). This model involved abandoning the story of a habitable solar surface; but it emphatically did not involve abandoning a picture of life in the Sun. Herschel quickly found a new resource: the observations by the great amateur James Nasmyth of what were called "willow leaves," patterned markings on the solar surface. Nasmyth was a close ally both of Herschel and of de la Rue, having introduced the latter to astronomy. De la Rue found these "willow leaves" useful because he reckoned that his photoheliscope could make them objective phenomena and win over those who simply could not make out the shapes Nasmyth saw. Herschel found them useful because they could be welded to his sunspot model and, above all, because they might be alive: he lectured in 1861 that "we cannot refuse to regard them as organisms of some peculiar and amazing kind . . . we do know that vital action is competent to develop both heat, light and electricity" (1867, 84; on the willow leaves, see Bartholomew 1976, 263–89; Rothermel 1993, 158–62).

Work with spots and "granulations," as Nasmyth's markings came to be called, delivered a number of lessons for the experimental astronomers. It showed the alliance between their work and the doctrine of universal habitability. Most spectroscopists were also pluralists, including Huggins, who urged from 1864 that the "unity of operation" here and elsewhere taught the truth of the plurality of worlds.[25] It reinforced the local importance of trials which linked terrestrial and solar life. The Kew sunspot patrol simultaneously investigated spot structure and movement, the vital role and putative evidence of organic form. The view was emphatically and remarkably reinforced in a manifesto of 1868 issued by Stewart and Lockyer: "The Sun as a Type of the Material Universe." The spectroscopists argued here that the Kew program was a highly desirable bulwark against materialism, because experimental astronomy taught lessons about "the place of life in a universe of energy" (1868, 257).[26] The strategy developed by the Kew managers and their allies therefore connected stories about energy and vitality with the sunspot patrol's project and local Victorian values. The exactly contemporary controversy about the source of the Sun's heat (see Burchfield 1975), in which Stewart, William Thomson, and Helmholtz were protagonists, was here glossed as a classroom in morals. Stewart announced in 1873 that "the Sun is in the position of a man whose expenditure exceeds his income. He is living upon his capital

25. William Huggins, "On the Spectra of Some of the Fixed Stars" (1864), in Huggins 1909, 60. See also Crowe 1986, 362.

26. The paper is discussed in Myers 1989, 322–23.

and is destined to share the fate of all who act in a similar manner" (cited in Myers 1989, 326). Such moralized solar physics was a Victorian commonplace, bolstered by Max Müller's "solar mythography" and the uses of the limits on the Sun's age against a Darwinian time scale. While Proctor reassured his many readers of the permanence of solar life, Thomson asserted the limited past and determinate future of our nearest star. The "death of the Sun" nicely linked Kew experimentation with its host culture (for solar death, see Beer 1989).

However, these links were always tendentious and never stable. Techniques such as solar photography often failed to settle dispute. The inferences from the tabletop trials of Lockyer and Huggins about the Sun's life needed careful handling to survive. Lockyer reflected on this lesson after the closure of the debate on solar markings: "The 'willow-leaf' controversy shows us the difficulty observers have, not in collecting facts, but in actually being sure of what they see" (1865b, 106–7; cited in Rothermel 1993, 162). Huggins told the Royal Astronomical Society the following year that solar spectroscopy and "the law of exchanges of Balfour Stewart" could not easily resolve the problem of solar structure. If the Sun were a body that obeyed Stewart's principle, then why were spots dark and why were there dark parts between the granules? Lab experiments could possibly help here. We should attend very closely, therefore, to the linguistic and experimental techniques these workers used to help stabilize solar phenomena. Huggins recognized clearly the links between general changes in "the doctrine of evolution in organic as well as in inorganic nature" and the status of his solar spectroscopy.[27] Lockyer and his allies used bolder techniques. Notably, they persistently spoke of the Sun's appearance using imagery drawn from the aesthetics of Victorian natural history. This is how Lockyer described the results of using a wider slit arrangement, as opposed to an oscillating but conventional slit, to deal with the rapid translations of the solar prominences, an arrangement he had learned from Huggins in 1869:

> The solar and atmospheric spectra being hidden, and the image of the wide slit alone being visible, the telescope or slit is moved slowly, and the strange shadow-forms flit past. Here one is reminded, by the fleecy, infinitely delicate cloud-films, of an English hedgerow with luxuriant elms; here of a densely intertwined tropical forest, the intimately interwoven branches threading in all directions, the prominences gradually expanding, and changing slowly, indeed almost imperceptibly. By this method the smallest

27. William Huggins, "On the Bright Granules of the Solar Surface" (1866), in Huggins 1909, 310; and Huggins, "On the Spectra of Some of the Fixed Stars," 60.

> details of the prominences and of the chromosphere itself are rendered perfectly visible and easy of observation. (1869, 355)

The move between unfamiliar experiment and bucolic familiarity was an important value of the new astronomy. It helped calibrate the work of the astronomical lab. Lockyer's reference to "perfect visibility" and "easy observation" was designed to embody a radical new technology of tabletop trials. That embodiment depended directly on connections between Earth and heaven, between his lab and other ranges of skills and disciplines. The course of specific efforts to calibrate troublesome instruments and make tabletop experiments on the Sun more widely credible shows us how this disciplinary order was established.

6. "We Can Take the Very Sun Itself to Pieces"

Two tasks dominated Victorian solar physics: the determination of sunspot behavior and the determination of the solar energy source. Both tasks involved laboratory experimentation. Here we select a couple of exemplary moments in the course of experimental work to illuminate the development of the term "experiment" in this setting. One is the debate of the mid-1860s on the motion of solar matter in sunspots; the other is the debate of the early 1880s on the restoration of solar activity. In both cases, domestic trials were held to provide quite decisive information about the Sun's structure. Thus, when Lockyer and Frankland worked at the Royal College of Chemistry on the behavior of the sunspot spectra, Lockyer announced that changes in line widths must be due to the change of pressure acting on the solar hydrogen in the chromosphere (fig. 10.4). "There is an experiment by which it is perfectly easy for us to reproduce this artificially, so that you see we can begin at the very outside of the Sun by means of hydrogen, and see the widening of the hydrogen lines as the Sun is approached; and then we can take the very Sun itself to pieces" (1873, 107).[28] How could Victorian experimenters claim that they could achieve this remarkable dissociation on the lab bench? The answer to this question hinges on astronomers' work to set up calibration trials. What lab devices resembled spectrometers? And what earthly phenomena resembled those in space? Lockyer sought to show that spectrometry was like barometry; Huggins reckoned spectroscopy was a means of detecting astral speeds.

28. Lockyer used the same phrase in a letter to Stokes, Cambridge University Library MSS Add 7656, L590: "we are slowly pulling the Sun to pieces." I thank Barbara Becker for pointing this out.

Fig. 10.4. Lockyer's apparatus for comparison between solar and laboratory spectra (Lockyer 1887, 212)

Lockyer's claim came in the course of detailed work with Frankland and the Kew sunspot patrol. In 1865 the patrol announced that a statistical analysis of de la Rue's photographs of 694 spots revealed that more than three-quarters showed the spots as cavities, not clouds as Kirchhoff held. Furthermore, the "faculae" (matter associated with spots) concentrated on average to the west of (i.e., behind) most spots, suggesting the cavity structure model was right. The Kew numerology was crucial here: de la Rue's photos did not compel assent, but careful statistics did. Furthermore, these numbers backed up Herschel's model. In sunspots the Kew patrol saw the downrush of cooler matter toward the dark solar core. Lockyer quickly added his testimony to those of his allies. He turned up at the Royal Astronomical Society to remind his colleagues of contradictions between the Herschel model of spots and willow leaves and the models of European observers, who held that spots were high clouds. Herschel believed that spots were holes and that the willow leaves and faculae were mere clouds condensing and moving over the spots. Lockyer confirmed this view with a report of observations of faculae moving over a spot and then of "a downward current into a spot." But Lockyer reckoned his own observations were weak: "it would be rash to say that one absolutely puts the right reading on what one saw." So he backed up his claim with French blowpipe trials with silver borax crystals, which successfully replicated faculae and sunspot behavior (Lockyer 1865a, 236–41; Rothermel 1993,

162–63). To win authority, Lockyer needed his own opportunity to deploy lab work. The opportunity soon came when the distinguished Paris astronomer Hervé Faye produced a controversial survey of sunspot theories. Faye summarized Kirchhoff's argument that the photosphere was liquid and that the sunspots were clouds above it:

> Kirchhoff has conjecturally carried this admirable laboratory arrangement to the Sun itself; he needed a source of continuous light: this would be the photosphere; he needed interposed metallic vapors: these would form the invisible solar atmosphere. . . . but incandescent solids and liquids are the only substances which give continuous spectra, while gases and vapors only provide a spectrum reduced to some bright rays: so the photosphere, far from being gaseous, as we believed and as Arago believed he had demonstrated experimentally, would be a molten solid or, rather, a liquid. (1865, 93)

Arago's lab work on the lack of polarization of sunlight and the hypotheses of Thomson and Helmholtz about the recruitment of solar energy provided Faye with resources to deny Kirchhoff's hypothesis and assert the wholly gaseous nature of the Sun. Faye's 1865 speech at the Académie des Sciences represents a coherent attempt by post-Laplacian physics to obviate solar spectroscopy and experimental astronomy. As Karl Hufbauer has argued, the details of Faye's application of the nebular cosmogony of Helmholtz and Thomson to solar physics were not entirely original, and several details of his model of the Sun as a system of circulating hot gases were immediately challenged, but his claims that gravitational collapse generated the Sun's heat, which then reached the surface by convection, long stayed on astronomers' agenda. Faye reckoned that he could now displace tendentious "analogies between laboratory experiments [*expériences de cabinet*] and the most inaccessible of celestial phenomena" (1865, 144). He rejected the Kew claim that sunspot cycles affected Earth's magnetism. Instead, the photosphere's activity relied on a circulation of heated gas from within the Sun's body. Central gases were so hot that they scarcely emitted visible light, and as the Sun cooled, sunspots would appear as holes through which this heated mass rose from the surface (see Clerke 1885, 193–96; Hufbauer 1991, 62–64). "It is not the solid, cold, and dark kernel of the sun which one sees but the surrounding internal gaseous mass, whose emissive power, at the temperature of the liveliest incandescence, is so weak compared with that of the luminous clouds of nongaseous particles that the difference of these powers suffices to explain the striking contrast of the two layers observed with our darkening telescopes." Eventually the Sun would cool so much that it would completely liquefy, "the definitive extinction, the geological phase" (Faye 1865, 146). By reviving views of Buffon and of Fourier on universal cooling and by connecting this sunspot model with modish doctrines of the

Sun's death and with laboratory thermometry, Faye directly challenged the Kew program on solar vitality. Lockyer's response was a decisive mobilization of his spectroscopic trials, using the spectroscope as a device for taking the sunspots' temperature.

However authoritative it seemed, Faye's work did not go down well with de la Rue or with Stewart. Stewart, the author of the law of exchanges, pointed out that a superheated gas incapable of emission would be transparent, and thus, were Faye right, sunspots should be bright, not dark. Lockyer adopted a much more experimental strategy against the Frenchman. In March 1866 he devised a new arrangement for solar spectroscopy, in which the image was cast onto a screen and a slit in the screen arranged so that light from a specific spot could be admitted to the spectroscope's eyepiece. With a fine Cooke equatorial on a clock drive, Lockyer was able to make the light from a spot's umbra display the same spectrum as that of the photosphere, but "greatly enfeebled in brilliancy." In the Kew model, this was just what would be expected; in Faye's story, however, bright emission lines should be detected, and Lockyer saw none (fig. 10.5). Troubles remained with this evidence: Lockyer had only seen sunspots under cloudy strata, and he reckoned that "the dispersive power of the spectroscope . . . was not sufficient to enable me to determine whether the decreased brilliancy of the spot-spectrum was due in any measure to a greater number of bands of absorption." The absorption bands in the umbral spectrum obsessed Lockyer, since, as he repeatedly pointed out, not only was the continuous spectrum dimmed there but the Fraunhofer lines were thickened. This latter claim was highly controversial. Many held this "thickening" might be an illusion due to the brightness of the background. Certainly it was very hard to reproduce. The problem of this selective absorption was a crucial topic for astrospectroscopists. In sunspots, the late Victorian textbooks agreed, astronomers could see many dark lines, some of which corresponded exactly to those visible at the surface. But other absorption lines in the umbra could not be seen in light from the surface, some lines were very weak, and yet others were completely reversed and appeared as bright emission lines. Lockyer and his colleagues worked with this recalcitrant material to make sense of the sunspot structure. "If the theory of absorption be true, we may suppose that in a deep spot rays might be absorbed which would escape absorption in the higher strata of the atmosphere" (Lockyer 1866, 256–58; Meadows 1970, 42). This made sunspot spectroscopy a key new site for solar work.

Lockyer's strategy in the later 1860s was to turn the spectroscope into a superior substitute for a range of more familiar terrestrial instruments, barometers in particular. This was the work which culminated in Lockyer's identification of "helium" in spectra of prominences in 1869. To

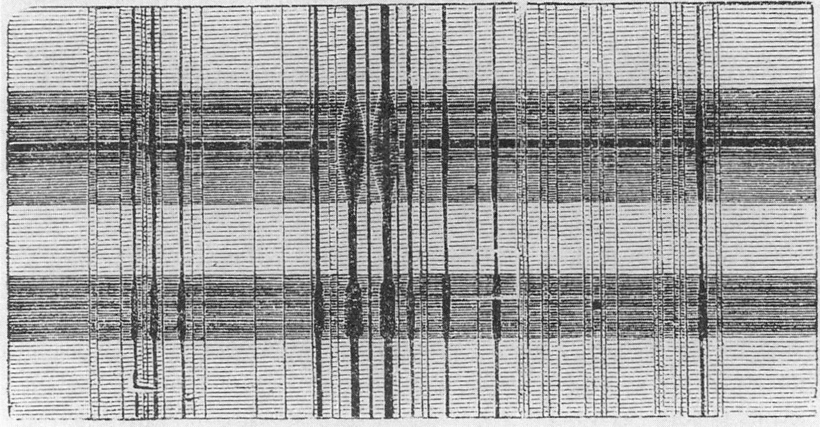

Fig. 10.5. Lockyer's representation of sunspots (*top*) and their spectra (*bottom*). He was keen to stress the line broadening and darkening of the spectrum near the spot. (*Top*) Lockyer 1873, 104; (*bottom*) Lockyer 1887, 100.

make his optical device count as a reliable transmitter of solar pressure, Lockyer had to link his skills with those of Kew Observatory and the Royal College of Chemistry. This was not easy. Huggins's comments show this well, because he was skeptical of Lockyer's inferences. In April 1866, soon after Lockyer began his observing program, Huggins wondered publicly whether the Kew-postulated mechanism for sunspot formation, a downrush of cool gas, might not also explain the granulation, which was interspersed by dark pores, and he reaffirmed that some combination of Stewart's law and the experimental differences of gaseous (line) and stellar (continuous) spectra would produce "a feasible explanation of solar phenomena." In August, Faye wrote to Huggins asking him to check whether the umbra spectrum was "compound, consisting of a continuous spectrum with dark lines and a second spectrum of bright lines."[29] The following spring, Huggins told the Royal Astronomical Society that "the spectra formed by the light from different parts of a solar spot has [*sic*] been examined. No certain modification of the solar spectrum has been detected" (1867). So Huggins's astrospectroscope, developed with the chemist William Miller for nebular and stellar trials, simply failed to register any changes of spot spectra.

Lockyer's analysis of selective absorption was therefore an important test case for making experimental spectroscopy into an effective way of probing the Sun. Yet at this point he lacked any experience of chemical spectroscopy inside the laboratory. So, "by the kindness of Dr Frankland," from 1868 Lockyer engaged in an attempt "to familiarize myself with the spectra of gases and vapours under previously untried conditions. . . . The experience I have gained at the College of Chemistry has guided me greatly in my observations at the telescope" (1869, 350). This attempt involved a subtle combination of eclipse observations in India, lab trials with Frankland's assistants, and Lockyer's own work with the spectroscopy of solar prominences. Alex Soojung-Kim Pang's recent studies of eclipse expeditions show that a fragile system of "small instruments, direct observations and drawing" characterized the work of such temporary field observatories. The Indian observations were highly unreliable. But many observers agreed that the hydrogen lines seen from solar prominences in eclipse broadened out nearer the Sun's surface. Lockyer and Frankland baptized the layer in which the prominences were visible the "chromosphere" and mapped it with the green hydrogen line. Lockyer's lab work then extended this model to the sodium lines seen from spots. Their broadening and that of the green line were attributed to the same cause, and this cause was demonstrated in the lab. The key apparatus here

29. See Huggins, "On the Bright Granules of the Solar Surface," 308–11.

was Frankland's iron chamber, in which a burning jet of hydrogen gas could be subjected to increasing pressure, and its spectrum observed directly. Thus, mapping the chromosphere's pressure became a matter of measuring line broadening: "we were in a position to determine the atmospheric pressure operating in a prominence, in which the red and green lines are nearly of equal width, and in the chromosphere, through which the green line gradually expands as the sun is approached." This made the green line the new calibration for barometric determination: "the green line expands in a more decided manner by pressure than does the red."[30]

Having established the spectroscope as a solar barometer, it was possible for Lockyer to link his measures to the claims of the Kew workers. They held that the photosphere was a zone of condensation and the spots a zone of downrush, where increasing darkness matched increasing pressure. Here was a compelling story for the spectroscope as barometer: Lockyer held that the thickening of the sodium D line in spot spectra must be a mark of a condensation. He performed a demonstration that was later to become a standard for his public lectures:

> Here I have an electric lamp, and by means of this slit I only permit a fine line of light to emerge from it; here the beam passes through a bisulphide of carbon prism, and there you see on the screen the glorious spectrum, due to the dismemberment of the fine line of polychromatic light. Mr Pedler [lab assistant at the Royal Institution] will now place a glass tube containing metallic sodium, sealed up with hydrogen, in front of the slit, and will heat it with a spirit lamp. As the sodium vapour rises you see the dark line of absorption make its appearance as an extremely fine line, and finally you see that the light which traverses the upper layer of the sodium scarcely suffers any absorption—the line is thin; while on the contrary the light which has traversed the lower, denser layers has suffered tremendous absorption; the line is inordinately thick, such as we see it in the spectrum of a spot. (1970, 1:96)

The establishment of this equivalence was a vital aim of Lockyer. Henceforth, he reckoned, chemists and experimental astronomers would be able to work together in an effort to replicate solar physics in the lab and map the Sun itself. No immediate consensus was formed in response to this message, however. Thus, George Stokes told Lockyer that "I see no evidence that for a given temperature the spectral appearance depends on density of active vapour present rather than quantity" (1907, 1:399).

30. For the Indian eclipse work, see Pang 1993, 270–72; and cf. Rothermel 1993, 149–51, on verbal description. For Lockyer, see Meadows 1972, 59; Frankland and Lockyer 1869, 290; and Lockyer 1970, 1:97.

Similarly, despite his concession that Lockyer's claims about line broadening were replicable, Huggins's work on sunspots in the mid-1860s is distinctly different from that of Lockyer. This difference hinges on the contrast in their alliances and their social positions: Huggins's alliance with Airy and with Miller pointed his attention toward nebular astronomy, and there his program depended directly on the criterion that gases gave line spectra. Lockyer and Huggins differed markedly, for example, in the lobbies they mounted for preparations for the eclipse expedition of 1868: Lockyer backed solar prominence observations with barometric spectroscopy, while Huggins tried to win money to compare solar and nebular spectra. His most famous achievement of 1864–65 was to show that Orion and other nebulae, allegedly resolved into stars, were really gaseous. Then questions of nebular motion and their difference from stars became important. From 1864 he and Miller were keen to detect line-of-sight motion by means of Doppler shifts of spectral lines from nebulae and stars. He consulted Clerk Maxwell and Stokes on these problems in 1867 and learned from them of the high-precision measures that would be necessary to establish reliable shifts above the level of background variation of line positions. Above all, they told him that Doppler effects would not change the apparent color of stars, because infrared and ultraviolet frequencies would be changed too, and so would only be detectable if Huggins could first recognize terrestrial elements in stellar and nebular spectra. So he worked hard with London electricians to make an experimental device which could perform this well.[31]

This device included a novel way of throwing the comparison spectrum onto the eyepiece to avoid air disturbance because of spark heat, and a powerful battery and induction coil built by Ladd with exchangeable plates. Unlike Lockyer, when Huggins used this new machine on sunspots in April 1868, he checked how much of the incoming light was from the umbra as opposed to Earth's atmosphere—about three-quarters, he estimated photometrically. Huggins judged there was no chemical difference between the umbra and the photosphere, but he differed from Lockyer in the barometric estimate. Lacking immediate contact with Frankland's team, Huggins was far less convinced of the allegedly self-evident link between broadening and increased pressure. He noted Stewart's argument that a temperature change deep in the Sun could explain it just as well. Lockyer's efforts to turn the spectrometer into a barometer did not win assent in Tulse Hill. Instead, disputes were endemic between Huggins and Lockyer, notably on the question of the gaseous nature of nebulae. Lock-

31. Huggins, "Further Observations on the Spectra of Some of the Stars and Nebulae," 205–10.

yer and Frankland reckoned Huggins was wrong to suppose his spectroscopy revealed details of nebular composition and believed that the differences observed might simply be due to pressure changes there. Differences in research goals and alliances set up different evidential contexts for instrumentation and experiment.[32]

7. THE END OF THE SUN AND THE ENDS OF EXPERIMENT

These differences could be resolved if Victorian astronomers could agree about the most important message their work might teach: the question of the Sun's life and death. The stories Huggins and Lockyer told from their observatories about suns, stars, and nebulae were interpreted by their audiences and their peers as messages about Earth's future (fig. 10.6). Stewart's notorious comparison between the Sun and a degenerate wastrel, cited previously, was an example of this commonplace. So too was the spectacular statement of the moral message of sunspots in the 1868 text of Stewart and Lockyer, "The Sun as a Type of the Material Universe," which made the tendentious similarity relation between the social and the physical quite explicit. Their comment that "as in the social world a man may degrade his energy so also in the physical world energy may be degraded" is an argument about the function of Kew Observatory, of human destiny, and of solar life (1868, 322). This was also the function of the influential work which Stewart produced with his Edinburgh ally Peter Guthrie Tait seven years later, *The Unseen Universe,* in which experimental physics was deployed against materialist determinism.[33]

Their sources lie with the established authorities of Victorian physics. Helmholtz (in his 1854 lecture "The Interaction of Natural Forces") and then Thomson (notably in his paper "On the Age of the Sun's Heat" in 1862) had developed a thermodynamic model of the source of solar energy in the gradual gravitational condensation of the Sun. For Helmholtz, this was a satisfactory reconciliation of the nebular hypothesis with the conservation of force. He told his Königsberg audience of the moral lesson of the Sun's exhaustion: "to our own race it permits a long but not an endless existence; it threatens it with a day of judgment, the dawn of which is still happily obscured" (1962, 90). In 1862 Thomson reckoned that "the sun has not illuminated the earth for 100,000,000 years" and that "unless sources now unknown to us are prepared in the great storehouses of creation" the course of life on earth would be very limited (cited in

32. William Huggins, "Observations of the Sun" (1868), in Huggins 1909, 315–16. For Lockyer's fights with Huggins, see Meadows 1972, 61.

33. For Stewart's later work, see Myers 1989, 322; and Heimann 1972.

FIG. 10.6. The death of the Sun and the end of the world (Camille Flammarion, *Le fin du monde,* 1893)

Smith and Wise 1989, 530–33; see also Burchfield 1975, 27–32; James 1982, 173–80; Hufbauer 1991, 56–57). But "our knowledge" was defined as that of William Thomson. A salient feature of the lengthy debates on the Thomson estimate of solar age was the imperialist claims made by the natural philosophers over evolutionists, geologists, and many astronomers. Thus Thomson was countered by the Cambridge authorities James Challis, who pointed out that Herschel had spoken of "vitality" in the Sun when discussing its granulations, and George Darwin, who cited Lockyer's astrochemistry to show that "the energy of concentration may be largely underestimated by computing simply from gravitation" (see Burchfield 1975, 122–28; Smith and Wise 1989, 536, 542). It was clear to most members of the Victorian scientific community that these issues hinged on rival models of the legitimate scope of physics. Orthodox astronomy eventually conceded the Helmholtz-Thomson model. The most influential historian of Victorian astronomy put the point bluntly: "There remains as the only intelligible rationale of solar sustentation, Helmholtz's shrinkage theory. And this has a very important bearing upon the nebular view of planetary formation; it may, in fact, be termed its complement" (Clerke 1885, 355).

The shrinkage theory and its consequences provided fertile grounds for testing out the scope of experimentation. This was because its defenders based their authority on their mastery of laboratory physics and conversion processes. In Thomson's case, notably, such processes were held to be those made familiar in the sphere of capitalist machinofacture and on show in his Glasgow lab. In a remarkable lecture at the Royal Institution in 1887, Thomson built what he called a "model mechanical sun," in which an imaginary assemblage of gear-driven paddle wheels was used to calculate the Sun's output: "it would be impossible, by any of the appliances of solar physics, to see the difference between our model mechanical sun and the true sun" (see Smith and Wise 1989, 538–41). Backed by the doyen of American astronomers, Simon Newcomb, Thomson urged a limit of about 20 million years on solar life and showed clearly how his grip on engineering mechanism could master the phenomena of solar spots and prominences, of spectroscopy, and of experimental astronomy. An ingenious mechanism, he reckoned, could explain away the spectroscopic claim that spots were sites of downward-moving pressurized gas. Glasgow engineering beat South Kensington solar chemistry. This claim to mastery was often contested, and these contests show rather well the variability in the meaning and extension of laboratory experiment. A good case here is the work of William Siemens, who in the early 1880s directly challenged Thomson with a rival engineering story about the Sun. Siemens's work is our final example of a set of tabletop experiments which Victorians assessed for their applicability to extraterrestrial events.

Siemens, manager of the British end of the great electrotechnology network, was easily Thomson's equivalent as a master engineer. He preceded Thomson as founder-president of the Society of Telegraph Engineers and was already distinguished for his work on submarine telegraphy, for which he designed the remarkable cable-laying ship the *Faraday* (Pole 1888, 207–10, 265–66). Obsessed by the basic issues of waste and economy, in 1856 he and his brother Frederick designed the revolutionary regenerative furnace for steel manufacture, and in 1869 he announced his classical theory of the conversion of electrical energy into dynamical force without the aid of permanent magnetism (Pole 1888, 97–104, 130–40, 224–48). These machines came to dominate iron and steel manufacture and thermal power generation throughout the European market. He developed a cosmology in which waste was the ultimate sin, and systematic efficiency a universal virtue. Thus he told Coventry science classes in 1882 of the "five kinds of waste": time, food, personal energy, mechanical energy, and material. He lectured on the domestic economies of the great manufacturers, while his own home in Kent was a temple to prudent economy, run on a steam dynamo which powered what he called, significantly, an "artificial Sun."[34]

Siemens's ideology of efficiency helped him define the relation between the insulation of the laboratory, occupied by what he called "the high priests of science," and the factory shop of the "rule-of-thumb" practitioner "who is guided by what comes nearer to instinct than to reason." There was a close link between his model of this division of labor and his argument that terrestrial processes could be transferred to solar physics. His domesticated "artificial Sun" was a fine case of this alliance of technology and translation promoted by a cadre of scientific engineers who could move from the lab and the workshop to the "great sources of power in nature."[35] So the techniques of the regenerative furnace were fundamental resources for his approach to the Sun. From the late 1840s, Siemens worked with British engineers to design a series of feedback processes in heat engines. The principle of his early "metallic respirator," which he introduced in 1847–50, was the passage of hot waste gas through a respirator on its way to the condenser. Then water from the condenser would be heated on the way to the boiler by passage through this respirator. The same reclamation principle was used in his double-chamber furnace design half a decade later: here hot combustion products were led into a brick chamber, whose resultant heat was then used to warm the furnace air supply before it entered the combustion chamber.[36] This principle, then, informed Siemens's riposte to Thomson in the early 1880s. In talks at the Royal Society and the Royal Institution given in the same year as his Coventry lecture on waste, he urged that the Sun could scarcely be less efficient than his own steelworks, and the steelworks technology provided him with the means to show this in the lab. The implication was that Thomson had simply underestimated the "sources of energy now known to us." Thomson's contraction model violated Siemens's aesthetic distaste for wastefulness: "the true solution will be furnished by a theory according to which radiant energy which is now supposed to be dissipated into space and irrecoverably lost to our solar system could be arrested, wholly or partly, and brought back in another form to the sun himself, there to continue the work of solar radiation." The "wanton dis-

34. William Siemens, "On Waste" (1882), in Siemens 1889, 3:360, 363–64.

35. William Siemens, "Address to the British Association" (1882), in ibid., 319.

36. William Siemens, "On a Regenerative Steam Engine" (1856), in ibid., 1:50–61; William Siemens, "On a Regenerative Gas Furnace" (1862), in ibid., 81–95.

Fig. 10.7. Siemens's model of the regeneration of the solar energy (*Nature* 25 [9 Mar. 1882]: 441)

sipation" imagined by Thomson and Helmholtz could in fact be compensated by a process "somewhat analogous to the action of the heat recuperator in the regenerative gas furnace."[37]

Siemens described the following process, taking care to draw as much authority as he could from thermal engineering and astrophysics: space was full of an attenuated gas mixture; these gases were drawn into the Sun at the poles under solar gravity; within the Sun pressure forced these gases into chemical combustion; finally combustion products were ejected at the equator (fig. 10.7). These combustion products, just like waste in the regenerative furnace, were then chemically dissociated under the action of sunlight and returned to the poles. The resources Siemens used here included the spectroscopy of Huggins, which showed the presence of water vapor and carbon compounds in comets and meteorites, and the Kew sunspot patrol, which provided evidence for gaseous compression and combustion. He also consulted Lockyer on the existence of oxygen gas in the

37. William Siemens, "On the Conservation of Solar Energy" (1882), in Siemens 1883a, 6, 21–22.

sun's atmosphere and backed Lockyer against Draper on the hotly debated question of the bright oxygen lines the American had seen.[38] Where these inferences broke down, Siemens made up with lab work. For example, he performed a tabletop experiment on the effect of sunlight on water vapor in which he filled a vacuum tube with water vapor and placed it between two platinum electrodes. When the tube was cooled, no discharge occurred across the electrodes, but a discharge was observed when sunlight was shone on the tube. This increased the plausibility of the mechanism whereby sunlight dissociated and thus reenergized the solar combustion products. These trials in what Siemens called "electrohorticulture" proved that "terrestrial sources of radiant energy were capable of imitating solar action." He reckoned that stellar space was "filled with rarefied matter analogous to that which we can actually produce in our vacuum tubes," another appeal to state-of-the-art lab physics.[39]

The key element in Siemens's solar model, however, as his appeal to the regenerative furnace indicated, was the account of solar temperature he envisaged. Victorian chemistry held that above about 3000°, hydrogen would be too heated to burn.[40] So for the solar furnace to work, the Sun's temperature must not be greater than this value. Siemens reprinted currently accepted graphs of the variation of radiation intensity with frequency and temperature to show that it was plausible that the temperature of the Sun's photosphere "could not materially exceed that of a powerful electric arc." Otherwise it would shine in the ultraviolet, extinguishing life on Earth.[41] Here Siemens was on authoritative ground and deployed much earlier work from his Bakerian lecture at the Royal Society in the spring of 1871. Since the resistance of a gas would be related to the velocities of its component molecules, it would be possible to construct a parabolic function linking resistance with temperature. This function was checked for low temperatures by Werner Siemens in his Berlin works. The immediate application was a method for high-temperature measurement of otherwise inaccessible hot systems, since it would be possible to change the temperature of some external known resistance until it balanced a resistor within the hot chamber. At balance point, the two temperatures would be equal. Alternatively, a differential voltmeter could be used to assess the different resistances. Siemens consulted his friend Clerk Maxwell on this question: Maxwell commented that while metals become more resistant

38. For Siemens's solar model, see Siemens 1970, 1:207–13; Siemens, "On the Conservation of Solar Energy," 10; Lockyer to Siemens, 28 Oct. 1873, in Siemens 1953, 49.

39. Siemens, "On the Conservation of Solar Energy," 13; William Siemens, "Reply to the Objections Presented by M. Faye" (1882), in Siemens 1883a, 51; Siemens 1970, 1:208.

40. G. D. Liveing, "Dissociation of Attenuated Compound Gases" (1882), in Siemens 1883a, 108–11.

41. Siemens 1970, 1:209.

when heated, gases become more conductive, and so was dubious whether Siemens's touted formulae would work (Siemens 1871; Maxwell to Siemens, 23 June 1871, in Siemens 1953, 23). In the early 1880s, however, Siemens used these formulae to compelling effect, or so he hoped. He performed a series of lab trials at the Royal Institution and elsewhere designed to test and establish a relation between radiation and temperature. He set up a circuit which allowed him to measure the electrical power developed across a resistance. Values of the resistance were matched against corresponding power inputs, and using his relation for temperature and resistance, he was able to derive a relation linking power input and temperature. From Thomson's estimate of the solar constant he was then able to derive the maximum solar temperature: around 2800° (Siemens 1883b, 1970, 1:212). Siemens reckoned that this combination of thermoelectric trials would provide sterling proof for his solar theory by demonstrating the precise equivalence of the Sun's work with that of a domestic electric arc and an industrial regenerative furnace.

Siemens soon found allies for his combination of electro-horticulture and thermoelectricity. Support came from the eccentric Astronomer Royal for Scotland, Charles Piazzi Smyth, who had developed similar models on spectroscopic grounds in the 1870s and now announced replication of the electrical trials on dissociation in sunlight. Equivocal support was also gained outside the astronomical community from the American scientists Charles Morris and Sterry Hunt. Morris backed Siemens's story about "chemical vigour" against Thomson, while Hunt found a Newtonian precedent for Siemens's argument that solar activity could be recruited from gases in space.[42] However, decisive criticism came from Thomson himself, who pointed out privately to Siemens that his argument violated the principle of dissipation, and from the great Irish physicist G. F. FitzGerald, who pointed out that if solar energy was completely recycled, then all stars should be invisible. Siemens vainly tried to argue for a differential absorption of high-frequency rays in space, but as an editorial in the influential amateur magazine *The Observatory* noted, this was a fatal confession. It was the conflict between Siemens's engineering and the received tenets of experimental astronomy that mattered. "This ingenious theory originated in the laboratory of one of the greatest living electricians, but it supposes a condition of things in and around the Sun of which the telescope and the spectroscope afford us no indication."[43] The disciplinary limits on

42. For electro-horticulture, see Pole 1888, 321–28; and Smyth to Siemens, 17 Mar. and 5 Apr. 1882, in Siemens 1953, 178–79. For the Americans, see Charles Morris, Sterry Hunt, and William Siemens, "Letters on the Conservation of Solar Energy" (1882), in Siemens 1883a, 27–37.

43. G. F. FitzGerald, "Dr Siemens's Solar Hypothesis" (1882), in Siemens 1883a, 42–44. The editorial is in *Observatory* 6 (1883): 293–95.

experimentation were heavily marked here. What counted as legitimate experimentation for Siemens was utterly illicit for the Victorian astronomers. Faye spoke for most of the astronomical community when he insisted, characteristically, that faced with a choice between the death of the Sun and the death of astronomical autonomy, his colleagues would choose the former: "It is not very likely that astronomers would adopt such hypotheses. They would doubtless be glad to think that there was some means by which the Sun could preserve its heat longer: but as its final cooling down is, in any case, a very distant catastrophe, they will console themselves in thinking that the things of this world, even the most lovely, do not seem to be made to last for ever."[44]

Work on sunspots in the 1860s and on solar energy in the 1880s shows the difficulty of defining the limits to experimental trials. In order to make their work count as relevant to the Sun, Lockyer and Siemens had to make standards and values that were effective in one limited setting work outside that setting. Their projects are part of the enterprise which Galison and Assmus (1989) identify as the mimetic impulse in the Victorian morphological sciences. In their study of the translation of C. T. R. Wilson's efforts to make clouds in the lab into the context of Cavendish particle physics, these historians tell us how Wilson's guru, John Aitken, following attendance at Thomson's Glasgow natural philosophy class, inaugurated key new processes within his laboratory: cyclones, tidal vortices, clouds. We should interpret the spectra made by Frankland and Lockyer and the incandescent resistors handled by Siemens in similar terms. Such techniques help lab manipulations translate between local milieux and the wider world and help the limited metrology of one set of experiments work as measures of the cosmos.

Commentators on experiment have often been impressed by the robust character of such local enterprises. It would be dangerous to conclude from this healthy localism that all we need to know is to be found on the tabletop of experimentation. We also need to understand how experiments get to work elsewhere. Changes in the work styles and topology of the sciences help make the world fit for experimental trial. Ian Hacking counsels that "we are convinced because instruments using entirely different physical principles lead us to observe pretty much the same structures in the same specimen" (1983, 209). So similarity judgments help make conviction. Similarity judgments are institutions. In this paper, I have been concerned with the work astronomers and other Victorians performed to set out criteria which allowed them to see the Sun and the laboratory as the same kind of place and to see the spectroscope, the barometer, and the

44. Hervé Faye, "On Dr Siemens's New Theory of the Sun" (1882), in Siemens 1883a, 48–49.

regenerative furnace as referring to the same kind of world. A very wide range of resources and techniques were used by astrophysicists to establish the similarity between earthly and stellar physics. These techniques hinged on practical tasks performed in the observatory: the claims that varying forms of light from varying sources affected photochemical agents in the same sort of way; that gases in stellar atmospheres absorbed and emitted radiation in the same way as terrestrial fluids did; that high pressures affected light absorption the same way in chemistry labs and in the solar chromosphere; and that furnaces in steelworks behaved the same way as the heart of the Sun did. In turn, these claims hinged on institutionalizing new social formations, such as the remarkable systems of personal discipline through which observatory managers calibrated observers' performance against galvanic clocks and through which workers sent out as delegates from metropolitan observatories could be trusted to respond reliably in distant and exotic sites. The claims also hinged on some idiosyncratic theological and natural-philosophical themes, such as the modish concern with the possibility of other inhabited worlds or with the divine warrant which maintained a continuous space-filling luminiferous ether. In these enterprises, "experiment" is best understood as a label applied to complex arrays of personnel, technique, hardware, and social institutions. The label does indeed refer to the production of phenomena and the organization of the techniques which enable this production. Because of this relation between phenomena, technique, and institutionalization, the term "experiment" carried varying senses and the variations depended on local problems of astrophysical inference.[45] In defining those places in the universe which could possibly be judged to be similar, these Victorians simultaneously redefined where their "experiments" ended.

Acknowledgments
I am grateful to Barbara Becker, Jed Buchwald, David DeVorkin, Ian Hacking, Karl Hufbauer, Holly Rothermel, and Sven Widmalm for their very generous help. Thanks to the University Library Cambridge for access to manuscripts in its collections.

45. The misfortunes of experiment in astrophysics continue. Ed Regis explains the low status of astrophysics at the Princeton Institute for Advanced Study between 1940 and 1970: "astronomy is not an experimental science in the same sense that the rest of physics is. In physics proper, theories are testable by laboratory experiments. . . . the same thing doesn't hold for astrophysics, because there's just no way to get out there and handle the objects themselves. . . . So observational astronomers have to be content with sitting at their telescopes and passively gazing off into the nether reaches." Regis claims that radio telescopes and computer simulations changed all this: "the computer gave astrophysicists a measure of experimental control over their subject matter." See Regis 1987, 170–71.

ELEVEN

THE STURDY PROTESTANTS OF SCIENCE: LARMOR, TROUTON, AND THE EARTH'S MOTION THROUGH THE ETHER

Andrew Warwick

1. INTRODUCTION

IN 1990 BRITAIN'S LEADING RESEARCHER IN ELECTROMAGNETIC THEORY, Joseph Larmor, published what was to become his most widely known work, *Aether and Matter* (Larmor 1900). The book marked the culmination of almost eight years of work for Larmor, during which he had constructed a new electrodynamics using the concept of discrete electric nuclei, or "electrons."[1] As the mature expression of his most original work, *Aether and Matter* carried much of Larmor's reputation as a mathematical physicist at Cambridge. Within a year of its publication, however, the experimental foundations of the new electrodynamics had been challenged by William Hicks, professor of mathematics and physics at Firth College (Sheffield).[2]

The dispute that developed between Larmor and Hicks concerned the legitimacy and interpretation of the null result obtained by Michelson and Morley in their now famous ether-drift experiment.[3] Larmor claimed that the null result provided firm empirical evidence that matter contracted when it moved through the ether (Larmor 1900, 64). Hicks made two counterclaims: first, that the null result reported by Michelson and Morley could not be trusted, because the experiment had been incorrectly analyzed; and, second, that even if the null result were ultimately confirmed,

1. For a detailed technical discussion of the development of Larmor's new electrodynamics of moving bodies, see Warwick 1991, 29–91. On Larmor's introduction of the electron see Buchwald 1985a, 133–73; and Hunt 1991, 209–39.

2. William Hicks was 7th Wrangler in the Cambridge Mathematical Tripos of 1873. For further biographical details, see Milner 1932–38.

3. On the Michelson-Morley experiment, see Swenson 1972; and Goldberg and Stuewer 1988.

it could not legitimately be interpreted as evidence that moving matter contracted. Larmor responded to Hicks's challenge by collaborating with Frederick Trouton to produce a new electrical method of investigating the effects of the earth's motion through the ether. The new method, Larmor hoped, would corroborate Michelson and Morley's null result and furnish independent evidence for the contraction of moving matter. In this essay I recount these events and their aftermath, focusing upon the following three issues.

The first issue is historiographic. Larmor's claim, that moving matter contracts in precisely the manner first suggested by George FitzGerald, is frequently portrayed as a classic example of the adoption of an ad hoc hypothesis in physics.[4] According to this view, Larmor adopted a theoretically unfounded hypothesis simply to evade an otherwise troublesome result. I have argued elsewhere that this view is untenable on the grounds that Larmor's new electrodynamics actually predicted that moving matter would contract in this manner (Warwick 1991, 59). My argument is further substantiated here in the following way. Larmor's response to Hicks's claims shows quite clearly that, far from being troubled by the outcome of the Michelson-Morley experiment, Larmor *required* the null result as evidence that his new electrodynamics was correct. Had Larmor believed his explanation of the null result to be somehow contrived, he would have welcomed Hicks's claims as evidence that the contraction hypothesis could be abandoned. I shall argue that acknowledgment of this fact is crucial to our understanding of the work of those British mathematical physicists who took up and developed Larmor's electromagnetic theory during the early 1900s.

The second issue concerns the relationship between experimental and theoretical work in late-Victorian physics. Larmor needed *experimental* evidence to refute Hicks's challenge, but Larmor was no experimenter. He had to rely upon the willingness of others to undertake experiments on his behalf. In this case Larmor turned to Trouton's laboratory in search of the new material technology he required, but this kind of technology alone could not fulfill his needs.[5] Larmor had to employ his considerable skill as a mathematical physicist to generate the *evidential context* in which the new experiment could act as a test for the contraction of moving matter.[6]

4. This problem is discussed in Hunt 1988 and Nersessian 1988.

5. On the role of material technologies in the establishment of scientific facts, see Shapin and Schaffer 1985, 25–60.

6. By "evidential context" I mean the range of research problems and debates in physics to which any given experiment is thought to be of relevance. For further discussion of the term, see Pinch 1985.

By studying the nexus of theoretical and experimental work from which Larmor's new test was forged, I shall show how the test was effectively coproduced.

The third issue raised concerns the operation and breakdown of what I shall call the "Maxwellian network."[7] It was not by chance that Larmor turned to Trouton's laboratory for help following Hicks's challenge. During the 1890s Larmor had become part of a network of physicists, the Maxwellians, who shared a common interest in certain theoretical and experimental questions raised in Maxwell's *Treatise* (Maxwell 1873).[8] Two leading members of this network, George FitzGerald and Oliver Lodge, helped to shape Larmor's electromagnetic theory during crucial phases of its development. When FitzGerald died in 1901 his assistant, Trouton, took over the running of his laboratory in Trinity College (Dublin) before moving to University College (London) as professor of physics. Trouton continued to experiment on the ether until 1908, yet his collaboration with Larmor in 1901 represented the final achievement of the Maxwellian network. I shall argue that by studying both the successful operation of the network in response to Hicks's challenge and the effects of the network's subsequent collapse, we can gain considerable insight into the nature and extent of the local economies within which Larmor's electromagnetic theory held currency.

2. Larmor and the Contraction of Moving Matter

In order fully to appreciate the potentially destructive nature of Hicks's challenge to Larmor, it is necessary to understand the importance of Michelson and Morley's null result to Larmor's electrodynamics. I begin, therefore, by explaining how the null result became incorporated into British electromagnetic theory during the 1890s. Michelson and Morley's announcement in 1887 that they had failed to detect the earth's motion through the ether did not cast doubt upon the ether's existence.[9] To the majority of practicing physicists, the null result suggested, if anything at all, merely that the ether was carried along at the earth's surface as it orbited the sun. For the small group of Maxwellians working on the electromagnetic theory of light, however, the result was rather puzzling on two counts.[10]

7. On some advantages of approaching scientific work in terms of networks of cooperation, see Latour 1987, 179–213.

8. For a discussion of the emergence of the Maxwellian group, see Hunt 1991.

9. On the initial impact of Michelson and Morley's null result, see Buchwald 1988 and Hunt 1988.

10. One other physicist, H. A. Lorentz, also found the null result problematic. Lorentz had made a thorough study of the optical effects of the earth's motion through the ether and

First, the most convincing explanation of the other optical phenomenon associated with the earth's motion—the aberration of starlight—was couched in terms of a nonconvected ether; that is, an ether that was not dragged along at the earth's surface.[11] Second, and perhaps more seriously, much of the mathematical simplicity of the emergent Maxwellian electrodynamics would be lost if the ether were dragged along at the surface of moving bodies (Larmor 1894, 477).[12] A worrisome contradiction therefore existed for these men between the most obvious interpretation of Michelson and Morley's null result (the ether is convected by moving matter) and their preferred explanations of stellar aberration and the electrodynamics of moving bodies (the ether is not convected by moving matter).[13]

George FitzGerald attempted to reconcile nonconvected ether theory and Michelson and Morley's null result in 1889 by postulating his now infamous contraction hypothesis (Hunt 1988). He suggested that moving matter contracted in its direction of motion through the ether by exactly the amount required to negate the effect that they had sought.[14] But, as I have shown elsewhere, this idea was not taken seriously by any of his British colleagues (Warwick 1991, 43–45). Oliver Lodge tried to resolve the problem experimentally. During 1891/92 he attempted to convect the ether artificially using his great whirling machine. Lodge passed the light beams of an interferometer in opposite directions between two rapidly rotating steel plates. He reasoned that if the ether were dragged along by the motion of the plates, the velocity of the beams would be differentially altered, and the interference fringes shifted. But the invariably null results obtained by Lodge only heightened the contradiction that he had tried to resolve. If the ether was not convected by moving matter, as Lodge's experiments seemed to confirm, why had Michelson and Morley failed to detect the earth's motion?[15]

---

concluded that nonconvected ether theory gave the best explanation of first-order optical effects. See Miller 1981, 18–25.

11. Most supporters of nonconvected ether theory (including H. A. Lorentz) also adhered to atomic theories of matter. They believed that the ether streamed freely between the atoms of which matter was composed. See, e.g., Larmor 1894, 479.

12. Lorentz had also argued that the most comprehensive convected ether theory, due to George Stokes, was inconsistent; see Miller 1981, 18–19.

13. The introduction of a convected ether would have generated formidable mathematical complications to the Maxwellian account of the electrodynamics of moving bodies. If the ether surrounding a moving charged body was itself subject to viscous flow, then it would be necessary to employ both electrodynamics and hydrodynamics in order to solve the simplest problems.

14. Hunt (1988) has shown that FitzGerald had some theoretical reasons for believing that moving matter might contract in the manner that he proposed.

15. Lodge's work on the whirling machine is discussed in Hunt 1986.

In 1894, Larmor began to develop a theory of optical and electromagnetic phenomena based on a new, rotationally elastic, ether. For reasons of mathematical simplicity, he too decided that his ether would have to be of the nonconvected variety. But although initially puzzled by Michelson and Morley's null result, Larmor soon became convinced that the result could be accommodated quite naturally within his electrodynamics (Warwick 1991, 56–60). By 1897, moreover, he actually *required* the null result as firm evidence that his new, electron-based *electronic theory of matter* (hereafter ETM) was correct. Since the events recounted below turn crucially upon this important shift in Larmor's electromagnetic theory, I shall briefly describe the two lines of research that led him to his explanation of the null result.

The first line of research concerned the phenomenon of light convection. When a bean of light passes into a moving, transparent medium, part of the medium's velocity is imparted to the light beam. Larmor constructed a theoretical model of this effect and used it to investigate the behavior of the light beams in Michelson's interferometer. He found he could explain the troublesome null result if he employed new space-time transformations to relate the stationary ether frame to the rest frame of the moving interferometer. He subsequently found that Maxwell's equations were invariant under these new space-time transformations provided that new electromagnetic vectors were introduced in the moving system (Warwick 1991, 83–91).[16] Larmor eventually explained this finding in the following way. He claimed that the electromagnetic fields measured by observers in moving systems (on the earth's surface, for example) were not the real fields that would be measured by observers who were stationary in the ether. Rather, they were subsidiary fields generated partly by the earth's motion through the ether. It just happened that these subsidiary fields were also subject to Maxwell's equations. Larmor noted, moreover, that the new space-time transformations had a strange property; they seemed to imply that moving electrical systems contracted in precisely the manner suggested by FitzGerald (Warwick 1991, 59).

The second line of research that led Larmor to his explanation of Michelson and Morley's null result concerned the potential surrounding of a moving electron. Larmor had introduced the electron into his theory in 1894 to solve a number of problems that had plagued his account of electrical conduction (Buchwald 1985a, 133–73). By 1897 he had come to believe that positive and negative electrons were the sole constituents of

16. Strictly speaking, Larmor only showed that Maxwell's equations were invariant under these transformations to second order. The notion of first- and second-order results is discussed in Appendix A.

gross matter. In that year, Larmor derived a general expression for the total electromagnetic potential surrounding a moving electron. This expression indicated that the potential surfaces surrounding the electron would contract by exactly the amount predicted by FitzGerald (see Appendix B). In order to remain in electromagnetic equilibrium, a stable group of electrons moving through the ether would therefore have to move closer together by the same amount. Since Larmor believed matter to be solely electronic in construction, it seemed reasonable to him to suppose that moving matter would also contract by this amount. He now had a physical explanation of the contraction implied by the new space-time transformations. Furthermore, the contraction explained why the Michelson-Morley experiment gave a null result. It was this comprehensive, electromagnetic theory of matter that eventually became known as the ETM. The important point to note is that, during 1897, the relationship between the ETM and the Michelson-Morley experiment changed dramatically. Larmor now claimed that the null result provided crucial evidence that matter was electrically constructed and did contract in the manner predicted by his theory.

Toward the end of 1897 Larmor began to collect and develop the most important results of the new electrodynamics for his 1899 Adams Prize essay "The Theory of the Aberration of Light."[17] In this essay he gave the first systematic account of the ETM and introduced the electromagnetic and space-time transformations that were later to be given by Lorentz and Einstein. It was a revised and expanded version of this essay that appeared in 1900 as *Aether and Matter.* The ETM thus took electrodynamics beyond the Maxwellian synthesis of the 1880s in three important ways.[18] First, it made the electron the natural unit of electric charge. Second, it resolved the problem of the relationship between ether and matter by asserting that all matter was constructed solely from electrons. Third, it introduced new electromagnetic and space-time transformations which correlated the electromagnetic fields measured in moving electrical systems with those of the stationary ether system.[19] But as we have seen, the credibility of much of the ETM rested heavily upon the claim that moving matter contracted, a claim guaranteed by Michelson and Morley's null result.

17. The contents of this essay are discussed in detail in Warwick 1991, 60–69.

18. The characteristics of the stable Maxwellian "synthesis" achieved during the late 1880s are discussed in Buchwald 1985b and Hunt 1991.

19. There were, or course, many more-subtle differences between Larmor's work and the earlier Maxwellian synthesis. Some of these are discussed in Buchwald 1985a, 168–73.

3. Larmor, Hicks, and the Michelson-Morley Experiment

The first, and completely unwitting, attack on Larmor's ETM came in September of 1897, shortly after he had begun work on his Adams Prize essay. In an article published in the *Philosophical Magazine,* the Australian physicist William Sutherland attempted to reconcile nonconvected ether theory with the Michelson-Morley experiment by casting doubt upon the null result reported by Michelson and Morley (Sutherland 1898). Working in isolation from the Maxwellian network and apparently unaware of Larmor's recent work, Sutherland continued to regard FitzGerald's explanation of the null result—the contraction hypothesis—as highly dubious. Sutherland claimed that the interference pattern produced by the recombination of a split beam is shifted due to the interferometer's motion through the ether. Michelson had not taken account of this effect and had, according to Sutherland, been looking in the wrong place for the maximum fringe shift. If the experiment were repeated, Sutherland argued, a positive result might yet be obtained and the contraction hypothesis rendered superfluous.

Larmor was not greatly perturbed by Sutherland's paper, as it aroused little general interest and was quickly attacked by Oliver Lodge. Lodge informed Sutherland of the work of FitzGerald, Lorentz, and Larmor and denied the necessity of a reanalysis of the interferometer's operation. Lodge claimed to have demonstrated conclusively that neither the path of a ray of light nor its journey time through the interferometer could be affected (to second order) by motion through the ether (Lodge 1898, 343). He concluded accordingly that his own earlier discussion of the apparatus—which acknowledged the correctness of the null result—was quite adequate. Lodge did suggest that it might be worthwhile to design a new experiment that would search for any "second-order electrodynamometric effect" of the earth's motion through the ether but added that any such effect "might quite possibly be compensated" (Lodge 1898, 344).[20] Sutherland let the matter drop.

Four years later, however, Larmor became very concerned when, 29 September 1901, he received the manuscript of a paper by William Hicks which criticized Michelson's analysis of the ether-drift experiment and denied the competence of the contraction hypothesis to explain a null result.[21] As a fellow of both the Royal *Astronomical* Society and Larmor's

20. As the expression "electrodynamometric" implies, Lodge believed that some second-order electrical (as opposed to optical) effect generated by the earth's motion through the ether might be detectable.

21. The manuscript was published in January of 1902 (Hicks 1902a).

own Cambridge College, St. John's, Hicks had to be taken seriously. It was therefore almost certainly with some alarm that Larmor read the letter that accompanied the manuscript. Hicks wrote:

> Herewith I send the [manuscript] of the Morley paper. It is a first part. In the continuation I propose to consider (1) the finite size of the source (there is a fringe at the first mirror and nowhere else). To show that FitzGerald's explanation (with your + Lorentz' evidence of the *fact*) will not explain the result, because the angle between the mirrors as well as their distance is affected by the contraction (or lateral expansion) and that even if it did (on the old theory) we should require an extension of a solid along the direction of motion, or a contraction perpendicular.[22]

Put simply, Hicks made two claims that contradicted Larmor's account of Michelson's null result. First, Hicks claimed, as had Sutherland, that the standard theory of the interferometer's operation was invalid in the ether-drift experiment. But unlike Sutherland, Hicks gave a specific physical reason for his objection. Hicks claimed that reflection of a light ray at the surface of a moving mirror would minutely alter both the wavelength and the angle of reflection of the departing ray. In consequence, the shift in the interference fringes (due to motion through the ether) as the interferometer was rotated would be quite different from that anticipated by Michelson and Morley (Hicks 1902a, 10). Hicks's analysis implied, moreover, that the averaging procedure adopted by Michelson and Morley in working up their data would actually have concealed the effect that, according to Hicks, they should have detected (Hicks 1902a, 36).[23] In short, Hicks claimed that the null result reported by Michelson and Morley could not be relied upon.

Hicks's second claim was even more remarkable. He claimed that even if the null result were to be confirmed, a contraction of the interferometer in the direction of its motion through the ether could not explain the result. Indeed, Hicks reckoned that such a contraction would actually *enhance* the positive result that Michelson had originally anticipated.[24] According to Hicks's analysis, only an extension of the interferometer in its direction of motion through the ether could account for a null result. Notice that Hicks did not challenge the sensitivity of the interferometer; rather, he argued, first, that the experiment would have to be

22. Hicks to Larmor, 29 Sept. 1901, Larmor Collection, St. John's College Library, Cambridge.

23. Hicks also believed that Michelson and Morley had failed to eliminate the effects of temperature fluctuations from their results.

24. Hicks in fact claimed that Michelson and Morley's published data did show a small positive effect when correctly analyzed (Hicks 1902a, 36–38).

repeated because Michelson had sought the wrong effect and, second, that even if the null result were to be confirmed, it could not be explained—assuming a nonconvected ether theory—by the contraction of moving matter along its line of motion.

Hicks did not at this point reveal his reasons for having undertaken such a detailed reanalysis of the Michelson-Morley experiment, but they can easily be inferred. Since graduating 7th wrangler in the Cambridge Mathematical Tripos of 1873, Hicks had devoted the greater part of his research effort to the mathematical analysis of vortex motion in hydrodynamical ether. In his most recent work on this topic, which had appeared in 1899, he sought to account for the periodic properties of atoms by relating them to a series of stable vortex aggregates in the ether (Hicks 1899).[25] Hicks had taken little part in the development of Maxwellian electrodynamics during the 1880s and 1890s and invariably conceptualized physical problems in terms of hydrodynamics. In January of 1902 he emphasized this point himself in a letter to Larmor:

> I have accustomed my mind to hydrodynamical concrete images. . . . Consequently, my concept is one in which when a solid moves through the ether, its vortex atoms set themselves on the whole to move thus—the solid is merely a swarm which goes on because its single atoms are moving, and is not itself a thing which carries the atoms. It seems to me likely that in this case the vortex atoms would contract perpendicular to the line of motion, or expand along it.[26]

Hicks, unlike Larmor, did not have a well-developed and comprehensive matter theory that made precise *quantitative* predictions concerning the dimensions of moving bodies. From a purely experimental point of view, it was, therefore, of little direct concern to Hicks whether the Michelson-Morley experiment gave a positive or a negative result. Hicks did, on the other hand, have a *qualitative* matter theory which implied that "swarms" of moving atomic vortices would tend either to elongate in the direction of their motion through the ether or to contract in the direction perpendicular to their motion. He would, therefore, have been very concerned by Larmor's claim that a null result automatically furnished evidence that the contraction of moving matter was a fact. By reanalyzing the operation of the Michelson-Morley experiment, Hicks showed that a null result, if con-

25. Hicks published a useful abstract outlining the main claims of this paper as Hicks 1898.

26. Hicks to Larmor, 7 Jan. 1902, Hicks Collection, St. John's College Library, Cambridge.

firmed, would actually furnish evidence for the *expansion* of moving matter (Hicks 1902a, 41).[27]

Larmor must have realized at once that if Hicks's claims were substantiated, his own ETM would be virtually demolished. It was, however, by no means obvious to Larmor how he could refute Hicks's arguments. Hicks's claims were based upon a series of complicated wave-front diagrams and several pages of laborious calculations in which all approximations were assiduously avoided. At first sight, both the premises of his argument and the details of his mathematical analysis appeared perfectly sound. The reliability of the null result would therefore remain in doubt until Michelson and Morley repeated their experiment according to Hicks's prescription.[28] And, crucially, Hicks's claim that the contraction hypothesis could not explain a null result could not be directly faulted. The evidential context in which Larmor had himself placed the Michelson-Morley experiment was now in danger of working against him.[29] If it became widely accepted that the Michelson-Morley experiment furnished evidence that moving matter expanded, Larmor's whole electrodynamics of moving bodies would be fatally undermined.

Larmor's problem was that he could offer no alternative theoretical account of the operation of the interferometer that would both support Michelson's claim that the null result was correct and guarantee the competence of the contraction hypothesis to explain the null result. Furthermore, as Larmor himself noted, it was widely believed that no other experiment could provide such a sensitive test for the earth's motion through the ether "on account of the unapproached precision of optical observations" (Larmor 1902a, lxi). But confronted by Hicks's potentially damaging challenge, Larmor began seriously to consider the possibility of constructing a new instrument that *could* measure some electromagnetic effect of the earth's motion through the ether.[30] Such an instrument would be extremely useful in two respects: first, it would give independent corroboration of Michelson and Morley's null result; and second, the theory of its operation would lay squarely within the purview of Larmor's elec-

27. Hicks did not give a precise prediction concerning the extent to which moving matter would change its directions. Initially, he set out to demonstrate only that a negative result (if confirmed) should not be taken as definite proof that moving matter contracted.

28. Hicks's paper cast sufficient doubt upon the alleged null result to send Michelson and Morley back to the drawing board; see Swenson 1972, 142–45.

29. This illustrates how the evidential context of an experimental result can change. It was Larmor who had made the Michelson-Morley experiment a test for the electronic construction and contraction of moving matter.

30. As Lodge's remarks above (Lodge 1898) show, the idea that some electromagnetic effect of the earth's motion might be detectable was common among the Maxwellians.

tromagnetic theory. It would therefore be extremely difficult for other physicists to challenge the validity of a null result once it had been established. To find the material technology and experimental skill that he now required, Larmor turned to the Maxwellian network.

4. The Maxwellian Network and the "Swinging" Capacitor

During the late 1870s and 1880s, much of the most important work in the development of Maxwellian electrodynamics was carried out beyond the traditional home of British mathematical physics, Cambridge.[31] George FitzGerald (Dublin), Oliver Lodge (Liverpool), and Oliver Heaviside (London) were chiefly responsible for developing and popularizing some of the most profound insights contained in Maxwell's *Treatise*. Through regular correspondence and personal meetings these men discussed, criticized, and refined each other's work, eventually producing a coherent and stable Maxwellian electromagnetic theory—the Maxwellian synthesis—during the late 1880s. Larmor came into contact with this Maxwellian network only in the early 1890s. Having graduated senior wrangler in the Cambridge Mathematical Tripos of 1880, he contributed occasional papers on electrodynamics but worked in a typically wrangler style, using analytical dynamics, rather than physical models, to develop lines of argument begun by Maxwell in the *Treatise*.[32]

Larmor's excursion into fundamental electromagnetic theory began in the early 1890s and was inspired by a paper of FitzGerald's. FitzGerald had noticed a remarkable formal similarity between the expression given by James MacCullagh in 1839 for the mechanical energy stored in his rotationally elastic ether and that given by Maxwell for the energy stored in the electromagnetic field. By replacing the mechanical symbols in MacCullagh's theory with appropriate electromagnetic symbols, and applying the principle of least action to the resulting Lagrangian, FitzGerald was able to follow MacCullagh's analysis to obtain an electromagnetic theory of the propagation, refraction, and reflection of light. Working through FitzGerald's paper, Larmor came to believe that MacCullagh's ether had far greater potential than FitzGerald had appreciated; Larmor

31. The only significant work on Maxwellian electromagnetic theory produced in Cambridge during the 1880s was undertaken by J. J. Thomson and R. T. Glazebrook. This work is discussed by Buchwald (1985a, 111–19, 269–77).

32. Larmor contrasted his own mathematical style with FitzGerald's more physical style by noting that "although an accomplished mathematician, [FitzGerald] preferred to reason in terms of direct images of the phenomena, and to reserve algebraic representations for purposes of calculation" (Larmor 1902a, lxii).

reckoned that it could become the common foundation for a dynamical theory of optical and electromagnetic phenomena.

Larmor's work on MacCullagh's ether quickly brought him into close contact with the other members of the Maxwellian network. His subsequent collaboration with FitzGerald on a range of theoretical issues played a crucial role in the emergence of the ETM.[33] There is, however, a further dimension to Larmor's membership in the Maxwellian network which is especially relevant to our present concerns. Larmor had very little practical experience in experimental physics. Employed by St. John's College as a lecturer in mathematics, he had neither the facilities nor the ability to follow up and check the experimental implications of his emergent theory. Furthermore workers at Cambridge's Cavendish Laboratory, less than half a mile from Larmor's rooms in St. John's, were neither engaged nor interested in the kind of research that could assist in the development of Larmor's electrodynamics. Regarding his conjecture that high-speed cathode rays were simply electrons that had spun out of their orbits, for example, Larmor informed Lodge that "I propounded this in the Cav[endish] Lab[oratory] years ago: but I ain't a prophet here."[34] Via the Maxwellian network, however, he was able to refer his theoretical work directly to the laboratories managed by FitzGerald and, especially, by Lodge.[35] The powerful use that Larmor made of this network is nicely illustrated by his rapid response to Hicks's attack.

Faced with the possible refutation of Michelson's null result in the early autumn of 1901, Larmor turned to FitzGerald's laboratory in search of the alternative experiment that would corroborate Michelson's original findings. On 24 October 1901, less than a month after receiving Hicks's manuscript, Larmor wrote to Oliver Lodge in Liverpool stating that the papers by Sutherland and Hicks were impressive pieces of work, but that they had "not upset [his] faith in the correctness of the negative result." He then went on to outline a new experimental arrangement about which he had corresponded with Frederick Trouton in Dublin:

> I have been worrying out with Trouton the rights of a vibration apparatus which he set up under FitzGerald's direction this time last year, and worked with last winter. FitzGerald does not seem to have explained himself, and

33. Larmor's collaboration with FitzGerald is discussed in detail by Buchwald (1985a, 162–73) and Hunt (1991, 209–39).

34. Larmor to Lodge, 29 Nov. 1901, UCL Archive, Lodge Collection, MS Add 89/65. Thomson had a different working model of atoms and expressly denied that the mass of his "corpuscles" (which Larmor claimed were "electrons") was completely electromagnetic in origin (Thomson 1899).

35. Larmor's collaboration with Lodge is discussed in detail in Hunt 1986.

> perhaps he had not quite fully thought it out. But so far as I can see it is going to be an electromagnetic second-order experiment to place along side Michelson's optical one. It only remains to make the apparatus about 100 times (or perhaps 20 times) more sensitive than it is, and there seems plenty of margin to work on.
>
> Hang up a condenser nearly at 45° to the aether's flow: there will be a (second order) couple on it: by timed charges and discharges this can set up a vibration. The sensitiveness is tested by tying on a magnet and testing with the alternating couple produced to an alternating magnetic field.[36]

FitzGerald had conceived the new "vibration apparatus" in an attempt to detect the earth's motion through the ether by electrical means. His reasoning was as follows. According to Maxwellian electromagnetic theory, when a charged parallel-plate capacitor moves "edgeways" (see fig. 11.3 in Appendix A) through the ether (with the motion of the earth), its plates generate a magnetic field (see Appendix A). When the capacitor is not charged, no such magnetic field exists. FitzGerald conjectured that as the moving capacitor was charged, the magnetic field was generated, not by the battery producing the electric field, but by an electromechanical interaction between the electric field and the ether.[37] According to FitzGerald's conjecture, the capacitor should receive a tiny mechanical impulse from the ether each time it is charged or discharged. Calculation of the magnitude of the predicted effect indicated that it might just be possible to measure the impulse received by a large capacitor moving edgeways through the ether if it were charged rapidly to a high potential (see Appendix A).

It is important to note at this point that although FitzGerald was very familiar with Larmor's ETM, he did not design the experiment simply to test or to refine Larmor's theory. In July 1900, just three months before the apparatus was constructed, FitzGerald wrote a review of Larmor's *Aether and Matter* for the weekly electrical trade journal *The Electrician.* In the review FitzGerald explained that Larmor's theory predicted that motion through the ether would be undetectable to second order and agreed with Larmor that "this seems to make it practically certain that Michelson and Morley's experiment is really a very accurate measurement of the change of length of a body as it moves in different directions through

36. Larmor to Lodge, 24 Oct. 1901, UCL Archive, Lodge Collection, MS Add 89/65. I shall use the more modern term "capacitor" rather than Larmor's term "condenser."

37. The energy supplied by the battery in charging the capacitor is QV, whereas the energy stored electrically on the capacitor is only one-half of this value. The battery could, therefore, supply the energy of the tiny magnetic field, but FitzGerald believed that the energy came from an electromechanical interaction between the building electric field and the ether.

the ether" (1900, 484). He nevertheless lamented the fact that motion through the ether appeared to be completely undetectable to second order:

> It is to be regretted that no effect other than aberration and the change of frequency of light waves, due to the relative motion of the earth and the ether, can be detected; for, if any considerable change in the electromagnetic forces due to this cause existed, there might be some hope of our being able to utilize this relative motion to drive engines, and thus obtain a source of power practically unlimited in amount, and confined to no one part of the earth. (1900, 484)

It was partly the dream of building a machine that could provide unlimited power at any point on the earth's surface that led FitzGerald, and later Trouton, to go on seeking an experiment that could detect an electromechanical interaction between the earth and the ether. In the autumn of 1900 he devised the experiment described above and directed his assistant, Frederick Trouton, to build the apparatus. A few preliminary runs indicated that no effect was detectable, but before the instrument could be calibrated with sufficient accuracy to put the result beyond doubt, FitzGerald, whose health had been failing for some time, became seriously ill. He did not recover from the illness and died in February 1901.[38]

Neither the exact form nor the precise purpose of the experiment originally envisaged by FitzGerald is known. As Larmor later remarked, FitzGerald either had failed to explain his experiment to Trouton thoroughly or else had not "sufficiently thought it out" before his unexpected death. Furthermore, Trouton did not describe the original apparatus constructed under FitzGerald's direction. It seems quite likely, however, that it consisted simply of a large capacitor suspended on the end of a long wire in the edgeways position. The idea would have been to set the capacitor swinging like a pendulum by charging and discharging it at the resonant frequency of oscillation. What is certain is that Trouton's problem with the apparatus was one of calibration. He began his research by causing a "small object" to strike against a mock-up of the proposed apparatus (Trouton 1902, 560). The "highly encouraging" results obtained by this crude calibration led him to build the actual experiment. But having obtained consistently null results, the problem of calibration became even more acute. In order to demonstrate that the predicted effect did not exist, Trouton had to show convincingly that the apparatus was capable of registering the sought effect (if it existed). In other words, he needed to

38. The impact of FitzGerald's early death on the other Maxwellians is discussed by Hunt (1991, 241–43).

find an acceptable surrogate against which to calibrate the apparatus.[39] Unable to solve this problem, Trouton redesigned the entire experiment.

As we have already seen, Trouton believed that a charged capacitor moving edgeways through the ether generated a net magnetic field in its vicinity. By a similar line of argument, a capacitor moving in the flatways orientation would not generate a net magnetic field (see Appendix A). It follows that if a charged capacitor is slowly rotated from the edgeways orientation to the flatways orientation, the magnetic field must, by Trouton's argument, vanish, while the electric field remains undiminished. The energy of the capacitor is therefore a minimum in the flatways position. Trouton inferred that if a charged capacitor were suspended by a fine thread at an angle of 45° to its direction of motion through the ether, a torque would be generated that would tend to turn the capacitor toward the flatways (minimum-energy) position. It was this torque that he now hoped to measure.

This form of the experiment has two distinct advantages over the previous arrangement. First, it is possible to calculate the value of the predicted torque precisely (see Appendix A) so that the degree of accuracy required to measure, or to rule out, the effect is known exactly. Second, the torsional—as opposed to linear—nature of the oscillation sought enabled Trouton to employ a modified form of a well-known technique to calibrate his apparatus. A standard technique for measuring the magnetic moment of a bar magnet was to suspend it by a fine thread and measure its period of oscillation in a magnetic field of known strength.[40] Trouton used a variation on this technique to calibrate the suspended-capacitor apparatus. He placed a small magnet of known magnetic moment on the suspended capacitor and used a magnetic field of known strength to deliver tiny regular impulses synchronously for half the period of oscillation of the apparatus.[41] By systematically reducing the strength of the field generating the impulses, he could ascertain the magnitude of the smallest regular impulse required to set the apparatus oscillating. By the late summer of 1901, Trouton had succeeded in making his new experiment register a synchronously applied torque of 7.5 ergs. Unfortunately, the torque generated by the suspended capacitor was only 1 erg (see Appendix A).

Larmor's letter to Lodge of 24 October reveals that this was more or less the situation that Trouton was in when he and Larmor began to correspond in the late summer of 1901. Trouton appears to have become

39. On the use of surrogates in the calibration of scientific instruments, see Collins 1985, 79–111.

40. This technique is described in detail in Maxwell 1873, vol. 2, pt. 3, chap. 7.

41. The small magnet and the applied magnetic field provide the "surrogate" for the sought interaction between the earth and the capacitor and the ether.

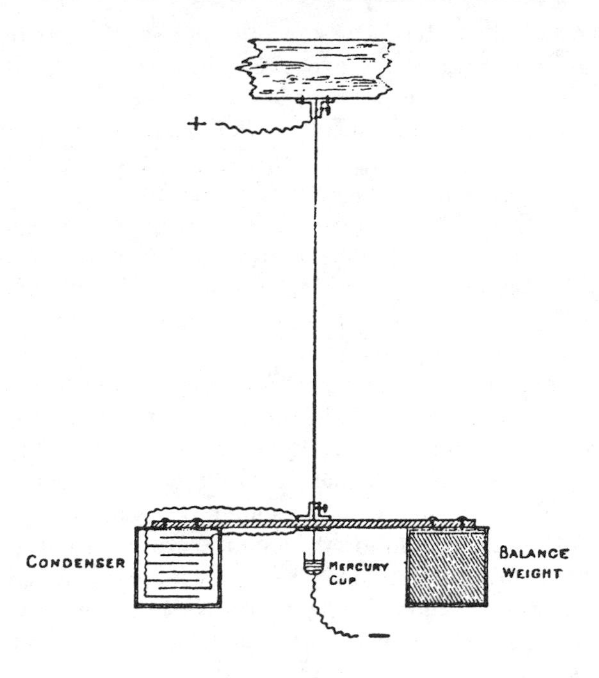

Fig. 11.1. The swinging capacitor apparatus (Trouton 1902)

disheartened by his inability to build an experiment that could either detect or rule out the effect he sought. But, urged on by Larmor, he devised yet another arrangement, which incorporated the most successful aspects of his previous efforts. Trouton now reverted to FitzGerald's original idea of seeking the impulse delivered to a capacitor when it is suddenly charged in the edgeways position. This time, however, he designed an experiment that would convert the postulated linear impulse into a rotatory oscillation.

Trouton suspended a rigid bar by its center using a long thin wire. To one end he attached the large capacitor; and to the other end, a counterweight to balance the bar (see fig. 11.1). A tiny impulse now delivered to the capacitor will set up a torsional oscillation of the whole apparatus. Attached to the center of the bar was a mirror that reflected a tiny spot of light onto a millimeter scale 1.860 m from the apparatus. The slightest rotational movement of the bar would now be revealed by the movement of the spot.[42] Finally, Trouton determined the period of free oscillation of the whole apparatus.

42. Maxwell also described in detail the technique of employing a reflected beam of light to magnify the effect of angular deflection (Maxwell 1873, vol 2, art. 450).

The idea of the experiment was to try to set the bar oscillating by charging and discharging the capacitor with exactly the resonant frequency of the apparatus (using a clockwork device). By synchronizing the impulses with the natural frequency of the apparatus, the sensitivity was greatly enhanced. The tiny regular impulses, if they existed, would accumulate constructively to produce a perceptible oscillation of the bar.

This time Trouton was able to use the surrogate developed in his previous experiment to calibrate the apparatus. He placed a small magnet on the bar and subjected it to systematically smaller impulses using a magnetic field. The strength of each impulse was measured in terms of the kinetic energy that it imparted to the apparatus. Trouton found that the cumulative effect of synchronized impulses of just 0.004 erg each was sufficient to produce detectable oscillations. Theory predicted that the capacitor would deliver an impulse 250 times larger than this each time it was charged or discharged in the edgeways position (see Appendix A). Trouton had finally produced an instrument that was capable of detecting the effects of the earth's motion through the ether with an accuracy that matched Michelson's interferometer (Trouton 1902, 562).[43] Unlike Michelson and Morley's interferometer, however, the theory of this instrument lay completely within the area of Larmor's greatest expertise—electromagnetism.

5. A New *Perpetuum Mobile*

While Trouton was perfecting the material technology of the new experiment, Larmor continued to work hard on its theoretical interpretation. Although he had confidence that Trouton would shortly achieve the required degree of accuracy, it was not at once clear to Larmor how the anticipated null result could act as positive proof of the contraction of moving matter. Elsewhere I have used the term "theoretical technology" to characterize the set of taken-for-granted physical principles and mathematical methods routinely employed by individual physicists (or groups of physicists) in constructing their arguments (Warwick 1992, 631–34). In this essay I shall employ the term to emphasize the mutually supportive nature of material and theoretical technologies in the coproduction of plausible arguments in physics. As we shall see, Larmor had to build an electromagnetic Carnot cycle into the theoretical technology of the

43. This method of calibrating the apparatus was not challenged, even by opponents of Larmor's ETM, because it was generally accepted that it represented an accurate surrogate for the phenomenon under investigation. For a discussion of the potentially controversial nature of the use of surrogate phenomena in calibration, see Collins 1985, 100–106.

ETM in order to create the required evidential context for Trouton's experiment.

On 10 November 1901, Larmor wrote to Lodge announcing his confidence in both the new apparatus (which had yet to be tested) and the line of argument that he was going to develop to lend significance to the anticipated null result:

> I think the swinging condenser experiment will work.
>
> After some worrying over it I think I now see this whole affair of negative second order results. If they were not null to *all* others, reversible cycles could be arranged for drawing off in unlimited amount—the energy of the earth's motion through the aether for the purposes of mechanical work. If the FitzG[erald] Lorentz contraction hypothesis is denied I can specify such a cycle with the rotating condenser.
>
> This is a new kind of *perpetuum mobile.* Although on abstract ground it seems as safe as Carnot's one. Yet I think it ought to be tested: and so I am trying to get Trouton to go on on the new tack.[44]

It had taken Larmor just over two weeks to come up with the new *perpetuum mobile* linking Trouton's experiment with the contraction hypothesis and the ETM.[45] This was, moreover, a new departure in Larmor's approach to the problem of the electrodynamics of moving bodies, and one that he never explicitly stated in his published writings. Larmor's remarks above imply that there can be *no* second-order electromechanical interaction between the moving earth and the ether since any such interaction would long since have brought the earth (and other celestial bodies) to rest in the ether (Larmor 1902a, lxi–lxii). It follows that any device that *was* capable of producing energy from an interaction between the moving earth and the ether would constitute a perpetual motion machine (because it could provide a limitless source of useful work).[46]

Within a week Larmor had written out the argument in full mathe-

44. Larmor to Lodge, 10 Nov. 1901, UCL Archive, Lodge Collection, MS Add 89/65.

45. Larmor frequently resorted to thermodynamical arguments when unable to give a complete dynamical or electrodynamical account of a phenomenon. See, e.g., Warwick 1991, 37.

46. Larmor never published his claim to Lodge—that attempts to measure motion through the ether must give null results to all orders of v/c—for two reasons. First, his claim that interaction between moving celestial bodies and the ether would soon bring them to rest was not plausible at fourth order (effects due to odd powers of v/c were eliminated on grounds of symmetry) and above because the energy exchange would be too small to have any major effect. Second, Larmor had good theoretical reasons to believe that effects of the earth's motion would be measurable at higher orders due to the finite radius of the electron. On the last point see Warwick 1991, 68–69.

matical detail (1902b). He sent the paper to Lodge on 17 November, enclosing the following remarks in a covering letter:

> I have written out a sketch of the rotating condenser argument which you might return. In the mechanical process of writing, new aspects arise in one's head. If the experiment is carried through with null result, its use will be to corroborate Michelson simply. Then either (i) the ether will travel with the earth, or (ii) the F[itzGerald] L[orentz] [contraction] state of affairs will exist which will bring things to the same pass as (i). This is excluding the alternative of the "perpetual motion."
>
> The only way to discriminate between (i) and (ii) is to say that (i) disagrees with the whole framework of our knowledge.[47]

Larmor claimed that the anticipated null result of Trouton's experiment could be explained in only two ways: *either* the ether was carried along with the earth at its surface, *or* the FitzGerald contraction must be a fact. He dismissed the first possibility by arguing that no comprehensive theory of the electrodynamics of moving bodies had been proposed that was consistent with the notion of a convected ether. Now he had only to show that any plausible electromagnetic theory based upon the notion of a nonconvected ether had to embody FitzGerald's contraction hypothesis.

Trouton, meanwhile, had successfully calibrated the new experiment and undertaken a series of experimental runs. He invariably obtained null results, confirming that no reaction between the capacitor and the ether was detectable. This was, of course, precisely the result upon which Larmor had been counting. Trouton himself did not believe this result to hold any special significance for electromagnetic theory. The energy of the magnetic field could straightforwardly be attributed to the battery with which the capacitor had been charged. As far as he was concerned, he had merely demonstrated that the impulse predicted by FitzGerald did not exist. But for Larmor the result was of far greater significance.

In his published account of Trouton's experiment, Larmor began with an oblique reference to Hicks's work. Larmor noted that the hitherto unique status of Michelson's ether-drift experiment had rendered its interpretation "somewhat ambiguous" (Larmor 1902b, 566). He then invited his reader to consider the following thought experiment. Two identical capacitors, AB and PQ (see fig. 11.2), are charged, wired in parallel, and traveling through the ether with the motion of the earth. The capacitor AB is suspended from a point O. Both capacitors initially carry the same

47. Larmor to Lodge, 17 Nov. 1901, UCL Archive, Lodge Collection, MS Add 89/65.

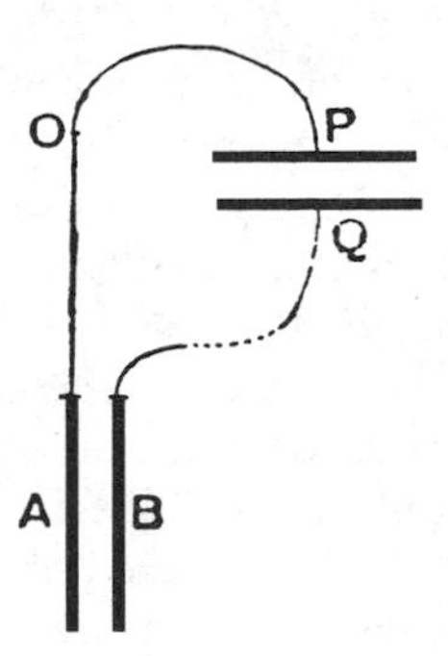

Fig. 11.2. The two-capacitor cycle (Larmor 1902b)

quantity of charge, but the distribution can be altered by (slowly) altering the distance between the plates of one or the other of the capacitors.[48]

Larmor now described the following reversible cycle: (1) the charge on AB is increased by separating the plates of PQ; (2) AB is disconnected and rotated into the edgeways position (raising its overall potential); (3) the plates of PQ are adjusted to bring them to the same potential difference as AB; (4) the capacitors are reconnected and AB is discharged back into PQ until both capacitors again have equal charge; (5) the capacitors are again disconnected, AB is rotated back into the flatways position, the plates of PQ are adjusted to bring them to the same potential as AB, and the capacitors are reconnected and the cycle is complete.

Larmor's point was that this cycle constitutes a perpetual motion machine.[49] The potential at which the capacitor AB is discharged is higher than the potential at which it was charged; useful work can be extracted from the cycle even though no net work is done on the system. This conclusion could be evaded within Maxwellian electrodynamics only by arguing that an electromechanical force acted on the capacitor (doing work) whenever its charge was altered in the edgeways orientation. But Trouton had shown that no such force exists. Larmor now exploited the two-

48. All of these operations are imagined to take place extremely slowly so that losses due to radiation can be made arbitrarily small. This ensures that the cycle remains, in principle, reversible.

49. Perpetual motion could now be avoided (assuming standard Maxwellian electromagnetic theory) only by arguing that a mechanical torque acts on the capacitor as it is rotated. Larmor considered this extremely unlikely because: (1) Trouton had already shown that no mechanical interaction occurred when the capacitor is charged in the edgeways position; (2) the ETM predicted that no such torque existed (because the capacitor's potential was independent of its orientation); and (3) if the energy of the earth's motion were available for useful work (at second order) the earth would long since have come to rest in the ether (Larmor 1902a, lxi).

capacitor cycle to show that the perpetual motion problem does not arise when the cycle is analyzed using the ETM. The reason the problem does not arise is as follows. The ETM predicts that moving matter will contract in precisely the way suggested by FitzGerald. If a moving capacitor contracts in this way, the potential between its plates can be made independent of the capacitor's orientation with respect to its direction of motion through the ether. The two-capacitor cycle does not then constitute a perpetual motion machine, because the potential across AB is the same in the flatways and edgeways orientations. Larmor could now argue that the contraction of moving matter, as predicted by the ETM, offered the only satisfactory solution to the perpetual motion problem raised by Trouton's null result. He could also claim, conversely, that Trouton's null result provided firm evidence that moving matter did contract in the manner predicted by the ETM. Thus Larmor concluded that "the energy of motion of the earth through the ether is available for mechanical work to an unlimited extent, unless the potential difference in the condenser is independent of its orientation; that is, by accepted electrodynamics, unless the FitzGerald-Lorentz shrinkage is a fact" (1902b, 568).

By "accepted electrodynamics" Larmor meant, of course, the ETM. The conviction carried by Larmor's argument derives in large measure from the skill with which he analyzed the moving-capacitor problem using the ETM. Unfortunately, an appreciation of the subtlety and originality of his argument requires considerable familiarity with early-twentieth-century electromagnetic theory. For this reason I have relegated discussion of the argument to Appendix B, where it is set out in full. It is sufficient to note here that Larmor's British contemporaries, insofar as they were competent to follow his argument, found it highly convincing.[50]

From Larmor's point of view, his brief collaboration with Trouton had been very successful; it had fully restored his confidence in the ETM. In August 1902, for example, when Lord Rayleigh announced that he had designed a new experiment that he hoped would detect the contraction of moving matter, Larmor responded, publicly, that no second-order effects of the earth's motion could be measured experimentally.[51] Three months

50. Larmor's argument was never challenged. On the development of Larmor's work in Cambridge, see Warwick 1989, chaps. 4–5.

51. This exchange took place at the 1902 meeting of the British Association for the Advancement of Science (Rayleigh 1902a). See also Cunningham 1914, 38. Cunningham remarks in a footnote, "the expectation of a positive result in this [double refraction] or any other experiment was demurred to by Larmor when this paper was read at the British Association meeting in Belfast (September 1902)." See also Robb 1921. Robb was also present at this meeting and remarked in the preface of his book that Larmor had predicted that Rayleigh would not be able to detect any double refraction.

later Larmor's prediction received further support when Rayleigh reported a null result (Rayleigh 1902b). Larmor had now succeeded in shifting the onus of proof back onto Hicks. In order to deny the contraction of moving matter convincingly, Hicks would now have to show either that Larmor's analysis of Trouton's experiment was fundamentally flawed or that the experiment had been incompetently performed. By January 1902, however, Hicks had very serious problems of his own. His paper on the Michelson-Morley experiment had appeared in print in the January issue of the *Philosophical Magazine,* but by 7 January one of its major claims had already been undermined. In the published version of his paper Hicks claimed that Michelson and Morley had failed to take proper account of the orientation of the interferometer (with respect to its direction of motion through the ether) in working up their data. Hicks reanalyzed the published data and claimed that it actually indicated a small positive effect. He was quickly informed by Morley, however, that appropriate allowance *had* been made for the orientation of the interferometer, but that it had been included in the published results without comment.[52] Hicks had to write a note to the *Philosophical Magazine* withdrawing his claim (Hicks 1902b, 256).[53]

Much worse was to follow. By the end of January one of Larmor's Cambridge colleagues, H. M. Macdonald, had located an important error in Hicks's paper.[54] Macdonald's own electromagnetic theory also embodied the contraction hypothesis, and he, like Larmor, was very concerned by Hicks's claims.[55] Macdonald worked painstakingly through the calculations of Hicks's paper, eventually locating an algebraic slip. Having corrected the error, Macdonald showed that Hicks's analysis agreed with Larmor's; only a contraction in one arm of the interferometer (in the direction of its motion through the ether) could account for the null result. Hicks wrote an apologetic letter to Larmor on 5 February acknowledging the error and ascribing it to "approaching senile decay."[56] He also asked Larmor's advice on how he should confess his mistake in public. Larmor counseled that, for speed of publication, he should write at once to the

52. Hicks to Larmor, 7 Jan. 1902 (see n. 26). Hicks also informed Larmor in this letter that he had asked to see Morley's original data but that Morley claimed to have lost the record.

53. The letter was dated 9 January 1902.

54. Hicks to Larmor, 5 Feb. 1902, Hicks Collection, St. John's College Library, Cambridge.

55. H. M. Macdonald was 4th wrangler in 1891. His derivation of the contraction hypothesis was published as Macdonald 1902, 185.

56. Hicks to Larmor, 5 Feb. 1902 (see n. 54). Hicks was clearly not senile; he won the Adams Prize some fifteen years later (Milner 1932–38).

weekly magazine *Nature* withdrawing his earlier claims. The letter appeared just eight days later, marking the end of the dispute (Hicks 1901/2, 343).[57] The following year Larmor succeeded George Stokes as Lucasian Professor of Mathematics at Cambridge University.

6. The Sturdy Protestants of Science

Larmor and Trouton each benefited from their brief collaboration: Larmor had reestablished the experimental foundations of the ETM; Trouton had embarked upon what he saw as a potentially fruitful line of experimental research. But their respective research programs had actually become incompatible in almost every respect. Larmor believed Trouton's null result provided evidence for the general proposition that second-order effects of the earth's motion could not be measured. Trouton, on the other hand, believed that he was on the trail of a whole new electromechanical technology that would revolutionize power generation. He was hardly aware that Larmor's ETM ruled out the very second-order effects he now sought to measure. Unlike his friend and mentor FitzGerald, Trouton was no mathematical physicist. He trained originally as an engineer and subsequently relied heavily upon FitzGerald to guide him in theoretical matters. It is unlikely that he would even have brought the swinging-capacitor experiments to fruition following FitzGerald's death had Larmor not intervened.[58] In this section I discuss the final marginalization of Trouton's research program and his gradual isolation from the remaining Maxwellians.

Toward the end of his 1902 paper on the swinging-capacitor experiment, Trouton made public his conjecture (discussed in section 4) that a torque is exerted on a charged capacitor whose plates make an angle of 45° with its direction of motion through the ether. He explained that if the effect were detected it would "open up a road leading to illimitable possibilities, for it would at once remove from the category of utter hopelessness the idea of mankind ever being able to utilize the vast store of energy in the Earth's motion through space" (1902, 564). Trouton's project remained the harnessing of the energy of the earth's motion through the ether. He went on to describe a machine he had designed to convert this energy into a usable mechanical form. A series of parallel-plate capacitors are arranged around a large rotatable cylinder with their plates tangential to the cylindrical surface. The axis of the cylinder is kept perpendicular to its direction of motion through the ether. As the cylinder rotates, each

57. The letter was dated 10 February 1902 and appeared in print on 13 February.
58. Trouton acknowledged Larmor's help at the end of his paper (1902 565).

capacitor is automatically charged and discharged every quarter of a revolution such that the couple exerted on it by the ether tends always to drive the cylinder round in the same direction (Trouton 1902, 564–65).[59] That Trouton could never produce the tiny couple that would render this machine practicable was, as we shall see, a major blow to his professional career.

In 1902 Trouton was appointed professor of physics at University College London (UCL). Having settled in his new post, he quickly resumed his experiments on the earth's motion through the ether. With the help of a research student, H. R. Noble, Trouton built a modified version of the capacitor apparatus with which he hoped to detect the torque that would drive his ether machine.[60] The bulk of his joint effort with Noble was expended in improving the sensitivity of the apparatus by systematically eliminating unwanted "disturbances" (Trouton and Noble 1904, 166). In the end, they established, experimentally, what Larmor had confidently asserted the previous year; no second-order torque acted on the charged capacitor. Trouton and Noble did not conclude, however, that they had provided further evidence for Larmor's ETM. They stated merely that the electrostatic energy of the capacitor must somehow diminish in the edgeways orientation (1904, 181).

Over the next four years Trouton collaborated with another of his research students, A. O. Rankine, in the design and construction of a new experiment to detect the earth's motion.[61] This time Trouton hoped to detect matter contraction indirectly by measuring its effect on the resistance of a current-carrying wire. Trouton reckoned that the resistance of a wire oriented parallel to its direction of motion would be decreased because it became physically shorter. By the same argument it followed that the resistance of the wire would be increased when oriented perpendicular to its direction of motion because its cross-sectional area would be reduced. By building a rotatable current balance, Trouton hoped to detect these changes in resistance (Trouton and Rankine 1908, 421). The reasoning behind the anticipated effect did not derive from the ETM (which would have predicted a null result); rather, it relied upon a little Maxwellian electromagnetic theory and a commonsense approach to the contraction of moving matter. Had he chosen to use the Maxwellian network by consulting Larmor on the proposed experiment, he would have been informed at once that the reasoning behind the prediction was quite falla-

59. Trouton even discussed the details of the mechanism by which the charge could most efficiently be switched from one capacitor to the next.

60. Noble took a B.Sc. at London University (University College) in 1901.

61. Rankine took a B.Sc. at London University (University College) in 1904. For further biographical details see Thomson 1956.

cious.[62] Trouton was now working in a theoretical vacuum. His faith in the existence of the sought effect was based upon a theoretical technology that had been abandoned by every leading theoretician in Europe.[63]

Trouton's young assistant, Rankine, went a step further. He reckoned he could prove that second-order effects of the earth's motion were, in principle, detectable. The reception afforded to Rankine's bold claim makes it worthy of further comment. In August 1907, before he and Trouton had completed their new experiment, Rankine delivered a paper to the British Association entitled "On a Theoretical Method of Attempting to Detect Relative Motion between the Ether and the Earth" (Rankine 1907). He described the following thought experiment to his audience. A thin bar with a large mass at each end is supported in the middle by a fine thread. The bar is now made to execute small rotatory oscillations. The period of these oscillations is measured (1) when the rod is aligned parallel to the direction of its motion through the ether and (2) when it is perpendicular to this motion. The frequency of oscillation, Rankine claimed, would be greater in position (1) than in position (2), because in position 1 the rod is subject to the FitzGerald contraction. This conclusion could be avoided, Rankine continued, only if the mass at each end of the bar were somehow reduced in position 2. The ETM predicts, however, that the mass will be greater in position 2, making the frequency difference still larger.[64] Thus, concluded Rankine, either it was, in principle, possible to measure the earth's motion through the ether or the ETM was badly in error.[65]

The reporter who covered this session of the British Association meeting for *Nature* correctly inferred that Rankine's argument was "in opposition to the view held by Larmor, Einstein, and others" that motion through the ether could not be measured. This anonymous reporter also noted, with some surprise, that Lodge, who attended the session, defended Larmor and Einstein by warning that Rankine's argument was probably

62. Trouton was by no means the only experimenter to make predictions concerning the observable consequences of the contraction of moving matter in almost total ignorance of the electromagnetic theory that embodied the effect. See, e.g., Brace 1904 and the response in Larmor 1904.

63. I do not mean to suggest that Trouton could not have obtained a positive result. But from the theoretical point of view, he might equally have expected a positive result from any other arbitrarily designed experiment.

64. In case 1 the effective mass would be the so-called transverse mass, whereas in case 2 it would be the longitudinal mass. For a simple modern explanation of these now outdated terms, see Rindler 1977, 90.

65. If the ETM is correctly applied, the frequency is the same in both directions. Many other objections to Rankine's argument were raised; see, e.g., Bumstead 1908.

erroneous (Anonymous 1907, 459). Expressing similar sentiments, the reporter from *Knowledge* recorded:

> Either the current theories are very wrong, or else it must be possible to obtain positive evidence of relative motion of the ether and the earth. The possibility of the latter is denied by one school of physicists amongst whom must be placed Larmor, Einstein, and (as it turned out in the short discussion which followed the reading of the paper) Sir O. Lodge. (Porter 1907, 210)

Lodge's objections to Rankine's paper prompted Trouton, who was also present, to retort that he and Rankine were engaged in a new set of experiments designed to detect motion through the ether and that they appeared to have observed a "small positive effect" (Anonymous 1907, 459). These new experiments were the electrical resistance ones described above. Unfortunately, the positive effect reported by Trouton could not be stabilized and had, ultimately, to be ascribed to systematic error. Eventually Trouton and Rankine had to acknowledge a completely null result. Once again they did not present their result as support for Larmor (or Einstein). They concluded merely that "in view of the very general acceptance of the FitzGerald-Lorentz shrinkage theory," they had to assume that some other compensating factor was at work (Trouton and Rankine 1908, 434).

The above exchanges at the British Association meeting reveal how isolated Trouton had become by failing to acknowledge recent work in electromagnetic theory. By 1907, reporters from the popular scientific press were aware that Rankine and Trouton were challenging the collective wisdom of the masters of theory in attempting to measure second-order electromagnetic effects of the earth's motion. Even Lodge, whose eloquent, often passionate, advocacy of the mechanical ether concept was well known, rallied to Larmor's defense on this issue. With the remaining members of the Maxwellian network aligned against him, Trouton must have realized that he could expect no further support for this line of experimentation. Having published an account of the null results obtained in the electrical resistance experiments, he abandoned his research on the earth's motion for good.

We should not conclude, however, that Trouton had been convinced of the impracticality of his ether machine. He preferred to believe that he had been betrayed by his more theoretically minded colleagues. Struck down by ill health, Trouton was forced to retire from UCL in 1912. As president of the mathematics and physical sciences section of the British Association in 1914, he took advantage of what would be his last major public platform to air his views on the physics that had occupied much of his professional career. He began his address by harking back 25 years to the speculative attempts made by his teacher, FitzGerald, at "unlocking and utilizing the internal energy of the atom." In that case, he reminded his audience, it had been the "brilliant work of Rutherford and others which had shown that no key was required to unlock the energy—the door lay open" (Trouton 1914, 285). But in the "analogous case of a hitherto untapped source of energy arising from our motion through the ether," he continued, similar progress had not been achieved (285). This was not, Trouton added reproachfully, a reason for adopting the notion of "Relativity," which sought merely to "remove the lion in the path by laying down the proposition that the existence of lions [was] an impossibility" (286).

Trouton was dismayed by the willingness with which many physicists appeared to accept that the earth's motion through the ether could never be measured. Some, following Einstein, were even prepared to elevate this acceptance to the status of a physical principle. Trouton reminded his audience that such a principle would not have been accepted "half-a-century ago when a purely dynamical basis was expected for a full explanation of all phenomena." The men of that generation, he continued, "were the sturdy Protestants of Science, to use an analogy, while we of the present day are much more catholic in our scientific beliefs, and in fact it would seem that nowadays to be used to anything is synonymous with understanding it" (1914, 285).

Trouton's sectarian analogy spoke forcefully of his Protestant roots in southern Ireland. In Dublin he had worked in a Protestant enclave, Trinity College, which had stubbornly resisted the increasingly vociferous and violent demands for Home Rule by the local Catholic majority.[66] Protestant intellectuals at Trinity College believed that the Catholic people of southern Ireland were too irrational to rule themselves effectively. FitzGerald, for example, claimed that superstition "prevented [Catholics] from rational activity almost as much as the fetish worship of an African savage."[67] And the following tirade was occasioned by the mere suggestion that the British government might grant Hole Rule to the Catholic majority in Ireland: "Woe to Britain if for the sake of saving trouble over Irish squabbles that are only now beginning, if for the sake of puny local

66. Trinity College was represented at Westminster by two of only three safe Unionist seats in southern Ireland. In the general election of November 1885, Trinity College returned the only two non-Home-Rule candidates outside Ulster. See Buckland 1988, 85; and Downing 1980, 60.

67. FitzGerald to Lodge, 21 Dec. 1896, UCL Archive, Lodge Collection, MS Add 89/65. These remarks were prompted by a conversation overheard by FitzGerald in which a Catholic "gentleman" stated that he would not willingly pass a graveyard.

ends, for the sake of some supposed principle of so called popular government, she hands over the intelligence, the industry, the whole people of Ireland to be a prey for these greedy vultures."[68] FitzGerald clearly attributed the intellectual and industrial achievements of Ireland solely to the Protestant ascendancy. It was from this familiar refrain that Trouton's analogy sprang. For him, the abandonment of the search for a method of harnessing the earth's energy of motion in favor of a "principle of relativity" was synonymous with the betrayal of sturdy Protestant values in favor of what he saw as Catholic superstition.

But Trouton's sense of betrayal was partly misplaced. Larmor was an Ulsterman who aligned himself very closely with the distinguished school of mathematical physicists—James MacCullagh, William Rowan Hamilton, and George FitzGerald—associated with Trinity College.[69] Indeed, he considered himself to be developing the tradition begun by them. Had Trouton remained abreast of developments in electromagnetic theory during the late 1890s and early 1900s, he would have realized that Larmor and Einstein actually held little in common when it came to the problem of the earth's motion through the ether. Einstein's "principle of relativity" ruled out the possibility of measuring the earth's motion at any order and made the ether concept superfluous to electrodynamics. Larmor, on the other hand, believed that the earth's motion had to be measurable at some higher order and insisted that the notion of a unifying dynamical ether was indispensable to any intelligible electromagnetic theory; he surely remained a "sturdy Protestant of Science."[70] It was Trouton's failure to familiarize himself with the theoretical technology of the ETM that led him to lump together Larmor, Einstein, and perhaps even Lodge as physicists who had lost faith in the utilitarian aspirations of the Maxwellian enterprise.

Despite their idiosyncratic nature, however, Trouton's remarks are informative in two respects. First, they capture an important change that had taken place in British physics over the course of his professional career. When Trouton had trained, during the 1880s, physics and electrical science were partly coextensive. In striving to understand and improve a range of practical electrical devices using Maxwell's *Treatise,* the Max-

68. FitzGerald to Lodge, 16 Mar. 1893, UCL Archive, Lodge Collection, MS Add 89/65. FitzGerald's invective against Catholic intentions continues for three pages. He also urged Lodge to publish the letter in "a Liverpool paper." I would like to thank Bruce Hunt for drawing this letter, and the one above, to my attention.

69. Larmor even represented Cambridge University as a Unionist member of Parliament between 1911 and 1922; see the obituary notice by Eddington (1944, 206).

70. On Larmor's reasons for believing that the earth's motion had to be measurable at some order, see Warwick 1991, 68.

wellians contributed directly to electrical engineering while remaining at the cutting edge of research in physics. Furthermore, in this quintessentially electrical age the relentless advance of the electrotechnology industry guaranteed their moral sense that the Maxwellian enterprise worked for the betterment of Victorian society.[71] For Trouton, the electromagnetic ether remained the invisible, yet omnipresent, workhorse of this industry, and his attempt to unleash the energy of the earth's motion in the service of humanity was a moral duty that could not be shirked.[72] By 1914, however, few of those listening to his address would have recognized the enterprise for which he pleaded. The rich legacy of the Maxwellians had already been divided between electrical engineering and an emergent theoretical physics. The laboratories that had flourished during the Maxwellian heyday had sunk into obscurity.[73] A new and increasingly international microphysics of electrons, X rays, Becquerel rays, and atomic spectra now defined new networks of cooperation for a different generation of physicists.

The second, and related, sense in which Trouton's remarks are informative concerns our own perception of the relationship between early-twentieth-century British physics and Einstein's theory of relativity. Trouton's lamentations regarding what he saw as the demise of a purely dynamical and directly utilitarian physics make Einstein's "Relativity" the villain of the piece. But, as we have seen, Larmor's ETM no more allowed the second-order effects sought by Trouton than did Einstein's relativity theory. This is worth stressing because the increasingly divergent traditions represented by Larmor and Trouton, respectively, tend to be conflated by historians of physics (see, e.g., Goldberg 1970). Relativity theory was indeed a direct affront to Trouton's plans to harness the earth's energy of motion, but it was initially welcomed by some of Larmor's students in Cambridge. It is, therefore, highly misleading to characterize all of British mathematical physics circa 1905 in terms of the common Maxwellian enterprise of the 1880s and 1890s. As I have argued elsewhere, explaining the range of responses accorded to relativity theory by British physicists during the first quarter of the twentieth century entails a much deeper

71. On the impact of electrotechnology on popular culture during the late nineteenth century, see, e.g., Marvin 1988.

72. FitzGerald expressed similar sentiments in 1888 in stating that "in our electromagnetic engines we are using as our mechanism the ether, the medium that fills all known space" (1902, 240).

73. The laboratory directors who succeeded FitzGerald in Dublin (W. E. Thrift) and Lodge in Liverpool (L. R. Wilberforce) carried out no significant research and were little interested in the work of their famous predecessors. See McDowell and Webb 1982, 456; and Rowlands 1990, 252.

appreciation of the different factions at work within the British scientific community (Warwick 1992).

7. Conclusion

In this essay I have focused upon the Maxwellian network in order to highlight the productive power, the locality, and the historically situated nature of that collective enterprise. The mixture of ignorance, indifference, and outright hostility shown to Larmor's electromagnetic theory by some of his Cambridge contemporaries—J. J. Thomson and William Hicks, for example—clearly illustrates that late-nineteenth-century British electromagnetic theory cannot be treated as homogeneous. Recognition of this fact is crucial, moreover, because the stability and power of Larmor's theory were both generated and bounded by the network within which the theory was actively practiced. Larmor could not have responded so effectively to Hicks's challenge, for example, had it not been for the range of resources and common currency provided by the Maxwellian network. Active interest in the emergent ETM was, as we have seen, distributed across a number of important centers of research in Britain. Indeed, it was largely the common interest in Larmor's electromagnetic theory that held this otherwise heterogeneous collection of theoretical and experimental resources together during the 1890s.[74]

Larmor himself supplied that unique combination of mathematical dexterity and dynamical insight peculiar to high Cambridge wranglers; FitzGerald's agile mind provided Larmor with a rich source of incisive criticism and physical inspiration at crucial stages in the theory's development; FitzGerald, Lodge, and Trouton made available the material resources of their laboratories and their considerable skill as experimenters to follow up the observable consequences of Larmor's latest speculations. In short, although Larmor became the acknowledged author of the ETM, the theory was actually the common product of the Maxwellian network. With the breakup of the network, Larmor's achievement ceased to refer to any active research program beyond Cambridge.

Hunt has described the combination of death, illness, and career changes that brought the most productive period of the Maxwellian network to a close around 1900 (1991, 240). Lodge retired from active research that year to become principal of Birmingham University. Fitz-

74. As the collected papers of these men reveal, they each had major interests beyond Maxwellian electromagnetic theory that they did not discuss directly with members of the Maxwellian network.

Gerald died in 1901. Larmor was appointed secretary of the Royal Society in 1901 and Lucasian Professor of Mathematics at Cambridge in 1903. Following these appointments, he devoted much of his time to writing semipopular lectures and fulfilling general administrational duties. He continued to publish papers on diverse topics in dynamical physics but, apart from a few short papers on radiation pressure, made no further contribution to fundamental electromagnetic theory. Between 1902 and 1908, Trouton was the only member of the old network who continued to investigate possible second-order effects of the earth's motion through the ether. But as we have seen, where Larmor had taken electromagnetic theory beyond the original Maxwellian synthesis, Trouton was neither able nor willing to follow his lead.

This effective localization in Cambridge had a marked effect on the practice of the ETM. Through the publication of *Aether and Matter* and Larmor's own professorial lecture courses, the theory became widely known to students of the Mathematical Tripos. These men did not see the ETM as moribund or ad hoc but as a new and exciting research program to which they could contribute using their recently acquired skills in mathematics (Warwick 1992). This local upsurge of interest in Larmor's work both established Cambridge as the foremost center in Britain for research in electromagnetic theory and shifted the focus of that research from physical to mathematical aspects of the ETM. Where the Maxwellians had been concerned with practical electrical devices, electromagnetic waves, and the dynamical properties of the ether, this new generation of young wranglers focused their interest upon the mathematical properties of the new space-time and electromagnetic transformations introduced by Larmor. They necessarily followed Larmor in believing all matter to be electronic in construction but were equally prepared to abandon the notion of a unique dynamical ether if it impeded their attempts to introduce more-general transformations of the electromagnetic field equations (Warwick 1992, 644–50).

This emphasis upon the formal mathematical structure of electron-based electromagnetic theory made it increasingly easy for mathematicians in Cambridge to see themselves as engaged in a common project with some of their European colleagues. Where Larmor had sharply distinguished his work from that of Lorentz, Abraham, and, especially, Poincaré, the new generation of Cambridge mathematical physicists had no qualms about drawing freely upon their work. Some even drew upon Einstein's early papers on relativity, believing him to be a supporter of an electromagnetic view of nature (Warwick 1992). As a result of this mathematical homogenization of European electromagnetic theory, the local Maxwellian enterprise from which Larmor's ETM had emerged during the 1890s was rapidly effaced during the first decade of the twentieth century.[75]

With the benefit of hindsight, it is clear that Hicks's challenge to Larmor in 1901 provoked the final collaborative effort of the Maxwellian network. Even in 1902, however, Larmor recognized the episode as an important milestone in the development of Maxwellian electromagnetic theory. At the time of his collaboration with Trouton, Larmor was collecting and editing the late FitzGerald's papers for publication. He took the unusual liberty of appending both Trouton's account of the swinging-capacitor experiment and his own paper on the two-capacitor cycle to the end of the volume. Larmor considered it a fitting monument to FitzGerald's memory that the contraction proposed by him in 1889 had finally been established beyond doubt (Larmor 1902a, lxii). In retrospect, the multiauthored volume itself stands as a fitting monument to the Maxwellian network.

Acknowledgements

I would like to thank St. John's College (Cambridge) and University College (London) for granting me permission to reproduce manuscript material held in their archives. I would also like to thank Theodore Arabatzis, Bruce Hunt, Simon Schaffer, and Otto Sibum for reading an earlier draft of this essay and making a number of very helpful comments and suggestions.

Appendix A
Trouton's Analysis of the Swinging-Capacitor Experiment

In this appendix I show how Trouton used Maxwellian electromagnetic theory to calculate the electric and magnetic field energies associated with a moving capacitor.[76] Trouton considered a parallel-plate capacitor (of capacitance C) with square plates of side a separated by a distance d (see fig. 11.3). The plates carry equal and opposite charges Q, and the capacitor is initially at rest in the ether. The electric field energy W of the stationary capacitor can now be straightforwardly calculated using the well-known expression $W_e = Q^2/2C$. Since $C = \epsilon_r \epsilon_0 A/d$ (where $A = a^2$, the

75. In the first three editions of Jeans, *The Mathematical Theory of Electricity and Magnetism* (1908, 1911, 1915), for example, Jeans discussed the ETM as a joint project being pursued by Larmor, Lorentz, Abraham, Bucherer, and Einstein. Jeans's book is discussed in Warwick 1989, chap. 4.

76. Here I am expanding upon Trouton's argument in Trouton 1902.

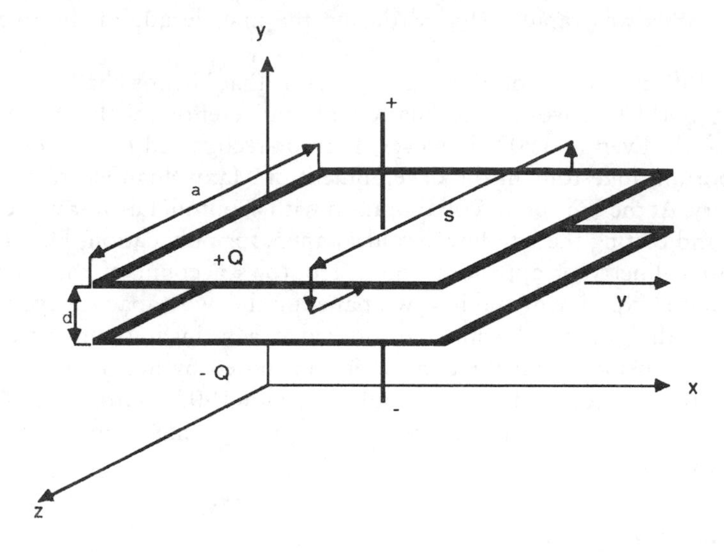

FIG. 11.3. Capacitor traveling edgeways

area of each plate) for a parallel-plate capacitor, W_e can be expressed in terms of A and d as follows:[77]

$$W_e = \frac{1}{2}\frac{Q^2 d}{\epsilon_r \epsilon_0 A}. \ldots \qquad (1)$$

If the capacitor now begins to move "edgeways" through the ether (see fig. 11.3) with velocity $v = (v, 0, 0)$, a magnetic field will be generated by the moving charge. Trouton explained that the equal and opposite charges carried along with the moving plates were equivalent to "currents running tangentially in opposite directions" (1902, 558).[78] Since the charge density on each plate is $Q/A = \sigma$, the current generated by each plate (with respect to the stationary ether) is $va\sigma$. This current generates a magnetic field, $\boldsymbol{B} = (B_x, B_y, B_z)$, which can be calculated using Ampere's law:

$$\int \boldsymbol{B} \cdot dl = \mu_0 I,$$

77. The terms ϵ_0 and ϵ_r represent the permittivity of free space and the relative permittivity of the dielectric between the capacitor plates, respectively.

78. For the purposes of this calculation it is convenient to imagine the equal and opposite charges moving in the same direction as equal and identical charges moving in opposite directions.

where $I = va\sigma$ (the total current passing through the closed loop around which the integral is taken).[79] Taking the line integral s around the surface of the positively charged plate (see fig. 11.3), we obtain $2aB_z = \mu_0 va\sigma$ so that

$$B_z = \frac{1}{2}\mu_0 v\sigma = \frac{\mu_0}{2}\frac{Qv}{A} \ldots \qquad (2)$$

in the close proximity of the plate. It is assumed here that $B_x = B_y = 0$ and that B_z is confined to the region between the plates.

These assumptions are justified in the following way. First, the motion of the plates (as shown in fig. 11.3) does not generate a magnetic field in the x-direction (hence $B_x = 0$ everywhere). Second, if the area of the plates is very large in comparison with the distance between them (i.e., $a \gg d$), then, to a good approximation, the B_y components of the magnetic fields generated by the plates will cancel each other out everywhere. Third, the B_z components of the magnetic fields will interfere constructively in the region between the plates and destructively outside this region. These arguments can be visualized if one imagines each moving plate as a collection of thin strips of current flowing parallel or antiparallel to the plate's direction of motion (the sense depending on the sign of the charge carried by the plate). Each current strip will be surrounded by circulating lines of magnetic force. Since the plates carry equal but opposite charges, the lines of magnetic force around the current strips of each plate will rotate in opposite senses. These lines interfere constructively between the plates (in the z-direction) but destructively everywhere else. Since the distance between the plates is very small, we can assume that the net magnetic field in this region is uniform.

The total uniform magnetic field between the plates (B_{zt}) is the sum of the fields due to the two plates; that is, exactly twice expression (2). Remembering also that $c^2 = (\mu_0\epsilon_0)^{-1}$, we can write

$$B_{zt} = \frac{1}{\epsilon_0 c^2}\frac{Qv}{A} \ldots . \qquad (3)$$

Since the magnetic energy per unit volume is given by

$$W_m = \frac{1}{2\mu_0}\int B_{zt}^2 dx\, dy\, dz \ldots, \qquad (4)$$

79. Here μ_0 is the magnetic permeability of a vacuum. Unlike the electric permittivity, the magnetic permeability is not appreciably affected by the presence of the dielectric medium.

and the volume between the plates is Ad, we obtain for the total magnetic energy in the edgeways position (W^e_m) due to the motion of the charged capacitor

$$W^e_m = \frac{Ad}{2\mu_0}\left(\frac{Qv}{\epsilon_0 Ac^2}\right)^2 = \frac{1}{2}\frac{Q^2 d}{\epsilon_0 A}\left(\frac{v}{c}\right)^2 = \frac{1}{2}\epsilon_r CV^2\left(\frac{v}{c}\right)^2 \ldots . \quad (5)$$

This is referred to as a "second-order" effect of the earth's motion. The "order" of such effects denotes simply the power to which the ratio v/c is raised in the expression. Since v/c is extremely small, its "order" will generally govern whether or not an effect is experimentally detectable. In Trouton's experiment, for example, the capacitor had a capacitance of 8 μF and a relative permittivity of 2 and was charged to a potential of 1200 V. He assumed the capacitor's velocity through the ether to be equal to the earth's orbital speed (19 miles per second) around the sun so that $v/c \approx 10^{-4}$. From the final term in equation (5) we therefore obtain

$$W^e_m = 11.5\left(\frac{v}{c}\right)^2 \simeq 10^{-7} \text{ joules.}$$

In cgs units this is about 1 erg.[80] FitzGerald postulated that the magnetic field generated when the capacitor was charged in the edgeways position derived its energy (W^e_m) through an electromechanical interaction between the capacitor and the ether. In the case of the swinging-capacitor experiment, Trouton always calibrated in terms of the energy delivered by each impulse. The actual forces in play would have been much more difficult to calculate. This means that each time the capacitor is charged or discharged, it should experience an impulse that delivers about 1 erg (Trouton 1902, 559).

When the capacitor is moving "flatways" through the ether (see fig. 11.4), the magnetic fields due to the charges on the plates will cancel each other everywhere (to a good approximation), and the magnetic energy will be zero. According to Trouton, this meant that the energy of the capacitor would be a function of its orientation with respect to its direction of motion through the ether. He claimed further, without demonstration, that since the magnetic energy was a function of B^2_{zt} (see expression [4]), the magnetic energy of a capacitor whose plates make an angle Ψ with its direction of motion through the ether must be of the form (Trouton 1902, 564)

80. Any effect that depended upon higher orders of v/c would be correspondingly harder to detect.

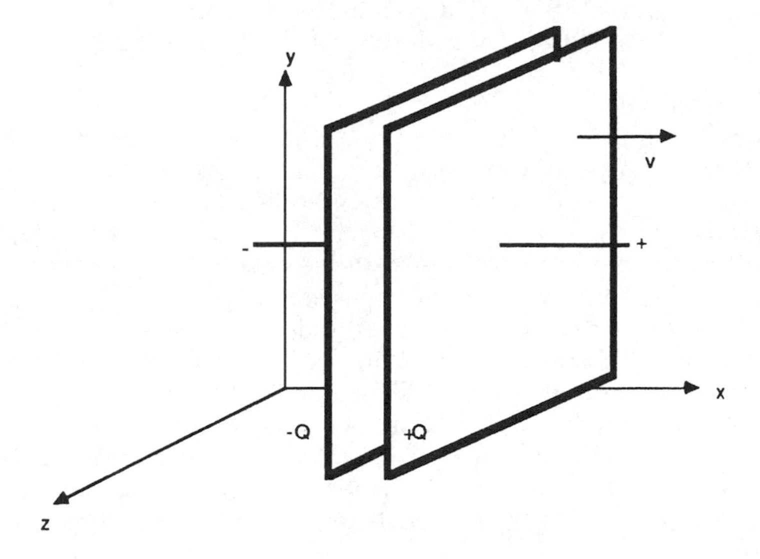

Fig. 11.4. Capacitor traveling flatways

$$W_{\Psi} = \left(\frac{v}{c}\right)^2 W^e_m \cos^2 \Psi.$$

The torque (T), tending to turn the capacitor to the position of minimum energy, may now be calculated as follows. The change in energy δW_{Ψ} due to a change in angle $\delta\Psi$ is $-Fr\delta\Psi$ (where F is the applied force and $r\delta\Psi$ is the distance through which the force acts).[81] But since rF is simply the torque acting on the capacitor, we have $T = \delta W_{\Psi}/\delta\Psi$. Hence, in the limit,

$$T = -\frac{\partial W_{\Psi}}{\partial\Psi} = \left(\frac{v}{c}\right)^2 W^e_m \sin 2\Psi \ldots, \qquad (6)$$

which is a maximum when $\Psi = 45°$ (approximately 7.5 dyne-cm). Trouton attempted to measure this torque by suspending the capacitor on the end of a long wire at an angle of 45° to its supposed direction of motion through the ether. Unable to make his apparatus sufficiently sensitive, he abandoned this experiment in the autumn of 1901. Trouton successfully demonstrated a null result for this effect in collaboration with H. R. Noble in 1903 (see section 6).

81. The minus sign appears because the energy decreases as the angle increases.

Appendix B
The Potential of a Moving, Charged Capacitor

B1. Introduction

I explained in the main text that Larmor did not solve the problem of perpetual motion in the two-capacitor cycle simply by asserting that moving matter contracted. Rather, he demonstrated that, according to the ETM, the difference in potential between the plates of a capacitor was independent of the capacitor's orientation with respect to its direction of motion through the ether. He did emphasize, however, that the contraction of moving matter was implicit in this demonstration. In this appendix I reconstruct Larmor's argument to show how he used his theoretical technology to generate the required evidential context for Trouton's experiment. Important points in Larmor's argument are sometimes highly compressed, usually because he assumed complete familiarity with contemporary electromagnetic theory on the part of the reader. I shall expand upon such points by supplying the relevant argument from Larmor's own writings.

The reader might find it helpful to refer to figures 11.3 and 11.4 in Appendix A while working through the next section. The factor β is defined as follows:

$$\beta = \frac{1}{\sqrt{1 - (v^2/c^2)}},$$

where v is the velocity of the electrical system through the ether, and c is the velocity of light.

B2. The Potential of a Moving, Charged Conductor

Having introduced his reader to the two-capacitor perpetual motion cycle described in section 5 above, Larmor set about calculating the difference in potential between the plates of a moving capacitor using the ETM. His procedure was based upon what he referred to as a "known electrodynamic result," which he simply stated:

> Compare an electrostatic conducting system at rest in the ether with the identically same system in motion with velocity v; suppose the charges to be the same in the two cases: to obtain the potential difference between the conductors in the latter case, find what they would be in the system when at rest, and elongated in the ratio β in the direction of motion, and multiply the result by β^2. (1902b, 567)[82]

82. Some 25 years after writing this paper Larmor edited it for inclusion in his *Mathematical and Physical Papers* (1929). As part of the editing process he "corrected" some of

This rule, which appears quite arbitrary as stated, in fact derives from the very heart of the ETM. Since the rule is central to the discussion that follows I shall explain its origin.

In 1897 Larmor derived an expression for the potential surrounding a charged conductor that moves through the ether with uniform velocity v.[83] He set up the problem in the following way. A moving charged conductor is surrounded by an electrostatic field ($\boldsymbol{E}$) and a magnetic field ($\boldsymbol{B}$) (generated solely by the motion of $\boldsymbol{E}$ through the ether). A unit test charge moving with the conductor would therefore experience a total force ($\boldsymbol{F}$) due partly to the electrostatic field of the charged conductor and partly to its motion through the magnetic field generated by the moving conductor. The total force, in vector notation, is $\boldsymbol{F} = \boldsymbol{E} + \boldsymbol{v} \wedge \boldsymbol{B}$.[84] Larmor assumed that this force could be expressed as the gradient of some general potential (V) such that

$$\boldsymbol{F} = -\left(\frac{\partial V}{\partial x} + \frac{\partial V}{\partial y} + \frac{\partial V}{\partial z}\right). \ . \ . \ . \qquad (1)$$

When the motion of the conductor is confined to the x-axis, $\boldsymbol{v} = (v, 0, 0)$, as in the case of the moving capacitor, the general expression found by Larmor reduces to

$$\beta^{-2} \frac{\partial^2 V}{\partial x^2} + \frac{\partial^2 V}{\partial y^2} + \frac{\partial^2 V}{\partial z^2} = 0 \qquad (2)$$

Larmor noted that equation (2) could be reduced to Laplace's equation by applying the transformations[85]

$$x_1 = \beta x, \qquad y_1 = y, \qquad z_1 = z, \ . \ . \ . \ , \qquad (3)$$

and the associated derivative transformations

$$\frac{\partial^2}{\partial x^2} = \beta^2 \frac{\partial^2}{\partial x_1^2}, \qquad \frac{\partial^2}{\partial y^2} = \frac{\partial^2}{\partial y_1^2}, \qquad \frac{\partial^2}{\partial z^2} = \frac{\partial^2}{\partial z_1^2}. \ . \ . \ . \qquad (4)$$

---

the definitions given in the paper to make them consistent with subsequent work. For this reason the reader is here referred to the original paper.

83. For a full derivation and discussion of this expression, see Warwick 1991, 81–83.

84. This expression, now known as the Lorentz force, was one of the fundamental equations of Larmor's ETM.

85. Although by 1900 Larmor was employing the Lorentz transformations, he continued to believe that the transformation $x_1 = \beta x$ was *always* valid because he assumed that clocks were always synchronized in the stationary ether frame. Hence $dx_1 = \beta\,(dx - vdt)$ could always be reduced to $dx_1 = \beta dx$ because dt (but not dt_1) could be set equal to zero for all dx. This is discussed more fully in Warwick 1991, 64–65.

Applying transformations (4) to equation (2) he obtained the following equality:

$$\beta^{-2}\frac{\partial^2 V}{\partial x^2} + \frac{\partial^2 V}{\partial y^2} + \frac{\partial^2 V}{\partial z^2} \equiv \frac{\partial^2 V}{\partial x_1^2} + \frac{\partial^2 V}{\partial y_1^2} + \frac{\partial^2 V}{\partial z_1^2} = 0. \ldots \qquad (5)$$

Expression (5) is important because it can be interpreted in two ways. Each way has different physical consequences.

B2.1 Mathematical Artifice. The equality in expression (5) can be thought of as a mathematical artifice that simplifies the calculation of the electrical potential surrounding a moving charged body. According to this interpretation the above transformation is made merely to reduce equation (2) to a recognizable mathematical form (Laplace's equation) in which it can more easily be solved. Having found the solution in terms of (x_1, y_1, z_1), the original variables are reinstated.

This interpretation attributes no physical significance to the fact that the potential surfaces surrounding a moving electrical system become contracted in their direction of motion by a factor of β. The technique is useful only because it reduces a difficult problem in the electrodynamics of moving bodies to a familiar problem in electrostatics. This procedure can therefore be expressed as follows: to find the potential difference between two points in a system of uniformly moving conductors, find the potential difference between the same two points when the system is at rest in the ether and elongated along the x-axis by a factor of β. This is precisely the first part of the rule given above by Larmor. For the remainder of this section I shall simply assume the second part of Larmor's rule (that the resulting potential has to be multiplied by a further β^2 factor), its explanation being given in section B.3 below. Larmor referred to the elongated stationary system constructed by this method as the "correlated" system.[86]

Larmor applied this rule to a moving capacitor as follows. Let the charge on a capacitor be Q and the potential between its plates V (when stationary in the ether). To find the potential between the capacitor's plates when it is moving flatways through the ether with velocity $\mathbf{v} = (v, 0, 0)$, we expand along the x-axis by a factor of β and multiply the resultant potential by a factor of β^2. Since the potential across a parallel-plate capacitor is given by[87]

$$V = \frac{Qd}{\epsilon_0 A} \ldots, \qquad (6)$$

86. On the origin of the "correlated-system" technique, see Warwick 1991, 60–69.

87. Larmor assumed for the purposes of this argument that $\epsilon_r = 1$. Any other choice renders the argument much more complicated.

the effect of expanding d by a factor of β is to raise the potential by the same factor. With the additional β^2 factor, the potential in the flatways orientation is $\beta^3 V$. Applying the same rule to the capacitor in the edgeways orientation, the elongation along the x-axis increases the length of the plates (and hence the area of the plates) by a factor of β. According to equation (6) this decreases the potential between the plates by a factor of β^{-1}. Multiplying this result by the additional β^2 factor, the potential in this orientation is βV. It follows that in rotating the moving capacitor from flatways to edgeways there is a second-order decrease in potential of[88]

$$\beta V - \beta^3 V = V\beta^3(\beta^{-2} - 1) \tag{7}$$

$$= -V\left(\frac{v^2}{c^2} + \frac{3}{2}\frac{v^4}{c^4} + \ldots\right)\ldots$$

Larmor made two comments on this result. First, energy is conserved when the capacitor is rotated from flatways to edgeways. This can be seen as follows. The potential energy stored on a parallel-plate capacitor is

$$W_{pot} = \frac{1}{2}QV = \frac{1}{2}\frac{Q^2 d}{\epsilon_0 A}.$$

Hence the decrease in potential given by (7) implies a decrease in potential energy (to second order) of

$$\delta W_{pot} = -\frac{1}{2}QV\frac{v^2}{c^2} = -\frac{1}{2}\frac{Q^2 d}{\epsilon_0 A}\frac{v^2}{c^2}.$$

But this does not take account of the magnetic energy that is stored kinetically in the ether in the edgeways position.[89] As was shown in Appendix A, the magnetic field that appears in the edgeways position has kinetic energy (assuming $\epsilon_x = 1$)

$$W_{kin} = \frac{1}{2}\frac{Q^2 d}{\epsilon_0 A}\frac{v^2}{c^2}\ldots \tag{8}$$

so that energy is conserved.

88. Larmor always approximated his calculations to second order because this was the limit of experimental accuracy. Since, for the earth's orbital velocity around the sun, $v/c \approx 10^{-4}$, an increase of order v/c increases the required sensitivity of the apparatus by four orders of magnitude. Equation (9) below shows that any residual mechanical effect between the capacitor and the ether would be of the order v^4/c^4, so that Trouton would have had to make his apparatus more sensitive by eight orders of magnitude to search for an effect.

89. Larmor continued to think of the ether as a dynamical object. The energy of the electric field was stored in potential form as some kind of strain; that of the magnetic field, in kinetic form as some kind of local motion.

Larmor's second comment was that although the energy of the electric and magnetic fields is conserved, the cycle still constitutes a perpetual motion machine. To see why this is so we have to consider precisely what equation (7) represents. Equation (2)—from which the rule is generated—is interesting precisely because it expresses the *complete* potential of a moving electrical system, including the contribution made by the magnetic field generated by the charge's motion. The work done in charging a moving capacitor is, therefore, done solely against this potential. When the capacitor is turned from edgeways to flatways, this electric potential alters by the amount given in equation (7). Hence the capacitor AB in the thought experiment could still be charged and discharged at different electric potentials (by rotating it from edgeways to flatways) without any external source of work being expended. For this reason, Larmor argued, the interpretation of equation (6) as a mathematical artifice was untenable (Larmor 1902b, 568).

B2.2 The Contraction of Moving Matter. According to Larmor's ETM, all matter was made solely of electrons and held together by purely electrical forces. Larmor claimed that the contraction of the potential surfaces surrounding moving electrons—predicted by expression (2)—would require the electrons of which matter was composed to move slightly closer together in order to remain in electric equilibrium.[90] An obvious implication of this claim is that matter will itself contract by a factor of β^{-1} in the direction of its motion through the ether. If matter did behave in this way, direct physical significance could be attributed to the equality in (5) by making (x_1, y_1, z_1) the physical space of the contracted moving system (all moving observers and measuring rods being similarly contracted). This contraction offered an explanation of the null result obtained in the Michelson-Morley experiment and, conversely, made the Michelson-Morley experiment evidence that this contraction actually occurred.

Following Hicks's challenge, Larmor used Trouton's experiment as fresh evidence that moving matter did contract. The argument runs as follows. We first use Larmor's rule to calculate the potential between the plates of the capacitor. In the flatways position the moving capacitor is shrunk so that the plates are only β^{-1} apart. If we now expand along the x-axis (the direction of d) by β and multiply the resulting potential by the additional factor of β^2, the final potential becomes $\beta^2 V$. This is now the potential as measured in the moving system (in this case, the surface of the earth). In the edgeways position the length of the plates is shrunk by a factor of β^{-1}. Expanding by a factor of β and multiplying the resulting potential by an additional β^2 factor, we obtain $\beta^2 V$. The potential is

90. For a detailed discussion of this point see Warwick 1991, 56–60.

now the same in both orientations. Seen from the stationary ether frame, energy is still conserved between the edgeways and flatways orientations. The change in total potential energy due to the change in the capacitor's dimensions as it rotates from flatways to edgeways (at constant potential) is

$$\delta W_{pot} = \frac{\epsilon_0}{2}\frac{A}{d}V^2(\beta^{-1} - \beta) = -\frac{1}{2}\frac{Q^2 d}{\epsilon_0 A}\frac{v^2}{c^2},$$

which is exactly the energy of the magnetic field in the edgeways orientation (see expression [8]). Energy is also conserved in the rest frame of the capacitor since in this frame no magnetic field appears to exist and the capacitor does not appear to alter its dimensions when rotated.[91]

According to this interpretation of equality (5), Larmor's two-capacitor cycle does not constitute a perpetual motion machine because the potential between the plates of AB is independent of its orientation. Larmor argued that this confirmed his claims that matter was wholly electronic in constitution and contracted when in motion through the ether. Since Trouton's experiment appeared to rule out any other account of the two-capacitor cycle, it effectively substantiated Larmor's claims.

B3. The Correlation of Electrical Potentials

In this section I explain why Larmor reckoned that the potentials found by expanding the correlated system had to be multiplied by an additional factor of β^2 to obtain the true potentials. Larmor's strategy in solving problems in the electrodynamics of *moving* bodies was to try to find a *stationary* (in the ether) electrical system that was surrounded by the same potential field as that which surrounded the moving body. It was this equivalent stationary system that Larmor referred to as the "correlated system." In deriving equation (2) above, Larmor showed that the relationship between the electric force $\boldsymbol{E}$ (that would surround a charged body when it was stationary in the ether) and the Lorentz force $\boldsymbol{F}$ (that would surround the same body when it moved uniformly through the ether) was[92]

$$\beta^{-2}(E_x, E_y, E_z) = (\beta^{-2}F_x, F_y, F_z). \; . \; . \qquad (9)$$

91. Larmor continued to believe that the description of events from the perspective of the stationary ether frame should be privileged because it described the real physical state of the ether.

92. For a detailed discussion of the origins of equation (9) see Warwick 1991, 81–83. The "electric force" and the "Lorentz force" are in this case the forces that would be experienced by a unit charge that was at rest with respect to the charged conductor.

In *Aether and Matter* he used this relationship to construct the correlated system for a charged system that moved through the ether with velocity v (1900, 150–53). His construction was based upon the fact that, according to Maxwellian electromagnetic theory, the charge density at a point in space is equal to the divergence of the electric field strength at that point. When stationary in the ether, the charge density ρ, electrical field strength $\boldsymbol{E}$, and electrical potential P of a charged conductor are related as follows:

$$\begin{aligned}\rho = \nabla \cdot \boldsymbol{E} &= \left(\frac{\partial E_x}{\partial x} + \frac{\partial E_y}{\partial y} + \frac{\partial E_z}{\partial z}\right) \\ &= -\left(\frac{\partial^2 P}{\partial x^2} + \frac{\partial^2 P}{\partial y^2} + \frac{\partial^2 P}{\partial z^2}\right) \ldots .\end{aligned} \tag{10}$$

In the rest frame of the contracted *moving* conductor, the Lorentz force $\boldsymbol{F}$ would be interpreted as a *purely electrostatic force* (because the charge in that frame is stationary and so does not generate a magnetic field). In the moving system, then, the equivalent of expression (10) will be

$$\begin{aligned}\rho_1 = \nabla \cdot \boldsymbol{F} &= \left(\frac{\partial F_{x1}}{\partial x_1} + \frac{\partial F_{y1}}{\partial y_1} + \frac{\partial F_{z1}}{\partial z_1}\right) \\ &= -\left(\frac{\partial^2 V}{\partial x_1^2} + \frac{\partial^2 V}{\partial y_1^2} + \frac{\partial^2 V}{\partial z_1^2}\right) \ldots ,\end{aligned} \tag{11}$$

where

$$(F_{x_1}, F_{y_1}, F_{z_1}) = -\left(\frac{\partial V}{\partial x_1} + \frac{\partial V}{\partial y_1} + \frac{\partial V}{\partial z_1}\right),$$

and ρ_1 is the charge density. Larmor sought to relate (10) and (11) using (9). He first took the divergence of both sides of (9) to obtain

$$\beta^{-2}\left(\frac{\partial E_x}{\partial x} + \frac{\partial E_y}{\partial y} + \frac{\partial E_z}{\partial z}\right) = \left(\beta^{-2}\frac{\partial F_x}{\partial x} + \frac{\partial F_y}{\partial y} + \frac{\partial F_z}{\partial z}\right) \ldots \tag{12}$$

and then transformed the right-hand side of (12) using transformations (3). In terms of the electrical potentials P and V, this transformation yields

$$\left(\frac{\partial^2 P}{\partial x^2} + \frac{\partial^2 P}{\partial y^2} + \frac{\partial^2 P}{\partial z^2}\right) = \beta^2\left(\frac{\partial^2 V}{\partial x_1^2} + \frac{\partial^2 V}{\partial y_1^2} + \frac{\partial^2 V}{\partial z_1^2}\right) \ldots . \tag{13a}$$

Expression (13) relates the electrical potential P that surrounds a charge ρ when at rest in the ether to the electrical potential V that surrounds a charge ρ_1 (measured in the rest frame of the moving charge) when moving through the ether with velocity v. According to (13a), the relationship between ρ and ρ_1 is $\rho = \beta^2\rho_1$. Larmor noted that if the potential P (on the left-hand side of [13]) is multiplied by β^2, then

$$\begin{aligned} \rho &= \left(\frac{\partial^2 P}{\partial x^2} + \frac{\partial^2 P}{\partial y^2} + \frac{\partial^2 P}{\partial z^2}\right) \\ &= \beta^2\left(\frac{\partial^2 V}{\partial x_1^2} + \frac{\partial^2 V}{\partial y_1^2} + \frac{\partial^2 V}{\partial z_1^2}\right) = \rho_1. \ . \ . \ . \end{aligned} \tag{13b}$$

We now have $\rho = \rho_1$ and $P = V$. In other words, the potential P surrounding a stationary charged conductor can be made the same as the potential V surrounding the same conductor (with the same charge) when it moves through the ether with velocity v by multiplying P by β^2. This was the source of the second part of Larmor's rule; namely, that the potentials surrounding the stationary correlated system needed to be multiplied by an additional factor of β^2.

One final point must be clarified. Larmor subsequently realized that in order for the *total* charge on the conductor to be the same in the stationary and moving cases, the charge *densities* needed to be related as $\rho = \beta\rho_1$ (rather than $\rho = \rho_1$) (Larmor 1900, 176).[93] It then followed that the potential P needed to be multiplied by β (rather than β^2) to make it equivalent to V. Larmor noted this correction in the corrigenda to *Aether and Matter* but then used a β^2 factor (presumably in error) a few months later when preparing his paper on Trouton's experiment.[94] The error does not affect Larmor's argument (which was limited to second order) and was corrected when the paper was reproduced in his *Mathematical and Physical Papers*.

93. On Larmor's reasons for investigating charge, see Warwick 1991, 64–69.

94. I would like to thank Michael Janssen for drawing my attention to Larmor's correction of the β^2 factor in the corrigenda to *Aether and Matter.*

CONCLUSION

Jed Z. Buchwald
and
Sylvan S. Schweber

THE ESSAYS IN THIS VOLUME share an image of science as an active process, one in which work in the laboratory and work on paper interact to forge a flexible, yet highly stable, structure. All the articles have been influenced in this by the insights that social constructivists have given historians and philosophers of science during the past two decades. Some of the authors nevertheless hold considerably different opinions from those of most constructivists. Others are significant contributors to constructivist understanding in their own rights. In this concluding essay, stimulated by the articles in this volume, we would like rather to emphasize what unites these various authors than what divides them. We feel that, different though they are, these essays nevertheless cohere in an important sense, namely, in their strong emphasis on science as a process for the making of stable, and hence reliable, knowledge. We want to use this unifying characteristic to suggest a place in contemporary philosophy of science for a pragmatic realism and to insist as well on a place in historical practice for the psychology of the individual scientist.

By "pragmatic realism" we (loosely) mean a commitment to the efficacy in practice of entities and effects that are engaged by, and that are not (merely) produced by, experiments. This naturally requires that we not yield to the ever-present temptations of absolute skepticism, while nevertheless insisting on the constructed character of scientific knowledge. By "constructed" we mean that scientific knowledge is actively produced rather than passively read off from a natural codex, as it were. Given our emphasis on pragmatic realism we do not of course mean to suggest that scientific knowledge lacks objectivity. On the contrary, we feel that it is highly objective, and that it becomes more so over time as novel effects are produced and as they are bound to past ones. Dewey's remarks on James and skepticism seem to us appropriate in this regard. Dewey noted that

James "consistently opposed absolute dogmatism in philosophy, and at the same time he has repudiated utter skepticism" (Dewey 1982, 20:220). For James "a skepticism which brings with it nothing that can contribute constructively to investigation" was "most distasteful." James, Dewey continued, "asks us to look for new truth in the results of our past experiments at the same time that we continue to experiment and to seek for a growing area of practical belief."

We want to place particular weight on a similar pragmatic skepticism. Although this is hardly a novel observation, we believe that the core of scientific practice lies precisely in its noncorrosive skepticism, out of which a "growing area of practical belief" emerges over time. Herein, we think, lies the proper source for a new kind of realism, one that speaks with historical verisimilitude to scientists' own beliefs. Although scientists, like all practitioners, have myths concerning their work and its history that rarely sustain critical inquiry, nevertheless we take very seriously their usual (if often nuanced and historically mutable) commitment to some form of realism. Indeed, we feel that the constructivists' own emphasis on the many different things—the "actants" in contemporary parlance—that work together to form the scientific enterprise coheres nicely with the generative quality that a commitment to certain forms of realism brings to both the science itself and to its historiography. Callon and Law have aptly written that "the actors themselves continuously seek to create links between such pairs as laboratory and consumers, scientific knowledge and methods of production, and science policy and economic growth" (1989). We believe further that neither the elements between which the links are created nor the links proper are any less real than the actors who construct and forge them. They are however evolved in time by practical activity; they are not to be found in an eternal world of abstractions divorced from human skills, wants, beliefs, and hopes.

Out of this pragmatic realism few if any universally applicable strictures on the practice of science are likely ever to emerge. As Morrison aptly remarks, "traditional realism (as well as some of its modern versions) attempts to impose the philosophical and logical conditions for truth and knowledge on a practice (science) that is ill equipped to meet these standards." Pragmatic realism seeks nothing like this; it looks instead to the immensely fertile seedbed of the laboratory, where scientific effects are made to happen and are used to make things happen, and to the quasi-stable models and representations of these effects and things. This space cannot be confined by the rigid strictures of historically dissociated philosophical norms. Norms may, and indeed do, operate, as Morrison insists, but their scope, nature, and efficacy are limited in space and time.

We want to open a place for the practice of the individual scientist in the making of scientific knowledge and indeed for psychological considerations that, we believe, must underpin any attempt to understand how knowledge is actually produced. Scientists *interact* with other scientists, and they *interact* with the knowledge and practices created by others; they also constantly think about how their work will be received. Groups do not think as undifferentiated units; individuals do, and we want to bind the constructivist emphasis on interaction to a deep understanding of individual practice. Hertz, Buchwald insists, worked in his cathode-ray laboratory with his German contemporaries and teachers ever in mind, and occasionally with some of them present. He constructed what we now regard as a false image of cathode rays; he also constructed knowledge about them that is not regarded as false at all. To understand what he did, and why he did it, requires understanding the unique facets of his own mental makeup, as well as the cultural features that he shared with other Berlin-trained physicists of the day.

This is not the place to elaborate a rigorous new philosophy of practice or to develop a scheme for the historical significance of the individual; nor would we want to do so in any case. To capture loosely the nexus of individual practice, group interaction, and constructivist production in a way that conduces to a renovated realism, we offer the metaphor of a scaffolding.[1] A scaffolding seems to us nicely to suggest the connectivity of conceptual schemes and practices that constitute scientific knowledge. It is especially significant that a scaffolding suggests issues of stability, and it is, we feel, stability that must be carefully emphasized in understanding how scientific knowledge can be objective and yet also be pragmatically constructed. The connections and linkages inherent in a scaffolding are precisely the sources of its stability, on the one hand, and they also form the resources that permit its extensions in novel directions; for our scaffolding indeed evolves over time. A metaphor of intertwined strands that form a sturdy cable invoked by C. S. Peirce for philosophy is accordingly apposite for us as well:

> Philosophy ought to imitate the successful sciences in its method, so far as to proceed only from tangible premisses which are subjected to careful scrutiny, and to trust rather to the multitude and variety of its arguments than to the conclusiveness of any one. Its reasoning should not form a chain which is stronger than its weakest link, but a cable whose fibers may be

1. John Herschel used the metaphor. See his *Preliminary Discourse* ([1832] 1987, 204). Herschel had in mind a scaffolding that was eventually to be taken down, leaving the edifice of science in place. For us the scaffolding is constantly being secured and partially rebuilt like a ship at sea.

ever so slender, provided they are sufficiently numerous and intimately connected.

We do not want to be overly precise or restrictive concerning the elements out of which the scaffolding is made. They include, at the very least, conceptual schemes, the networks connecting them, metaphysical commitments, revered and newly honed skills, long-working devices, new instruments, social institutions, indeed all of the many things that inevitably go into human production of anything. Most such constructions are not highly stable over time or, if stable, do not easily lend themselves to growth or novel articulation. The scientific scaffolding does, and it is precisely its stability under extension that makes it a uniquely objective form of knowledge. One might say that creative partaking in the scientific enterprise consists in the extension of the scaffolding to fit new circumstances[2] rather than in its nearly complete dismantling and replacement.

As noted, parts of the scaffolding are conceptual schemes and networks. When sufficiently stable and secure, these become maps for the further exploration of domains of inquiry. Such explorations can at times be thought experiments, whose results may restructure part of the scaffolding or make new linkages.[3] But the scientific scaffolding is not limited to conceptual linkages. New instruments, new practices, new institutions, etc., can likewise restructure it. Extensive restructurings may involve highly limited regions or subregions, or they may involve larger regions which reach across many disciplines and institutions.

We may think of the scaffolding as part of the equipment that scientists acquire during their training. It is subsequently secured and extended as a result of further research activities. The scaffolding delineates, connects, and amalgamates the constraints *and* the resources that Galison and Pickering discuss in their essays.[4] Each scientist has a particular mental map of the scaffolding, but some regions are essentially identical for all the members of a given scientific community. The scaffolding that individual scientists construct and sustain allows them to adumbrate blueprints for possible extensions (or restructurings) that they may try to effect. This private structure outlines and suggests the possibilities. Moreover, it is individual scientists who construct and sustain the novelties that enable these kinds of changes.

The individual's work involves many different kinds of knowledge,

2. We are here paraphrasing Pickering 1992, 6.

3. See Kuhn's essay "A Function for Thought Experiments" (1977b).

4. Using the metaphor of the scaffolding, Pickering would stress the difficult but ever-open process of forging connections between its elements. Galison, on the other hand, would point to the rigidity of the structure and therefore to the constraints imposed by it on possible extensions and mutations.

and we do not want to draw invidious distinctions among these kinds. We do however want to insist that cognition, and not only the types of skill involved in learning, for example, to make uniform and homogeneous pieces of glass or to ride a bicycle, necessarily enters into the production of science. Warwick's detailed analysis of Trouton's care in constructing and stabilizing a novel piece of equipment emphasizes this tight connection between knowledge and practice. Cognitive structures do bind tightly to instrumentalities and so to the objectification of knowledge through laboratory devices. Radder's contribution provides insight into mechanisms for objectification by discussing the differences between replication of effects and reproduction of experiments. In this respect his analysis complements Schaffer's article, which narrates the history of a transformation in the "topology," as he terms it, of the nineteenth-century astronomical workplace, that is, the transformation of the astronomical observatory into a laboratory.

Scientists and historians of science in years past had little doubt that scientific knowledge was knowledge *of* something that exists independently of the knower. We are not so convinced that even scientific knowledge can thoroughly transcend the sites of its production. Nevertheless, we continue to find it difficult to understand the stability of scientific practice—including its perennial adaptation of previous results to new ways of doing and thinking—without the active participation of the "entities" (the French school and Hacking lately speak of "things") that the particular discipline studies.[5] Our sense that knowledge has powerful local roots even given our belief in the independent existence of scientific entities is based on the "constructive" practices of physicists and chemists.

At the end of the last century the chemist Walther Nernst applied the adjective "constructive" to physics and chemistry. He had in mind the production of new worlds through a combination of cognitive endeavor and instrumental manipulation. In his inaugural lecture for the dedication of the Institute for Physical Chemistry and Electrochemistry at the University of Göttingen delivered on 2 June 1896, Walther Nernst suggested that the founding of this new institute "furnishes a new important demonstration of the fact that between two rather separate sciences [physics and chemistry] an intimate reunion had begun." He continued:

> Let us ask ourselves about the difference between physics and chemistry; one says frequently that chemistry is concerned with the construction of molecules from atoms, while physics has to do with the completed mole-

5. At times the confluence—the intertwining—of a quasi-stable ontology, a quasi-stable theory, a quasi-stable laboratory practice, and particular, quasi-stable styles of reasoning (as in the case of the domain described by nonrelativistic quantum mechanics) gives special robustness to certain scientific disciplines.

> cules; this definition is simply rejected *ad limine* since it uses an entirely specific hypothesis of these sciences, the so-called atomic hypothesis, which of course one uses with success very frequently, but which in no way forms such exclusive foundation for research that one may use it for a *general* unifying principle.
>
> However, one finds quite easily a highly characteristic feature that physics and chemistry have *together* in opposition to *all* other branches of the natural sciences; namely, as we want shortly to show, the other natural sciences find the objects of their investigations, their systems, ready in the external world; thus the astronomer is concerned with the properties of the solar system, the zoologist with the multiplicity of the species, the physiologist with the life processes of animals and plant organisms, etc. Only the physicists and chemists themselves create their systems for themselves, and ever according to theoretical representations, which they test and between which they want to distinguish. They themselves construct for themselves from the unformed raw material of the external world that one system which, from exhaustive research, appears to them quite especially worthy from their current viewpoint. We, therefore, can briefly designate these sciences as "*constructive sciences*." (1896, emphasis added)

As Nernst saw it, physicists and chemists are different from other scientists because they build their own worlds in the laboratory according to "theoretical representations." He felt that ever-changing knowledge is used by physicists and chemists to build new and ever-evolving worlds. They construct their own material worlds, which they then manipulate. This, he felt, had long been the distinguishing characteristic of these two groups, which in other respects were quite different from one another. Indeed, he went on to characterize the differences between them. As those who dedicate themselves to the constructive sciences, chemists "investigate [matter] in its individual characteristics in particulars, led by the wish to win in this way a clear picture of the whole." Physicists, on the other hand, "seek after the laws, which are as much as possible removed from the nature of matter; the appearances of the material world are to them . . . a more special example or image of their basically much more general laws and formulae."

Nernst was however writing at a time when even this division seemed to be breaking down. It had not previously been possible to unite the chemist and physicist in a single person because "the auxiliary means of both sciences were entirely too dissimilar for their simultaneous operation to have been able to become the rule." In his day, Nernst continued, unification looked to be possible because of the discovery of "a series of general, and thereby on account of their simplicity, especially important and useful natural laws which enable the inquiring intellect to unify an immense amount of experimental data in a few words or formulae." Nernst had in mind thermodynamics, Maxwell's electrodynamics, and the kinetic theory of gases. The hidden constitution of matter was now coming to the forefront of physicists' own research. This entailed a drastic reordering of the relation between physics and chemistry because the microworld, which had *not* been a major preoccupation of the working physicist throughout the century, was rapidly becoming a central concern. Many of the experiments discussed in this volume are part of the establishment of this new "microscopic" world, either as attempts to manipulate it or as attempts to do altogether without it.

Scientific revolutions restructure the scaffolding. They may involve but one domain or subdomain, or they may involve many disciplines, many institutions, and thus much larger regions of the scaffolding. It would be interesting, we believe, to analyze the dramatic shifts that occurred between the 1870s and the early 1920s in terms of the changes that occurred in the scientific community's scaffolding and thus flesh out what for the time being remains a suggestion.

References

Abraham, M.

1902a. Dynamik des Elektrons. *Nachr. Gesell. Wiss., Math.-Phys.* 4:20–41.

1902b. Prinzipien der Dynamik des Elektrons. *Phys. Zeit.* 4:57–63.

1904. Die Grundhypothesen der Elektronentheorie. *Phys. Zeit.* 5:576–79.

1905. *Ions, Electrons, Corpuscules: Mémoires réunis et publés.* 2 vols. Paris: Cauthiers-Villars.

Ackermann, Robert

1985. *Data, Instruments, and Theory.* Princeton: Princeton University Press.

1988. "Experiment as the Motor of Scientific Progress." *Social Epistemology* 2: 327–35.

Ackland, B.; R. Dickenson; J. Ensor; P. Gabbe; T. Lollaritsch; C. London; P. Subrahmanyam; and H. Watanabe

1985. *CADRE—A System of Cooperating VLSI Design Experts. Proceedings IEEE International Conference on Computer Design: VLSI in Computers.* Port Chester: IEEE.

Airy, G. B.

1833. Report on the Progress of Astronomy during the Present Century. In *Report of the Second Meeting of the British Association for the Advancement of Science,* 125–89. London: John Murray.

1859. Remarks on the Application of Photography. *Monthly Notices of the Royal Astronomical Society* 18:16–18.

Akin, W. E.

1977. *Technocracy and the American Dream: The Technocrat Movement, 1901–1941.* Berkeley and Los Angeles: University of California Press.

Amsterdamska, O.

1990. Surely You Are Joking Monsieur Latour. *Science, Technology, and Human Values* 15, no. 4:495–504.

Anonymous

1907. Mathematics and Physics at the British Association. *Nature* 76:457–62.

Apel, K.-O.

1980. *Towards a Transformation of Philosophy.* London: Routledge & Kegan Paul.

1990. *Penser avec Habermas contre Habermas.* Paris: L'Éclat.

Arago, D. J. F.
1859. Mémoire sur un moyen très-simple de s'affranchir des erreurs personelles. In M. J. A. Barral, ed., *Oeuvres complètes de François Arago,* 5:233–44. Paris: Gide.

Bachelard, G.
1972. *La formation de l'esprit scientifique.* 1938. Reprint, Paris: Vrin.
1975. *Le rationalisme appliqué.* 1949. Reprint, Paris: Presses Universitaires de France.

Baigrie, Brian S.
1991. The Justification of Kepler's Ellipse. *Stud. Hist. Phil. Sci.* 21:633–64.
1994. Social Epistemology, Scientific Practice, and the Elusive Social. *Argumentation* 8:125–44.

Baigrie, Brian S.; and J. N. Hattiangadi
1992. On Consensus and Stability in Science. *Brit. J. Phil. Sci.* 43:435–58.

Baillet, Adrien
1691. *La vie de Monsieur Des-Cartes.* 2 vols. in 1. Paris: Daniel Horthemels.

Barnes, B.
1974. *Scientific Knowledge and Sociological Theory.* London and Boston: Routledge & Kegan Paul.
1982. *T. S. Kuhn and Social Science.* London: Macmillan; New York: Columbia University Press.
1988. *The Nature of Power.* Urbana: University of Illinois Press.

Bartholomew, C. F.
1976. Discovery of the Solar Granulation. *Quarterly Journal of the Royal Astronomical Society* 17:263–89.

Bazerman, Charles
1988. *Shaping Written Knowledge.* Madison: University of Wisconsin Press.

Becker, Barbara J.
1993. Eclecticism, Opportunism and the Evolution of a New Research Agenda: William and Margaret Huggins and the Origins of Astrophysics. Ph.D. diss., Johns Hopkins University.

Becker, Howard S.
1982. *Art Worlds.* Berkeley and Los Angeles: University of California Press.

Beer, Gillian
1989. "The Death of the Sun": Victorian Solar Physics and Solar Myth. In J. B. Bullen, ed., *The Sun Is God: Painting, Literature, and Mythology in the Nineteenth Century,* 159–80. Oxford: Clarendon Press.

Ben-David, J.
1971. *The Scientists' Role in Society.* Englewood Cliffs, N.J.: Prentice Hall.

Bennett, J. A.
1984. *The Celebrated Phaenomena of Colours: A History of the Spectroscope in the Nineteenth Century.* Cambridge: Whipple Museum.

Bestelmeyer, A. C. W.
1907. Spezifische Ladung und Geschwindigkeit der durch Röntgenstrahlen erzeugten Kathodenstrahlen. *Ann. d. Phys.* 22:429–47.

Bhaskar, R.
1978. *A Realist Theory of Science.* Hassocks: Harvester Press.

Bloor, D.
1976. *Knowledge and Social Imagery.* London and Boston: Routledge & Kegan Paul.

Bogen, J.; and J. Woodward
1988. Saving the Phenomena. *Philosophical Review* 97:303–52.

Boltanski, L.; and L. Thévenot
1987. *Les économies de la grandeur.* Paris: Presses Universitaires de France.

Boorse, H. A.; and L. Motz; eds.
1966. *The World of the Atom.* Vol. 1. New York: Basic Books.

Borgmann, A.
1984. *Technology and the Character of Contemporary Life.* Chicago: University of Chicago Press.

Born, M.
1956. *Physics in My Generation.* London: Pergamon Press.
1965. *Einstein's Theory of Relativity.* New York: Dover.
1971. *The Born-Einstein Letters.* London: Macmillan.

Bourdieu, P.
1971. Genèse et structure du champ religieux. *Revue française de sociologie* 12: 295–334.
1979. Les trois états du capital culturel. *Actes de la recherche en sciences sociales* 30:3–6.
1981. Men and Machines. In K. Knorr-Cetina and A. V. Cicourel, eds., *Advances in Social Theory and Methodology: Toward an Integration of Micro- and Macro-Sociologies,* 304–17. Boston: Routledge & Kegan Paul.
1987. L'institutionnalisation de l'anomie. *Cahiers du Musée national d'art moderne* 19–20:6–19.
1989. *La noblesse d'Etat: Grandes écoles et esprit de corps.* Paris: Minuit.
1991. The Peculiar History of Scientific Reason. *Sociological Forum* 6, no. 1: 3–26.

Brace, D. B.
1904. On Double Refraction in Matter Moving through the Ether. *Phil. Mag.* 7: 317–29.

Braudel, Fernand
1950. Pour une économie historique. *Revue économique,* May, 37–44.
1958. La longue durée. *Annales,* Oct.–Dec.
1972. *The Mediterranean and the Mediterranean World in the Age of Phillip II.* New York: Harper & Row.
1973. *Capitalism and Material Life, 1400–1800.* New York: Harper & Row.

Bridgman, P. W.
1927. *The Logic of Modern Physics.* New York: Macmillan.

Brown, Theodore
1968. The Mechanical Philosophy and the "Animal Oeconomy." Ph.D. diss., Princeton University. Published 1981 by Arno Press.

Brush, Stephen
1976. *The Kind of Motion We Call Heat.* Amsterdam: North Holland.

Bucherer, A. H.
1904. *Mathematische Einführung in die Elektronentheorie.* Leipzig: Teubner.
1905. Das deformierte Elektron und die Theorie des Elektromagnetismus. *Phys. Zeit.* 6:833–34.

Buchwald, Jed
1985a. *From Maxwell to Microphysics: Aspects of Electromagnetic Theory in the Last Quarter of the Nineteenth Century.* Chicago: University of Chicago Press.
1985b. Modifying the Continuum: Methods of Maxwellian Electrodynamics. In Harman 1985, 225–41.
1988. The Michelson Experiment in the Light of Electromagnetic Theory before 1900. In Goldberg and Stuewer 1988.
1990. The Background to Heinrich Hertz's Experiments in Electrodynamics. In William Shea and Trevor Levere, eds., *Nature, Experiment, and the Sciences.* Dordrecht: Kluwer.
1992. Design for Experimenting. In J. Thompson and P. Horwich, eds., *Essays in Honor of T. S. Kuhn.* Cambridge: MIT Press.
1994. *The Creation of Scientific Effects: Heinrich Hertz and Electric Waves.* Chicago: University of Chicago Press.

Buckland, P.
1988. Irish Unionism and the New Ireland. In D. G. Boyce, ed., *The Revolution in Ireland, 1879–1923,* 71–90. Houndmills, Basingstoke, Hampshire: Macmillan Education.

Bumstead, H. A.
1908. Applications of the Lorentz-FitzGerald Hypothesis to Dynamical and Gravitational Problems. *American Journal of Science* 26:493–508.

Burchfield, Joe D.
1975. *Lord Kelvin and the Age of the Earth.* London: Macmillan.

Callon, M.
1987. Society in the Making: The Study of Technology as a Tool for Sociological Analysis. In W. E. Bijker, T. P. Hughes, and T. J. Pinch, eds., *The Social Construction of Technological Systems,* 83–103. Cambridge: MIT Press.
1991. Réseaux technico-économiques et irréversibilités. In *Les figures de l'irréversibilité en économie,* 95–230. Paris: Editions de l'Ecole des Hautes Études en Sciences Sociales.

Callon, M.; and B. Latour
1992. Don't Throw the Baby with the Bath School. In Pickering 1992, 343–68.

Callon, M.; and J. Law
1989. On the Construction of Sociotechnical Networks: Content and Context Revisited. *Knowledge and Society. Studies in the Sociology of Science Past and Present* 8:57–83.

Cambrosio et al.
1991. Analyzing Science Policy-Making: Political Ontology or Ethnography?: A Reply to Kleinman. *Social Studies of Science* 21, no. 4:775–81.

Campbell, N. R.
1907. *Modern Electrical Theory.* Cambridge: Cambridge University Press.

Cannon, S. F.
1978. *Science in Culture: The Early Victorian Period.* New York: Science History.

Carnap, Rudolf
1952. Empiricism, Semantics, and Ontology. In L. Linsky, ed., *Semantics and the Philosophy of Language,* 208–28. Urbana: University of Illinois Press.

Carneiro, P.; and P. Arnaud; eds.
1970. *Auguste Comte: Ecrits de jeunesse, 1816–1828.* Paris: Mouton.

Carrington, R. C.
1858. On the Distribution of the Solar Spots in Latitude. *Monthly Notices of the Royal Astronomical Society* 19:1–3.

Cartwright, N.
1983. *How the Laws of Physics Lie.* Oxford: Clarendon Press.

Cavendish, Henry
1798. Experiments to Determine the Density of the Earth. Philosophical Transactions 88:469–526.

Cherniak, C.
1986. *Minimal Rationality.* Cambridge: MIT Press.

Churchland, P.; and C. Hooker
1985. *Images of Science.* Chicago: University of Chicago Press.

Clarke, Peter
1976. Atomism versus Thermodynamics. In C. Howson, ed., *Method and Appraisal in the Physical Sciences,* 41–105. Cambridge: Cambridge University Press.

Clerke, Agnes M.
1885. *A Popular History of Astronomy during the Nineteenth Century.* Edinburgh: Black.

Cohen, E. R.; K. M. Crowe; and J. W. M. DuMond
1957. *The Fundamental Constants of Physics.* New York: Interscience Publishers.

Coleman, Sidney
1985. Secret Symmetry: An Introduction to Spontaneously Broken Symmetry and Gauge Theory. In his *Aspects of Symmetry.* New York: Cambridge University Press.

Collins, H. M.
1975. The Seven Sexes: A Study in the Sociology of a Phenomenon, or the Replication of Experiments in Physics. *Sociology* 9:205–24.
1985. *Changing Order: Replication and Induction in Scientific Practice.* Beverly Hills: Sage Publications.

Collins, H. M.; and T. J. Pinch
1982. *Frames of Meaning: The Social Construction of Extraordinary Science.* London: Routledge & Kegan Paul.

Collins, H. M.; and S. Yearly
1992. Journey into Space. In Pickering 1992, 369–89.

Common, A. A.
1970. Photography as an Aid to Astronomy. 1886. Reprinted in Bernard Lovell, ed., *The Royal Institution Library of Science: Astronomy,* 1:243–51. Dordrecht: Elsevier Publishing.

Comte, Auguste
1985. *Traité philosophique d'astronomie populaire.* 1844. Reprint, Paris: Fayard.

Creath, Richard
1987. Some Remarks on "Protocol Sentences." *Noûs* 21:471–75.
1990. *Dear Carnap, Dear Van: The Quine-Carnap Correspondence and Related Work.* Berkeley and Los Angeles: University of California Press.

Crowe, M. J.
1986. *The Extraterrestrial Life Debate, 1750–1900.* Cambridge: Cambridge University Press.

Cunningham, E.
1914. *The Principle of Relativity.* Cambridge: Cambridge University Press.

Cushing, J. T.
1981. Electromagnetic Mass, Relativity, and the Kaufmann Experiments. *Amer. J. Phys.* 49:1133–49.

Daumas, M., ed.
1969. *A History of Invention.* Vol. 2. New York: Crown Publishers. Originally published in French in 1964.

Davidson, Arnold
1987. Sex and the Emergence of Sexuality. *Critical Inquiry* 14:16–48.
n.d. Styles of Reasoning and the History of Concepts. In P. Galison and D. Stump, eds., *The Disunity of Science: Boundaries, Contexts, and Power.* Stanford: Stanford University Press. In press.

Dear, P.
1985. *Totius in Verba:* Rhetoric and Authority in the Early Royal Society. *Isis* 76: 145–61.

de la Rue, Warren
1866. Comparison of the Sun's Spots Observed at Kew. *Astronomical Register* 4: 40.

de Mey, Marc

1981. The Interaction between Theory and Data in Science. In K. D. Knorr, R. Krohn, and R. Whitley, eds., *The Social Process of Scientific Investigation.* Dordrecht: D. Reidel.

Dewey, J.

1982. *The Middle Works, 1899–1924.* Carbondale: Southern Illinois University Press.

Dingler, H.

1952. *Ueber die Geschichte und das Wesen des Experimentes.* Munich: Eidos Verlag.

Dirac, P. A. M.

1982. The Early Years of Relativity. In G. Holton and Y. Elkana, eds., *Albert Einstein: Historical and Cultural Perspectives,* 79–90. Princeton: Princeton University Press.

Downing, T., ed.

1980. *The Troubles.* London: Thames Macdonald.

Drake, Stillman

1977. Tartaglia's Squadra and Galileo's Compasso. *Annali dell'Instituto e museo di storia della scienza di Firenze* 2:35–54.

Dreyer, J. L. E.; and H. H. Turner

1923. *History of the Royal Astronomical Society, 1820–1920.* London: Royal Astronomical Society.

Duhem, Pierre

[1906] 1954. *Aim and Structure of Physical Theory.* Trans. Phillip Weiner. Princeton: Princeton University Press.

Durand-Delga, M.

1990. L'affaire Deprat. *Travaux du Comité français d'histoire de la géologie,* 3d ser., 4:117–215.

1991. L'affaire Deprat, l'honneur retrouvé d'un géologue. *La recherche* 237 (Nov.): 1342–46.

Durkheim, Emile

[1893] 1984. *The Division of Labor in Society.* Trans. W. D. Halls. Reprint, New York: Free Press.

[1895] 1964. *The Rules of Sociological Method.* Reprint, New York: Free Press.

[1897] 1930. *Suicide.* Reprint, Paris: F. Alcan.

[1912] 1976. *The Elementary Forms of Religious Life.* Reprint, London: Allen & Unwin.

1973. *On Morality and Society.* Ed. R. N. Bellah. Chicago: University of Chicago Press.

Eddington, A. S.

1944. Joseph Larmor. *Obituary Notices of Fellows of the Royal Society* 4: 197–207.

Einstein, A.
1905. Zur Elektrodynamik bewegter Körper. In Lorentz et al. 1952, 37–65.
1907a. Bemerkungen zu der Notiz von Hrn. Paul Ehrenfest: "Die Translation deformierbarer Elektronen und der Flächensatz. *Ann. d. Phys.* 23:206–8.
1907b. Über das Relativitätsprinzip und die aus demselben gezogenen Folgerungen. *Jahrb. Radio. u. Elek.* 4:411–62.
1950. *Out of My Later Years.* London: Thames & Hudson.
1969. Autobiographical Notes. 1949. Reprinted in P. A. Schilpp, ed., *Albert Einstein: Philosopher-Scientist,* 3–94. 3d ed. La Salle, Ill.: Open Court.
1981. What Is the Theory of Relativity? 1919. Reprinted in his *Ideas and Opinions,* 222–27. New York: Dell.

Falconer, I.
1985. Theory and Experiment in J. J. Thomson's Work on Gaseous Discharge. Ph.D. diss., Bath University.

Faragó, P. S.; and L. Janossy
1957. Review of the Experimental Evidence for the Law of Variation of the Electron Mass with Velocity. *Nuo. Cim.,* ser. 10, 5:1411–36.

Faye, Hervé
1865. Sur la constitution physique du soleil. *Comptes rendus des séances de l'Académie des sciences* 60:89–96, 138–50.

Feyerabend, Paul
1975. *Against Method.* London: New Left Books.

Fischer, Martin
1990. Reasoning about CAD Data. Paper presented at CIFE Expert System Workshop.

FitzGerald, George
1888. Electromagnetic Phenomena Due to the Action of an Intervening Medium. *Brit. Assoc. Report,* pp. 557–62.
1896. Hertz's Miscellaneous Papers. In *Writings,* 433–42.
1900. Aether and Matter. *Electrician* 45:483–84.
1902. *The Scientific Writings of the Late George Francis FitzGerald.* Ed. J. Larmor. Dublin: Hodges, Figgis, & Co.

Fleck, L.
1979. *Genesis and Development of a Scientific Fact.* Chicago: University of Chicago Press.

Forbes, G.
1916. *David Gill: Man and Astronomer.* London: John Murray.

Frankland, Edward; and J. Norman Lockyer
1869. Preliminary Note of Researches on Gaseous Spectra in Relation to the Physical Constitution of the Sun. *Proceedings of the Royal Society* 17:288–91.

Franklin, A.
1986. *The Neglect of Experiment.* Cambridge: Cambridge University Press.
1989. The Epistemology of Experiment. In Gooding, Pinch, and Schaffer 1989, 437–60.

Friedman, Michael
1987. Carnap's Aufbau Reconsidered. *Noûs* 21:521–25.

Fuller, Steve
1989. *Philosophy of Science and Its Discontents.* Boulder: Westview Press.

Galison, P.
1987. *How Experiments End.* Chicago: University of Chicago Press.
1988a. History, Philosophy, and the Central Metaphor. *Science in Context* 2: 197–212.
1988b. Multiple Constraints, Simultaneous Solutions. In A. Fine and J. Leplin, eds., *PSA 1988: Proceedings of the 1988 Biennial Meeting of the Philosophy of Science Association,* 157–63. East Lansing: Philosophy of Science Association.
1988c. Philosophy in the Laboratory. *Journal of Philosophy* 85:525–27.
1989. The Trading Zone: The Coordination of Action and Belief. Paper presented at TECH-KNOW Workshops on Places of Knowledge, Their Technologies, and Economies. UCLA Center for Cultural History of Science and Technology, Dec.
1990a. Aufbau/Bauhaus: Logical Positivism and Architectural Modernism. *Critical Inquiry* 16:709–52.
1990b. The Role of Experiment in Scientific Change. Paper presented at Virginia Polytechnic Institute and State University Department of Philosophy Conference.
1993. The Cultural Meaning of *Aufbau.* In F. Stadler, ed., *Scientific Philosophy: Origins and Developments,* 75–93. Dordrecht: Kluwer Academic Publishers.
n.d. *Image and Logic: The Material Culture of Twentieth-Century Physics.* Forthcoming.

Galison, Peter; and Alexi Assmus
1989. Artificial Clouds, Real Particles. In Gooding, Pinch, and Schaffer 1989, 225–74.

Gardner, M. R.
1979. Realism and Instrumentalism in 19th Century Atomism. *Philosophy of Science* 46:1–34.

Georgi, Howard
1982. *Lie Algebras in Particle Physics: From Isospin to Unified Theories.* Reading, Mass.: Benjamin/Cummings Publishing Co.

Gerhardt, C. I.
1875–90. *Leibniz: Die philosophischen Schriften.* 7 vols. Berlin.

Giere, R. N.
1988. *Explaining Science: A Cognitive Approach.* Chicago: University of Chicago Press.

Gilbert, G. Nigel; and Michael Mulkay
1984. *Opening Pandora's Box: A Sociological Analysis of Scientists' Discourse.* Cambridge: Cambridge University Press.

Gill, David

1970a. The Applications of Photography in Astronomy. 1887. Reprinted in Bernard Lovell, ed., *The Royal Institution Library of Science: Astronomy,* 314–28. New York: Elsevier Publishing Co.

1970b. An Astronomer's Work in a Modern Observatory. 1891. Reprinted in Bernard Lovell, ed., *The Royal Institution Library of Science: Astronomy,* 1: 362–77. New York: Elsevier Publishing Co.

Gingras, Y.

1991. *Physics and the Rise of Scientific Research in Canada.* Montreal: McGill-Queen's University Press.

Gingras, Y.; and R. Gagnon

1988. Engineering Education and Research in Montreal: Social Constraints and Opportunities. *Minerva* 26, no. 1 (spring): 53–65.

Gingras, Y.; and S. S. Schweber

1986. Constraints on Construction. *Social Studies of Science* 16:372–83.

Gingras, Y.; and M. Trépanier

1993. Constructing a Tokamak: Political, Economic, and Technical Factors as Constraints and Resources. *Social Studies of Science* 23, no. 1 (Feb.): 3–36.

Ginzburg, C.

1980. *The Cheese and the Worms: The Cosmos of a Sixteenth-Century Miller.* New York: Penguin Books.

Glashow, Sheldon L.

1982. Introduction. In Georgi 1982.

Gleick, J.

1987. *Chaos.* New York: Penguin Books.

Goldberg, S.

1967. Henri Poincaré and Einstein's Theory of Relativity. *Amer. J. Phys.* 35: 934–44.

1970. In Defence of Ether: The British Response to Einstein's Special Theory of Relativity, 1905–1911. *Hist. Stud. Phys. Sci.* 2:89–125.

1970–71a. The Abraham Theory of the Electron: The Symbiosis of Experiment and Theory. *Arch. Hist. Ex. Sci.* 7:7–25.

1970–71b. Poincaré's Silence and Einstein's Relativity. *Brit. J. Hist. Sci.* 5:73–84.

Goldberg, S.; and R. Stuewer; eds.

1988. *The Michelson Era in American Science, 1870–1930.* New York: American Institute of Physics.

Goldstein, E.

1880. On the Electric Discharge in Rarefied Gases. Part 1. *Phil. Mag.* 10:173–90.

1881. Ueber die Entladung der Electricität in verdünnten Gasen. *Ann. Phys. Chem.* 12:249–79.

Golinski, Jan

1990. The Theory of Practice and the Practice of Theory: Sociological Approaches in the History of Science. *Isis* 81:492–505.

Gooday, Graeme
1989. Precision Measurement and the Establishment of Physics Laboratories in 19th Century Britain. Ph.D. diss., University of Kent at Canterbury.

Gooding, David
1985. "In Nature's School": Faraday as an Experimentalist. In D. Gooding and F. A. J. L. James, eds., *Faraday Rediscovered: Essays on the Life and Work of Michael Faraday,* 105–35. New York: Stockton Press.
1989. "Magnetic curves" and the Magnetic Field: Experimentation and Representation in the History of a Theory. In Gooding, Pinch, and Schaffer 1989, 183–223.

Gooding, David; T. Pinch; and S. Schaffer; eds.
1989. *The Uses of Experiment.* Cambridge: Cambridge University Press.

Habermas, J.
1973. Wahrheitstheorien. In H. Fahrenbach, ed., *Wirklichkeit und Reflexion,* 211–65. Pfullingen: Neske.
1978. *Knowledge and Human Interests.* 2d ed. London: Heinemann.

Hacking, Ian
1983. *Representing and Intervening: Introductory Topics in the Philosophy of Natural Science.* Cambridge: Cambridge University Press.
1986. Language, Truth, and Reason. In M. Hollis and S. Lukes, eds., *Rationality and Relativism.* Cambridge: MIT Press.
1988a. On the Stability of the Laboratory Sciences. *Journal of Philosophy* 85: 507–14.
1988b. Philosophers of Experiment. In A. Fine and J. Leplin, eds., *PSA 1988: Proceedings of the 1988 Biennial Meeting of the Philosophy of Science Association,* 147–56. East Lansing: Philosophy of Science Association.
1989a. Extragalactic Reality: The Case of Gravitational Lensing. *Philosophy of Science* 56:555–81.
1989b. The Life of Instruments. *Studies in History and Philosophy of Science* 20: 265–70.

Hackmann, W. D.
1989. Scientific Instruments: Models of Brass and Aids to Discovery. In Gooding, Pinch, and Schaffer 1989, 31–65.

Hanson, H. R.
1963. *The Concept of the Positron.* Cambridge: Cambridge University Press.

Haraway, Donna Jeanne
1976. *Crystals, Fabrics, and Fields: Metaphors of Organicism in Twentieth-Century Developmental Biology.* New Haven: Yale University Press.

Harding, Sandra
1986. *The Science Question in Feminism.* Ithaca: Cornell University Press.

Harman, P. M., ed.
1985. *Wranglers and Physicists.* Manchester: Manchester University Press.

Harvey, B.

1981. Plausibility and the Evaluation of Knowledge: A Case Study of Experimental Quantum Mechanics. *Social Studies of Science* 11:95–130.

Hattiangadi, J. N.

1989. Kuhn Studies. In Fred D'Agostino and I. C. Jarvie, eds., *Freedom and Rationality: Essays in Honor of John Watkins from His Friends and Colleagues.* Boston: Kluwer Academic.

Heaviside, O.

1889. On the Electromagnetic Effects due to the Motion of Electrification through a Dielectric. *Phil. Mag.* 27:324–39.

Heimann, P. M.

1972. *The Unseen Universe:* Physics and the Philosophy of Nature in Victorian Britain. *Brit. J. Hist. Sci.* 6:73–79.

Helmholtz, Hermann von

1882. *Wissenschaftliche Abhandlungen.* Vol. 1. Leipzig.

1962. *Popular Scientific Lectures.* 1881. Reprint. Ed. Morris Kline. New York: Dover.

Henderson, D. K.

1990. On the Sociology of Science and the Continuing Importance of Epistemologically Couched Accounts. *Social Studies of Science* 20:113–48.

Herschel, John

1820. *Address of the Astronomical Society.* London.

1832. *Preliminary Discourse of Natural Philosophy.* London. Reprint, 1987.

1846. Address of the President. In *Report of the Fifteenth Meeting of the British Association for the Advancement of Science,* xxvii–xliv. London: John Murray.

1851. *Outlines of Astronomy.* 4th ed. London: Longman.

1867. *Familiar Lectures on Scientific Subjects.* London: Strahan.

Hertz, H.

1883. Versuche über die Glimmentladung. *Ann. Phys. Chem.* 19:782–816. In *Werke,* 1:242–76. Translated as "Experiments on the Cathode Discharge," in *Papers,* 224–54.

1895. *Gesammelte Werke.* 3 vols. Leipzig: J. A. Barth.

1898. *Miscellaneous Papers.* Trans. D. E. Jones and G. A. Schott. Leipzig: J. A. Barth.

1956. *The Principles of Mechanics.* Trans. D. E. Jones and J. T. Walley. New York: Dover Books.

1962. *Electric Waves Being Researches on the Properties of Electric Action with a Finite Velocity through Science.* Trans. D. E. Jones. New York: Dover Books.

Hicks, W. M.

1898. Researches in Vortex Motion. Part 3, On Spiral Gyrostatic Vortex Aggregates. *Proceedings of the Royal Society* 62:332–38.

1899. Researches in Vortex Motion. Part 3, On Spiral or Girostatic Vortex Aggregates. *Philosophical Transactions of the Royal Society* 192:33–100.

1901/2. The FitzGerald-Lorentz Effect. *Nature* 65:343.

1902a. On the Michelson-Morley Experiment Relating to the Drift of Ether. *Phil. Mag.* 3:9–42.
1902b. The Michelson-Morley Experiment. *Phil. Mag.* 3:256.

Hiebert, Erwin
1970. The Genesis of Mach's Early Views on Atomism. In R. Cohen and R. Seeger, eds., *Ernst Mach: Physicist and Philosopher,* 79–106. Dordrecht: D. Reidel.

Hirosige, T.
1968. Theory of Relativity and the Ether. *Jap. Stud. Hist. Sci.* 7:37–53.
1976. The Ether Problem, the Mechanistic World View, and the Origins of the Theory of Relativity. *Hist. Stud. Phys. Sci.* 7:3–82.

Holmes, F. L.
1985. *Lavoisier and the Chemistry of Life.* Madison: University of Wisconsin Press.

Holton, G.
1975. *Thematic Origins of Scientific Thought: Kepler to Einstein.* Cambridge: Harvard University Press.

Hon, Giora
1985. Kaufmann's Experiment and Its Reception. In On the Concept of Experimental Error. Ph.D. diss., King's College, London University.
1987a. H. Hertz: "The Electrostatic and Electromagnetic Properties of the Cathode Rays Are Either Nil or Very Feeble" (1883): A Case-Study of an Experimental Error. *Stud. Hist. Phil. Sci.* 18:367–82.
1987b. On Kepler's Awareness of the Problem of Experimental Error. *Annals of Science* 44:545–91.
1989. Towards a Typology of Experimental Error: An Epistemological View. *Stud. Hist. Phil. Sci.* 20:469–504.

Hones, M. J.
1990. Reproducibility as a Methodological Imperative in Experimental Research. In A. Fine, M. Forbes, and L. Wessels, eds., *PSA 1990,* 1:585–99. East Lansing: Philosophy of Science Association.

Hufbauer, Karl
1986. Amateurs and the Rise of Astrophysics. *Berichte zur Wissenschaftsgeschichte* 9:183–90.
1991. *Exploring the Sun: Solar Science since Galileo.* Baltimore: Johns Hopkins University Press.

Huggins, William
1866. Results of Some Observations of Bright Granules on the Surface of the Sun. *Astronomical Register* 4:159–61.
1867. [Report]. *Monthly Notices of the Royal Astronomical Society* 27:131.
1909. *Scientific Papers.* Ed. W. Huggins and M. Huggins. London: William Wesley.
1970a. On Some Further Results of Spectrum Analysis as Applied to the Heavenly Bodies. 1869. Reprinted in Bernard Lovell, ed., *The Royal Institution Library of Science: Astronomy,* 1:83–86. New York: Elsevier.

1970b. On the Physical and Chemical Constitution of the Fixed Stars and Nebulae. 1865. Reprinted in Bernard Lovell, ed., *The Royal Institution Library of Science: Astronomy,* 1:42–50. New York: Elsevier.
1970c. The Photographic Spectra of the Stars. 1880. Reprinted in Bernard Lovell, ed., *The Royal Institution Library of Science: Astronomy,* 1:172–84. New York: Elsevier.

Hughes, T. P.
1833. *Networks of Power: Electrification in Western Society, 1880–1930.* Baltimore: Johns Hopkins University Press.

Hull, D.
1988. *Science as a Process: An Evolutionary Account of the Social and Conceptual Development of Science.* Chicago: University of Chicago Press.

Hund, John
1990. Sociologism and Philosophy. *British Journal of Sociology* 41:197–224.

Hunt, B. J.
1986. Experimenting on the Ether: Oliver Lodge and the Great Whirling Machine. *Hist. Stud. Phys. Sci.* 16:111–34.
1988. The Origins of the FitzGerald Contraction. *Brit. J. Hist. Sci.* 21:67–76.
1991. *The Maxwellians.* Ithaca: Cornell University Press.
n.d. To Rule the Waves: British Cable Telegraphy and the Making of Maxwell's Equations. Forthcoming.

James, F. A. J. L.
1982. Thermodynamics and Sources of Solar Heat. *Brit. J. Hist. Sci.* 15:155–81.
1985. The Creation of a Victorian Myth: The Historiography of Spectroscopy. *History of Science* 23:1–24.

James, William
1907. *Pragmatism: A New Name for Some Old Ways of Thinking.* New York: Longman's Green & Co.

Jammer, M.
1961. *The Concepts of Mass.* Cambridge: Harvard University Press.

Janich, P.
1978. Physics—Natural Science or Technology? In W. Krohn, E. T. Layton, and P. Weingart, eds., *The Dynamics of Science and Technology,* 3–27. Dordrecht: Reidel.

Jardine, Nicholas
1991. *The Scenes of Inquiry.* Oxford: Clarendon Press.

Jeans, J.
1908. *The Mathematical Theory of Electricity and Magnetism.* Cambridge: Cambridge University Press.

Kaufmann, W.
1897. Die magnetische Ablenkbarkeit der Kathodenstrahlen und ihre Abhängigkeit vom Entladungspotential. *Ann. d. Phys.* 61:544–52.
1898a. Die magnetische Ablenkbarkeit elektrostatisch beeinflusster Kathodenstrahlen. *Ann. d. Phys.* 65:431–39.

1898b. Bemerkungen zu der Mittheilung von A. Schuster: "Die magnetische Ablenkung der Kathodenstrahlen." *Ann. d. Phys.* 66:649–51.
1901a. Die Entwicklung des Elektronenbegriffs. *Verh. Gesell. Dtsch. Naturf. Ärzte,* pt. 1:115–26.
1901b. Die magnetische und elektrische Ablenkbarkeit der Becquerelstrahlen und die scheinbare Masse der Elektron. *Nachr. Gesell. Wiss. Göttingen, Math.-Phys.* 3:143–55.
1901c. Methode zur exakten Bestimmung von Ladung und Geschwindigkeit der Becquerelstrahlen. *Phys. Zeit.* 2:602–3.
1902a. Die elektromagnetische Masse des Elektrons. *Phys. Zeit.* 4:54–57.
1902b. Über die Elektromagnetische Masse des Elektrons. *Nachr. Gesell. Wiss. Göttingen, Math.-Phys.* 4:291–96.
1903. Über die "Elektromagnetische Masse" der Elektronen. *Nachr. Gesell. Wiss. Göttingen, Math.–Phys.* 5:90–103.
1905a. Eine rotierende Quecksilberluftpumpe. *Zeit. Inst.* 25:129–33.
1905b. Über die Konstitution des Elektrons. *Preuss. Akad. Wiss. Berlin, Sitz. Ber.* 45:949–56.
1906a. Über die Konstitution des Elektrons. *Ann. d. Phys.* 19:487–553.
1906b. *Phys. Zeit.* 7:759–60, 761.
1907. Bemerkungen zu Herrn Plancks: "Nachtrag zu der Besprechung der Kaufmannschen Ablenkungsmessungen." *Verh. d. D. Phys. Gesell.* 9:667–73.

Kaye, G. W. C.
1927. *High Vacua.* London: Longmans.

Kim, K.-M.
1991. On the Reception of Johannsen's Pure Line Theory: Toward a Sociology of Scientific Validity. *Social Studies of Science* 21, no. 4 (Nov.): 649–79.

King, Henry C.
1955. *The History of the Telescope.* High Wycombe: Charles Griffin.

Kirchhoff, G.
1860. On the Relations between the Radiating and Absorbing Powers of Different Bodies for Light and Heat. *Phil. Mag.* 20:1–21.
1861. On the Chemical Analysis of the Solar Atmosphere. *Phil. Mag.* 21:185–88.

Klein, M. J.
1967. Thermodynamics in Einstein's Thought. *Science* 157:509–16.

Knorr-Cetina, K.
1981. *The Manufacture of Knowledge: An Essay on the Constructivist and Contextual Nature of Science.* Oxford: Pergamon.

Kuhn, T. S.
1970. *The Structure of Scientific Revolutions.* 2d ed. Chicago: University of Chicago Press. Originally published 1962.
1976. Reflections on My Critics. In I. Lakatos and A. Musgrave, eds., *Criticism and the Growth of Knowledge.* Cambridge: Cambridge University Press.
1977a. The Essential Tension: Tradition and Innovation in Scientific Research. 1959. Reprinted in his *The Essential Tension.* Chicago: University of Chicago Press.

1977b. A Function for Thought Experiments. In his *The Essential Tension*, 240–65. Chicago: University of Chicago Press.

1978. *Black-Body Theory and Quantum Discontinuity, 1894–1912*. New York: Oxford University Press.

Kunneman, H.; and W. Hullegie

1989. Een tripolair model van lichamelijkheid. Amsterdam. Unpublished manuscript.

Lakatos, Imre

1970. Falsification and the Methodology of Scientific Research Programmes. In I. Lakatos and A. Musgrave, eds., *Criticism and the Growth of Knowledge*, 91–195. Cambridge: Cambridge University Press.

Langevin, P.

1905. La physique des électrons. *Rev. Gen. Sci. Pure Appl.* 16:257–76.

Lankford, John

1981. Amateurs and Astrophysics: A Neglected Aspect in the Development of a Scientific Specialty. *Social Studies of Science* 11:275–303.

1984. The Impact of Photography on Astronomy. In Owen Gingerich, ed., *The General History of Astronomy*. Vol. 4, *Astrophysics and Twentieth-Century Astronomy to 1950*, Part A, 16–39. Cambridge: Cambridge University Press.

Laplace, Pierre Simon

1984. *Exposition du système du monde*. 1835. Reprint, Paris: Fayard.

Larmor, Joseph

1894. A Dynamical Theory of the Electric and Luminiferous Medium (Part 1). In Larmor 1929, 1:414–535.

1900. *Aether and Matter*. Cambridge: Cambridge University Press.

1902a. Introductory and Biographical. In FitzGerald 1902, xix–lxiv.

1902b. Can Convection through the Aether be Detected Electrically? In FitzGerald 1902, 566–69.

1904. On the Ascertained Absence of Effects of Motion through the Aether, in Relation to the Constitution of Matter, and on the FitzGerald-Lorentz Hypothesis. In Larmor 1929, 2:274–80.

1929. *Mathematical and Physical Papers*. 2 vols. Cambridge: Cambridge University Press.

Latour, B.

1983. Give Me a Laboratory and I Will Raise the World. In K. D. Knorr-Cetina and M. Mulkay, eds., *Science Observed*, 141–70. London: Sage.

1987. *Science in Action*. Milton Keynes: Open University Press.

1988a. *The Pasteurization of France*. Cambridge: Harvard University Press.

1988b. The Politics of Explanation: An Alternative. In Steve Woolgar, ed., *Knowledge and Reflexivity*, 155–76. London: Sage.

1993. *We Have Never Been Modern*. Cambridge: Harvard University Press.

Latour, B.; and S. Woolgar

1979. *Laboratory Life: The Social Construction of Scientific Facts*. Beverly Hills: Sage.

1986. *Laboratory Life: The Construction of Scientific Facts.* Enl. ed. Princeton: Princeton University Press.

Laub, J.
1910. Über die experimentellen Grundlagen des Relativitätsprinzips. *Jahr. d. Radio. u. Elek.* 7:405 ff.

Laudan, R.; L. Laudan; and A. Donovan
1988. Testing Theories of Scientific Change. In A. Donovan et al., eds., *Scrutinizing Science,* 3–44. Dordrecht: Kluwer.

Law, J.
1987a. On the Social Explanation of Technical Change: The Case of the Portuguese Maritime Expansion. *Technology and Culture* 28, no. 2:227–52.
1987b. Technology and Heterogeneous Engineering: The Case of Portuguese Expansion. In W. E. Bijker, T. P. Hughes, and T. J. Pinch, eds., *The Social Construction of Technical Systems,* 111–34. Cambridge: MIT Press.

Law, J.; and M. Callon
1988. Engineering and Sociology in a Military Aircraft Project: A Network Analysis of Technological Change. *Social Problems* 35, no. 3:284–97.

Lévi-Strauss, C.
1967. *Tristes tropiques.* Trans. John Russell. New York: Atheneum.

Lewis, G. N.
1908. A Revision of the Fundamental Laws of Matter and Energy. *Phil. Mag.* 16:705–17.

Lloyd, G. E. R.
1966. *Polarity and Analogy.* Cambridge: Cambridge University Press.
1979. *Science, Magic, and Experience.* Cambridge: Cambridge University Press.

Lockyer, J. Norman
1865a. Observations of a Sun Spot. *Monthly Notices of the Royal Astronomical Society* 25:236–41.
1865b. On the Physical Constitution of the Sun. *The Reader* 5:106–7.
1866. Spectroscopic Observations of the Sun. *Proceedings of the Royal Society* 15:256–58.
1869. Spectroscopic Observations of the Sun. *Proceedings of the Royal Society* 17:350–56.
1873. *The Spectroscope and Its Applications.* London: Macmillan.
1887. *The Chemistry of the Sun.*
1970. On Recent Discoveries in Solar Physics Made by Means of the Spectroscope. 1869. Reprinted in Bernard Lovell, ed., *The Royal Institution Library of Science: Astronomy,* 1:87–102. New York: Elsevier.

Lodge, O. J.
1898. Note on Mr. Sutherland's Objection to the Conclusiveness of the Michelson-Morley Aether Experiment. *Phil. Mag.* 46:343–44.
1907. *Electrons, or The Nature and Properties of Negative Electricity.* London: Bell.

Lorentz, H. A.
1895. Michelson's Interference Experiment. In Lorentz et al. 1952, 3–7.
1904. Electromagnetic Phenomena in a System Moving with Any Velocity Less than That of Light. In Lorentz et al. 1952, 11–34.
1924. The Radiation of Light. *Nature* 113:608–11.
1931. *Lectures on Theoretical Physics: Lectures Delivered at the University of Leiden in 1922.* Vol. 3. London: Macmillan.
1952. *The Theory of Electrons and Its Applications to the Phenomena of Light and Radiant Heat: Lectures Delivered at Columbia University.* 2d ed. New York: Dover.

Lorentz, H. A.; A. Einstein; H. Minkowski; and H. Weyl
1952. *The Principle of Relativity.* With notes by A. Sommerfeld. New York: Dover.

Lukasiewicz, J.
1991. Le principe de contradiction chez Aristote. *Rue Descartes,* nos. 1–2 (Apr.): 9–32. Originally published as Über den Satz des Widerspruchs bei Aristoteles. *Bulletin international de l'Académie des sciences de Cracovie, Classe d'histoire et de philosophie,* 1910, 15–38.

Lynch, M.
1985. *Art and Artifact in Laboratory Science: A Study of Shop Work and Shop Talk in a Research Laboratory.* London: Routledge & Kegan Paul.

Lynch, M.; E. Livingston; and H. Garfinkel
1983. Temporal Order in Laboratory Work. In K. D. Knorr-Cetina and M. Mulkay, eds., *Science Observed,* 205–38. London: Sage.

Macdonald, H. M.
1902. *Electric Waves.* Cambridge: Cambridge University Press.

Mach, Ernst
1863. *Compendium der Physik für Mediziner.* Vienna.
[1872] 1941. *History and Root of the Principle of the Conservation of Energy.* Trans. P. Jourdain. Chicago: Open Court.
[1883] 1942. *The Science of Mechanics.* La Salle: Open Court.
1890. The Analysis of Sensations—Antimetaphysical. *The Monist* 1:48.
1892. Facts and Mental Symbols. *The Monist* 2:198.
1976. *Knowledge and Error: Sketches on the Psychology of Enquiry.* 1905. Trans. T. J. McCormack and P. Foulkes. Boston: D. Reidel.

Mackenzie, D.
1987. Missile Accuracy: A Case Study in the Social Processes of Technical Change. In W. E. Bijker, T. P. Hughes, and T. Pinch, eds., *The Social Construction of Technological Systems,* 195–222. Cambridge: MIT Press.

Mackenzie, D.; and G. Spinardi
1988. The Shaping of Nuclear Weapon System Technology: US Fleet Ballistic Missile Guidance and Navigation II: "Going Broke"—The Path to Trident II. *Social Studies of Science* 18, no. 4:581–624.

Mann, A. K.
1990. Introductory Comments. In *Proceedings of the 1983 DPF Workshop on Collider Detectors: Present Capabilities and Future Possibilities.* Berkeley: Lawrence Berkeley Laboratory.

Marvin, C.
1988. *When Old Technologies Were New.* Oxford: Oxford University Press.

Maunder, E. Walter
1900. *The Royal Observatory, Greenwich.* London: Religious Tract Society.
1912. *Sir William Huggins and Spectroscopic Astronomy.* London.

Maxwell, James Clerk
1873. *Treatise on Electricity and Magnetism.* 2 vols. Oxford: Clarendon.
1890. *Scientific Papers.* 2 vols. Ed. W. D. Niven. Cambridge: Cambridge University Press.

May, R. M.
1976. Simple Mathematical Models with Very Complicated Dynamics. *Nature* 201 (10 June): 459–67.

McCormmach, R.
1970a. Einstein, Lorentz, and the Electron Theory. *Hist. Stud. Phys. Sci.* 2:41–87.
1970b. H. A. Lorentz and the Electromagnetic View of Nature. *Isis* 61:459–97.

McDowell, R. B.; and D. A. Webb; eds.
1982. *Trinity College, Dublin, 1592–1952.* Cambridge: Cambridge University Press.

McGucken, William
1969. *Nineteenth-Century Spectroscopy: Development of the Understanding of Spectra, 1802–1897.* Baltimore: Johns Hopkins University Press.

Meadows, A. J.
1970. *Early Solar Physics.* Oxford: Pergamon.
1972. *Science and Controversy: A Biography of Sir Norman Lockyer.* London: Macmillan.
1975. *Greenwich Observatory.* Vol. 2. London: Taylor & Francis.

Merleau-Ponty, Jacques
1983. *La science de l'univers à l'âge du positivisme.* Paris: Vrin.

Miller, A. I.
1973. A Study of Henri Poincaré's "Sur la dynamique de l'electron." *Arch. Hist. Ex. Sci.* 10:207–328.
1981. *Albert Einstein's Special Theory of Relativity: Emergence (1905) and Early Interpretation (1905–1911).* Reading, Mass.: Addison-Wesley.

Millikan, Robert Andrew
1917. A Re-determination of the Value of the Electron and of Related Constants. *Proc. Nat. Acad. Sci. U.S.A.* 3:231–36.

Milner, S. R.

1932–38. Frederick Trouton. *Obituary Notices of Fellows of the Royal Society* 1:393–99.

Misa, T. J.

1992. Review of D. Mackenzie's book on Missile Accuracy. *Science, Technology, and Human Values* 17, no. 1:127–29.

Mitchel, O. M.

1860. *Popular Astronomy.* London: Routledge.

Morrell, Jack; and Arnold Thackray; eds.

1984. *Gentlemen of Science: Early Correspondence of the British Association for the Advancement of Science.* Camden Society 4th ser., vol. 30. London: Royal Historical Society.

Morrison, Margaret

1986. More on the Relationship between Technically Good and Conceptually Important Experiments. *Brit. J. Phil. Sci.* 37:101–22.

1990. Theory, Intervention and Realism. *Synthese* 82:1–22.

Myers, Greg

1989. Nineteenth Century Popularizations of Thermodynamics and the Rhetoric of Social Prophecy. In P. Brantlinger, ed., *Energy and Entropy: Science and Culture in Victorian Britain,* 307–38. Bloomington: Indiana University Press.

Nagel, Ernest

1961. *The Structure of Science.* New York: Harcourt, Brace & World.

Nernst, W.

1896. *Die Ziele der physikalischen Chemie: Festrede gehalten am 2. Juni 1896 zur Einweihung des Instituts für physikalische Chemie und Elektrochemie der Georgia Augusta zu Göttingen.* Göttingen: Vandenhoeck & Ruprecht.

Nersessian, N.

1988. "Ad hoc" Is Not a Four-Letter Word: H. A. Lorentz and the Michelson-Morley Experiment. In Goldberg and Stuewer 1988,

Newton, Isaac

1934. *Sir Isaac Newton's Mathematical Principles of Natural Philosophy and His System of the World.* Trans. A. Motte. Revised by F. Cajori. Berkeley: University of California Press.

Nickles, Thomas

1978. Scientific Problems and Constraints. *PSA* 1:134–48.

1980. Can Scientific Constraints Be Violated Rationally? In T. Nickles, ed., *Scientific Discovery, Logic, and Rationality,* 285–315. Dordrecht: D. Reidel Publishing Co.

1986. Remarks on the Use of History as Evidence. *Synthese* 69:253–66.

1988. Reconstructing Science: Discovery and Experiment. In D. Batens and J. P. van Bendegem, eds., *Theory and Experiment,* 33–53. Dordrecht: D. Reidel Publishing Co.

Norman, Daniel
1938. The Development of Astronomical Photography. *Osiris* 5:560–94.

Nye, Mary Jo
1976. The Nineteenth Century Atomic Debates and the Dilemma of an Indifferent Hypothesis. *Stud. Hist. Phil. Sci.* 7:245–68.
1972. *Molecular Reality.* New York: American Elsevier.

Nygren, David R.; et al.
1990. Particle Identification. In *Proceedings of the 1983 DPF Workshop on Collider Detectors: Present Capabilities and Future Possibilities.* Berkeley: Lawrence Berkeley Laboratory.

Nyhoff, John
1988. Philosophical Objections to the Kinetic Theory. *Brit. J. Phil. Sci.* 39: 81–109.

O'Hara, J. G.; and W. Pricha
1987. *Hertz and the Maxwellians.* London: Peter Peregrinus in association with the Science Museum.

Ostwald, W.
1900. *The Scientific Foundations of Analytic Chemistry.* 2d English ed. Trans. George M'Gowan. London and New York: Macmillan.
1907. The Modern Theory of Energetics. *The Monist* 17:481–515.
1927. *Lebenslinien II.* Berlin: Klasing.

Pais, A.
1972. The Early History of the Theory of the Electron: 1897–1947. In A. Salam and E. P. Wigner, eds., *Aspects of Quantum Theory,* 79–92. Cambridge: Cambridge University Press.
1983. *"Subtle Is the Lord . . .": The Science and the Life of Albert Einstein.* Oxford: Oxford University Press.

Pang, Alex Soojung-Kim
1993. The Social Event of the Season: Solar Eclipse Expeditions and Victorian Culture. *Isis* 84:252–77.

Pauli, W.
1967. *Theory of Relativity.* Elmsford, N.Y.: Pergamon Press. Originally published as Relativitätstheorie. *Encyk. d. math. Wiss.* (Leipzig) 19 (1921).

Perrin, J.
1895. Nouvelles propriétés des rayons cathodiques. *Comptes Rendus* 121:1130. Oeuvres, 3–5.
1950. *Oeuvres scientifiques.* Paris: CNRS.

Piaget, Jean
1950. *Introduction à l'épistémologie génétique.* 3 vols. Paris: Presses Universitaires de France.
1985. *The Equilibrium of Cognitive Structures: The Central Problem of Intellectual Development.* Trans. Terrance Brown and Kishore Julian Thampy. Chicago: University of Chicago Press.

Pickering, A.
1981a. Constraints on Controversy: The Case of the Magnetic Monopole. In H. M. Collins, ed., *Knowledge and Controversy: Studies of Modern Natural Science.* Special issue of *Social Studies of Science* 11, no. 1:63–93.
1981b. The Hunting of the Quark. *Isis* 72:216–36.
1981c. The Role of Interests in High-Energy Physics: The Choice between Charm and Colour. In K. D. Knorr, R. Krohn, and R. D. Whitley, eds., *The Social Process of Scientific Investigation: Sociology of the Sciences,* 4:107–38. Dordrecht: Reidel.
1984a. Against Putting the Phenomena First: The Discovery of the Weak Neutral Current. *Stud. Hist. Phil. Sci.* 15:85–117.
1984b. *Constructing Quarks: A Sociological History of Particle Physics.* Chicago: University of Chicago Press.
1986. Positivism/Holism/Constructivism. Typescript.
1987a. Forms of Life: Science, Contingency, and Harry Collins. *Brit. J. Hist. Sci.* 20:213–21.
1987b. Models in/of Scientific Practice. *Philosophy and Social Action* 13:69–77.
1989a. Editing and Epistemology: Three Accounts of the Discovery of the Weak Neutral Current. In L. Hargens, R. A. Jones, and A. Pickering, eds., *Knowledge and Society: Studies in the Sociology of Science, Past and Present,* 8:217–32. Greenwich, Conn.: JAI Press.
1989b. Living in the Material World: On Realism and Experimental Practice. In Gooding, Pinch, and Schaffer,1989, 275–97.
1990a. Knowledge, Practice, and Mere Construction. *Social Studies of Science* 20:682–729.
1990b. Openness and Closure: On the Goals of Scientific Practice. In H. Le Grand, ed., *Experimental Inquiries: Historical, Philosophical, and Social Studies of Experimentation in Science,* 215–39. Dordrecht: Kluwer.
1991. Objectivity and the Mangle of Practice. In A. Megill, ed., *Rethinking Objectivity: Analysis, Research, Reflection.* Special issue of *Annals of Scholarship* 8:409–25.
1992. From Science as Knowledge to Science as Practice. In A. Pickering, ed., *Science as Practice and Culture,* 1–26. Chicago: University of Chicago Press.
1993. The Mangle of Practice: Agency and Emergence in the Sociology of Science. *American Journal of Sociology* 99:559–89.
1995. *The Mangle of Practice: Time, Agency, and Science.* Chicago: University of Chicago Press.

Pickering, A.; and A. Stephanides
1992. Constructing Quaternions: On the Analysis of Conceptual Practice. In Pickering, ed., 1992:139–67.

Pinch, Trevor
1985. Towards an Analysis of Scientific Observation: The Externality and Evidential Significance of Observation Reports in Physics. *Social Studies of Science* 15:3–35.

1986a. *Confronting Nature: The Sociology of Solar Neutrino Detection.* Dordrecht: Reidel.
1986b. Strata Various. *Social Studies of Science* 16:705–13.

Pinch, T. J.; and W. E. Bijker
1984. The Social Construction of Facts and Artefacts: Or How the Sociology of Science and the Sociology of Technology Might Benefit Each Other. *Social Studies of Science* 14:399–441.

Planck, M.
1906. Die Kaufmannschen Messungen der Ablenkbarkeit der β-strahlen in ihrer Bedeutung für die Dynamik der Elektronen. *Phys. Zeit.* 7:753–59.
1907. Nachtrag zu der Besprechung der Kaufmannschen Ablenkungsmessungen. *Verh. d. D. Phys. Gesell.* 9:301–5.

Plotkin, Howard
1977. Henry Draper: The Discovery of Oxygen in the Sun and the Dilemma of Interpreting the Solar Spectrum. *Journal for the History of Astronomy* 8:44–51.

Poincaré, H.
1904. Les rayons N existent-ils? Opinion de M. Poincaré. *Rev. Sci.,* 2d ser., 5:682.
1905. Sur la dynamique de l'électron. *C. R. Acad. Sci.,* (Paris) 140:1504–8.
1914. *Science and Method.* Trans. F. Maitland. London: Nelson.
1946. *The Foundations of Science.* Trans. G. B. Halsted, with an introduction by J. Royce. Lancaster, Pa.: Science Press. (This is a collection of three works: *Science and Hypothesis,* 9–197; *The Value of Science,* 201–355; *Science and Method,* 359–546.)
1952. *Science and Hypothesis.* Preface by J. Larmor. 1902. Reprint, New York: Dover.

Polanyi, M.
1964. *Science, Faith, and Society.* Chicago: University of Chicago Press, Phoenix Books.
1969. *Knowing and Being.* Chicago: University of Chicago Press.

Pole, William
1888. *The Life of Sir William Siemens.* London: John Murray.

Popper, K. R.
1965. *The Logic of Scientific Discovery.* New York: Harper & Row.

Porter, A. W.
1907. The Drift of the Ether. *Knowledge and Scientific News* 4:210.

Porter, Theodore M.
1990. Quantification as a Social Technology. UCLA Tech-Know Workshop, 24 February.
1992a. Objectivity as Standardization: The Rhetoric of Impersonality in Measurement, Statistics, and Cost-Benefit Analysis. In Allan Megill, ed., *Rethinking Objectivity II: Annals of Scholarship* 9, nos. 1/2:19–59.

1992b. Quantification and the Accounting Ideal in Science. *Social Studies of Science* 22:633–52.

Poynting, J. H.

1894. *The Mean Density of the Earth*. London: Charles Griffon & Co.

Pyenson, L.

1979. Physics in the Shadow of Mathematics: The Göttingen Electron-Theory Seminar of 1905. *Arch. Hist. Ex. Sci.* 21:55–89.

Radder, H.

1986. Experiment, Technology and the Intrinsic Connection between Knowledge and Power. *Social Studies of Science* 16:663–83.

1988. *The Material Realization of Science*. Assen: van Gorcum. Originally published as *De materiële realisering van wetenschap*. Amsterdam: VU-uitgeverij, 1984.

1992a. Experimental Reproducibility and the Experimenters' Regress. In D. Hull, M. Forbes, and K. Okruhlik, eds., *PSA 1992*, 1:63–73. East Lansing: Philosophy of Science Association.

1992b. Normative Reflexions on Constructivist Approaches to Science and Technology. *Social Studies of Science* 22:141–73.

1993. Science, Realization, and Reality: The Fundamental Issues. *Stud. Hist. Phil. Sci.* 24:327–49.

Rankine, A. O.

1907. On a Theoretical Method of Attempting to Detect Relative Motion between the Ether and Earth. *British Association Report*, 454–55.

Ravetz, J. R.

1973. *Scientific Knowledge and Its Social Problems*. Harmondsworth, England: Penguin Books.

Rayleigh, Lord

1902a. Does Motion through the Ether Cause Double Refraction? *British Association Report*, 546 (Only the title is given.)

1902b. Does Motion through the Aether Cause Double Refraction? *Phil. Mag.* 4:678–83.

1969. *The Life of Sir J. J. Thomson*. London: Dawson's.

Regis, Ed

1987. *Who Got Einstein's Office? Eccentricity and Genius at the Institute for Advanced Study*. London: Penguin Books.

Reisch, George A.

1991. Did Kuhn Kill Logical Empiricism? *Philosophy of Science* 58, no. 2: 264–77.

Richards, R. J.

1987. *Darwin and the Emergence of Evolutionary Theories of Mind and Behavior*. Chicago: University of Chicago Press.

Richardson, Alan W.

1992. Logical Idealism and Carnap's Construction of the World. *Synthèse* 93:59.

Richardson, O. W.
1914. *The Electron Theory of Matter.* Cambridge: Cambridge University Press.

Richter, S.
1980. Die "Deutsche Physik." In H. Mehrtens and S. Richter, eds., *Naturwissenschaft, Technik und NS-Ideologie,* 116–41. Frankfurt am Main: Suhrkamp.

Rindler, W.
1977. *Essential Relativity.* 2d ed. New York: Springer-Verlag.

Robb, A. A.
1921. *The Absolute Relations of Time and Space.* Cambridge: Cambridge University Press.

Ronald, Francis
1845. Report concerning the Observatory of the British Association at Kew. In *Report of the Fourteenth Meeting of the British Association for the Advancement of Science,* 120–42. London: John Murray.

Rossi, Paolo
1970. *Philosophy, Technology, and the Arts in the Early Modern Era.* New York: Harper & Row.

Roth, P. A.; and R. B. Barrett
1990. Deconstructing Quarks: Rethinking Sociological Constructions of Science. *Social Studies of Science* 20:579–632.

Rothermel, Holly
1993. Images of the Sun: Warren de la Rue, George Biddell Airy, and Celestial Photography. *Brit. J. Hist. Sci.* 26:137–69.

Rouse, Joseph
1987. *Knowledge and Power: Toward a Political Philosophy of Science.* Ithaca: Cornell University Press.

Rowlands, P.
1990. *Oliver Lodge and the Liverpool Physical Society.* Liverpool: Liverpool University Press.

Rudwick, M. J. S.
1985. *The Great Devonian Controversy: The Shaping of Scientific Knowledge among Gentlemanly Specialists.* Chicago: University of Chicago Press.
1988. The Closure of the Great Devonian Controversy. In Program, Papers, and Abstracts for the Joint Conference, British Society for the History of Science and History of Science Society, Manchester, England, 11–15 July 1988, pp. 155–59.

Rutherford, E.
1899. Uranium Radiation and the Electrical Conduction Produced by It. *Phil. Mag.* 47:109–63.

Salmon, W. C.
1984. *Scientific Explanation and the Causal Structure of the World.* Princeton: Princeton University Press.

Schaffer, Simon

1980. "The Great Laboratories of the Universe": William Herschel on Matter Theory and Planetary Life. *Journal for the History of Astronomy* 11:81–110.

1988. Astronomers Mark Time: Discipline and the Personal Equation. *Science in Context* 2:115–46.

1989a. Glass Works: Newton's Prisms and the Uses of Experiment. In Gooding, Pinch, and Schaffer 1989, 67–104.

1989b. The Nebular Hypothesis and the Science of Progress. In J. R. Moore, ed., *History, Humanity, and Evolution,* 131–64. Cambridge: Cambridge University Press.

1991. The Eighteenth Brumaire of Bruno Latour. *Stud. Hist. Phil. Sci.* 22: 174–192.

Schoenfeld, Eugen; and Stjepan G. Mestrovic

1991. From the Sacred Collectivity to the Sacred Individual: The Misunderstood Durkheimian Legacy. *Sociological Focus* 24:83 ff.

Schwartz, Heinrich

1987. *Art and Photography: Forerunners and Influences.* Chicago: University of Chicago Press.

Schweber, S. S.

1990. Auguste Comte and the Nebular Hypothesis. In R. T. Bienvenu and M. Feingold, eds., *In the Presence of the Past: Essays in Honour of Frank Manuel,* 131–91. Dordrecht: Kluwer.

Scott, R. H.

1886. The History of Kew Observatory. *Proceedings of the Royal Society of London* 39:37–86.

Searle, G. F.

1897. On the Motion of an Electrified Ellipsoid. *Phil. Mag.* 44:323–41.

Shapin, S.

1982. History of Science and Its Sociological Reconstructions. *History of Science* 20:157–211.

1988a. Following Scientists Around. *Social Studies of Science* 18, no. 3:533–50.

1988b. The House of Experiment in Seventeenth-Century England. *Isis* 79: 373–404.

1989. The Invisible Technician. *American Scientist* 77:554–63.

Shapin, S.; and S. Schaffer

1985. *Leviathan and the Air-Pump: Hobbes, Boyle, and the Experimental Life.* Princeton: Princeton University Press.

Shrum, W.

1988. Review Essay: The Labyrinth of Science. *American Journal of Sociology* 94, no. 2:396–403.

Siegel, Daniel M.

1976. Balfour Stewart and Gustav Robert Kirchhoff: Two Independent Approaches to "Kirchhoff's Radiation Law," *Isis* 67:565–600.

Siemens, C. W.
1871. On the Increase of Electrical Resistance in Conductors with Rise in Temperature. *Proceedings of the Royal Society* 19:443–45.
1883a. *On The Conservation of Solar Energy.* London: Macmillan.
1883b. On the Dependence of Radiation on Temperature. *Proceedings of the Royal Society* 35:166–77.
1889. *Scientific Works.* 3 vols. Ed. E. F. Bamber. London: John Murray.
1953. *Letters to Sir Charles William Siemens,* Ed. W. H. Kennet. London: English Electric Co.
1970. Some of the Questions Involved in Solar Physics. 1883. Reprinted in Bernard Lovell, ed., *The Royal Institution Library of Science: Astronomy,* 1:207–13. New York: Elsevier.

Simon, S.
1899. Über das Verhältnis der elektrischen Ladung zur Masse der Kathodenstrahlen. *Ann. d. Phys.* 69:589–611.

Smith, B. H.
1988. *Contingencies of Value.* Cambridge: Harvard University Press.

Smith, Crosbie; and M. Norton Wise
1989. *Energy and Empire: A Biographical Study of Lord Kelvin.* Cambridge: Cambridge University Press.

Smith, Robert W.
1991. A National Observatory Transformed: Greenwich in the Nineteenth Century. *Journal for the History of Astronomy* 22:5–20.

Smyth, W. H.
1864. *Sidereal Chromatics.* London: privately printed.

Sorensen, K. H.; and N. Levold
1992. Tacit Networks, Heterogeneous Engineers, and Embodied Technology. *Science, Technology, and Human Values* 17, no. 1:13–35.

Starke, H.
1903. Über die elektrische und magnetische Ablenkung schneller Kathodenstrahlen. *Verh. d. D. Phys. Gesell.* 5:241–50.

Stein, S. T.
1886. *Das Licht.* Vol. 1.

Stewart, Balfour; and J. Norman Lockyer
1868. The Sun as a Type of the Material Universe. *Macmillan's Magazine* 18:246–57, 319–27.

Stokes, G. G.
1860. On the Simultaneous Emission and Absorption of Rays of the Same Definite Refrangibility. *Phil. Mag.* 19:193–97.
1907. *Memoir and Scientific Correspondence.* 2 vols. Ed. Joseph Larmor. Cambridge: Cambridge University Press.
1966. *Mathematical and Physical Papers.* Vol. 1. Reprint, New York: Johnson.

Stump, D.
1988. The Role of Skill in Experimentation: Reading Ludwik Fleck's Study of the Wasserman Reaction as an Example of Ian Hacking's Experimental Realism. In A. Fine and J. Leplin, eds., *PSA 1988: Proceedings of the 1988 Biennial Meeting of the Philosophy of Science Association*, 1:302–8. East Lansing: Philosophy of Science Association.

Sutherland, W.
1898. Relative Motion of the Earth and Aether. *Phil. Mag.* 45:23–31.

Swenson, L.
1972. *The Ethereal Ether.* Austin: University of Texas Press.

Tagg, John
1988. *The Burden of Representation: Essays on Photographies and Histories.* London: Macmillan.

Tatum, C. B.; and M. Fischer
1990. Constructibility Improvement during Preliminary Design of Reinforced Concrete Structures. Paper presented at CIFE Expert System Workshop, 31 Aug.

Taylor, C.
1990. To Follow a Rule . . . . *Cahiers d'épistémologie,* no. 9019, UQAM. Reprinted in Craig Calhoun, Edward LiPuma, and Moishe Postone, eds., *Bourdieu: Critical Perspectives,* 45–60. Chicago: University of Chicago Press, 1993.

Tetens, H.
1987. *Experimentelle Erfahrung.* Hamburg: Felix Meiner Verlag.

Thomson, G. P.
1956. A. O. Rankine. *Biographical Memories of Fellows of the Royal Society* 2:249–55.
1964. *J. J. Thomson: Discoverer of the Electron.* New York: Doubleday.

Thomson, J. J.
1881. On the Electric and Magnetic Effects Produced by the Motion of Electrified Bodies. *Phil. Mag.* 11:229–49.
1893. *Notes on Recent Researches in Electricity and Magnetism.* Oxford: Clarendon Press. Reprinted, London: Dawson's of Pall Mall, 1968.
1894. On the Velocity of Cathode-Rays. *Phil. Mag.* 38:358–65.
1897a. Cathode Rays. (Discourse delivered at the Royal Institution, 30 Apr). *The Electrician* 39 (21 May): 104–9.
1897b. Cathode Rays. *Phil. Mag.* 44:293–316.
1899. Note on Mr. Sutherland's Paper on the Cathode Rays. *Phil. Mag.* 47: 415–16.
1906. *Conduction of Electricity through Gases.* 2d ed. Cambridge: Cambridge University Press.
1907. *The Corpuscular Theory of Matter.* 2d impression. London: A. Constable.
1937. *Recollections and Reflections.* New York: Macmillan.

Thomson, W.; and P. G. Tait
1888. *Treatise on Natural Philosophy.* Cambridge: At the University Press.

Tiles, M.
1984. *Bachelard: Science and Objectivity.* Cambridge: Cambridge University Press.

Treitel, Jonathan
1986. A Structural Analysis of the History of Science: The Discovery of the Tau Lepton. Ph.D. diss., Stanford University.
1987. Confirmation with Technology: The Discovery of the Tau Lepton. *Centaurus* 30:140–80.

Trouton, F. T.
1902. The Results of an Electrical Experiment, Involving the Relative Motion of the Earth and Ether, Suggested by the Late Professor FitzGerald. Reprinted in FitzGerald 1902, 557–65.
1914. Presidential Address (Mathematical and Physical Sciences Section). *British Association Report,* 285–90.

Trouton, F. T.; and H. R. Noble
1904. The Mechanical Forces Acting on a Charged Electric Condenser Moving through Space. *Philosophical Transactions of the Royal Society* 202A: 165–81.

Trouton, F. T.; and A. O. Rankine
1908. On the Electrical Resistance of Moving Matter. *Proceedings of the Royal Society* 80:420–34.

Tweeney, Ryan
1985. Faraday's Discovery of Induction: A Cognitive Approach. In D. Gooding and F. J. L. James, eds., *Faraday Rediscovered: Essays on the Life and Work of Michael Faraday, 1791–1867,* 189–209. New York: Stockton Press.

van Fraassen, Bas C.
1980. *The Scientific Image.* New York: Oxford University Press.
1990. *Laws and Symmetry.* New York: Oxford University Press.

Viala, A.
1985. *Naissance de l'écrivain.* Paris: Minuit.

Wacquant, L. D.
1989. For a Socio-analysis of Intellectuals: On Homo Academicus. An Interview with Pierre Bourdieu. *BJS Berkeley Journal of Sociology* 34:1–29.

Warwick, A. C.
1989. The Electrodynamics of Moving Bodies and the Principle of Relativity in British Physics, 1894–1919. Ph.D. diss., Cambridge University.
1991. On the Role of the FitzGerald-Lorentz Contraction Hypothesis in the Development of Joseph Larmor's Electronic Theory of Matter. *Arch. Hist. Ex. Sci.* 43:29–91.
1992. Cambridge Mathematics and Cavendish Physics: Cunningham, Campbell,

and Einstein's Relativity. Pt. 1, The Uses of Theory. *Stud. Hist. Phil. Sci.* 23:625–56.
1993. Cambridge Mathematics and Cavendish Physics: Cunningham, Campbell, and Einstein's Relativity. Pt. 2, Comparing Traditions in Cambridge Physics. *Stud. Hist. Phil. Sci.* 24:1–25.

Weber, Max
1949. "Objectivity" in Social Science and Science Policy. 1904. Reprinted in E. A. Shils and H. A. Finch, eds., *The Methodology of the Social Sciences.* New York: Free Press.

Weinberg, Steven
1980. Conceptual Foundations of the Unified Theory of Weak and Electromagnetic Interactions. *Science* 210:1212–18.

Westrum, R.
1989. The Social Construction of Technological Systems. *Social Studies of Science* 19, no. 1:189–91.

Whewell, William
1857. *History of the Inductive Sciences.* 3d ed. 3 vols. London: Pickering.

Whitrow, G. J.
1973. Time, Gravitation, and the Universe: The Evolution of Relativistic Theories. Inaugural lecture, 22 May 1973, Imperial College of Science and Technology, University of London.

Whittaker, E. T.
1953. *A History of the Theories of Aether and Electricity: The Modern Theories, 1900–1926.* Vol. 2. London: Nelson.

Wien, W.
1901. Über die Möglichkeit einer elektromagnetischen Begründung der Mechanik. *Ann. d. Phys.* 5:501–13.

Williams, Mari E. W.
1984. Beyond the Planets: Early Nineteenth Century Studies of Double Stars. *Brit. J. Hist. Sci.* 17:295–309.
1987. Astronomy in London, 1860–1900. *Quarterly Journal of the Royal Astronomical Society* 28:10–26.
1989. Astronomical Observatories as Practical Space: The Case of Pulkowa. In F. A. J. L. James, ed., *The Development of the Laboratory,* 118–36. London: Macmillan.

Wilson, George
1851. *The Life of Henry Cavendish.* London: Cavendish Society.

Woodruff, A. E.
1968. The Contributions of Hermann von Helmholtz to Electrodynamics. *Isis* 59:300–311.

Woolgar, S.
1988. *Science: The Very Idea.* London: Ellis Horwood/Tavistock.

Zahar, E.
1973. Why Did Einstein's Programme Supersede Lorentz's? *Brit. J. Phil. Sci.* 24: 95–123, 223–62.
1978. "Crucial" Experiment: A Case Study. In G. Radnitzky and G. Andersson, eds., *Progress and Rationality in Science,* 71–97. Bos. Stud. Phil. Sci., vol. 58. Dordrecht: Reidel.
1989. *Einstein's Revolution: A Study in Heuristic.* La Salle, Ill.: Open Court.

Zahn, C. T.; and A. H. Spees
1938. A Critical Analysis of the Classical Experiments on the Relativistic Variation of Electron Mass. *Phys. Rev.* 53:511–21.

Contributors

Brian S. Baigrie
Associate Professor
Institute for the History and Philosophy of Science and Technology
University of Toronto

Jed Z. Buchwald
Director of the Dibner Institute for the History of Science and Technology
Bern Dibner Professor of the History of Science
Massachusetts Institute of Technology

Peter Galison
Mallinckrodt Professor of the History of Science and Physics
Department of the History of Science
Harvard University

Yves Gingras
Associate Professor of History of Science
Département d'Histoire
Université du Québec à Montréal

Ian Hacking
University Professor
Institute for the History and Philosophy of Science and Technology
University of Toronto

Giora Hon
Senior Lecturer of Philosophy
Department of Philosophy
Haifa University

Margaret Morrison
Associate Professor of Philosophy
Department of Philosophy
University of Toronto

ANDREW PICKERING
Professor of Sociology
Department of Sociology
University of Illinois at Urbana-Champaign

HANS RADDER
Universitair Docent in Philosophy
Faculteit der Wijsbegeerte
Vrije Universiteit, Amsterdam

SIMON SCHAFFER
Reader in History and Philosophy of Science
Department of History and Philosophy of Science
Cambridge University

SILVAN S. SCHWEBER
Professor of Physics
Department of Physics
Brandeis University

ANDREW WARWICK
Lecturer in History of Science
Science and Technology Studies
Imperial College, London

Index